QSAR in
Environmental Toxicology

QSAR in Environmental Toxicology

Proceedings of the Workshop on Quantitative Structure-Activity Relationships (QSAR) in Environmental Toxicology held at McMaster University, Hamilton, Ontario, Canada, August 16-18, 1983.

Edited by

Klaus L. E. Kaiser

National Water Research Institute,
Canada Center for Inland Waters, Burlington, Ontario, Canada

D. Reidel Publishing Company

A MEMBER OF THE KLUWER ACADEMIC PUBLISHERS GROUP

Dordrecht / Boston / Lancaster

Library of Congress Cataloging in Publication Data

CIP

Workshop on quantitative structure-activity relationships
(QSAR) in Environmental Toxicology (1983: McMaster University)
Quantitative Structure-Activity Relations in Environmental Toxicology

Bibliography: p.
Includes index.
1. Pollutants–structure–activity relationships–Congresses. 2. Pollutants–toxicology–Congresses. I. Kaiser, Klaus L. E., 1941– II. Title.
QH545.A1W67 1983 574.2′4 84-9820
ISBN 90-277-1776-1

Published by D. Reidel Publishing Company
P.O. Box 17, 3300 AA Dordrecht, Holland

Sold and distributed in the U.S.A. and Canada
by Kluwer Academic Publishers,
190 Old Derby Street, Hingham, MA 02043, U.S.A.

In all other countries, sold and distributed
by Kluwer Academic Publishers Group,
P.O. Box 322, 3300 AH Dordrecht, Holland

Printed in The Netherlands

TABLE OF CONTENTS

PREFACE

Ever since Rachel Carson's Silent Spring, we have generally become aware of environmental contaminants and their effects on the ecosystem. The finding of PCB's in fish by Soren Jensen in Sweden, the recognition of mirex as contaminant in fish from Lake Ontario, and the discoveries of contaminant laden leachates from dumpsites such as the Love Canal have become milestones in the search for and characterization of contaminants in our environment.

At this time, the problem no longer is so much the identification of contaminants and their sources. Rather, we are now faced with solving questions on the fates and effects of such compounds. This includes the search for mechanisms to deal effectively with the large number of chemicals already found in water, air and biota. One of such time and cost saving scientific avenues is the field of quantitative structure-activity correlations for the prediction of the environmental behavior and effects of compounds.

This volume presents the papers and ensuing discussions at the Workshop on Quantitative Structure-Activity Relations in Environmental Toxicology at McMaster University, Hamilton, Ontario, August 16 to 18, 1983. In contrast to other conferences on QSAR which generally emphasize QSAR as a tool for the development of new drugs, biocides and so forth, this workshop addressed specifically the environmental implications of existing chemicals. Therefore, one of its primary objectives was to stimulate collaboration and discussion between biologists, chemists, physicists, and engineers, and between experimentalists and modellers. Consequently, not every paper in this volume is dealing with structure-activity correlations in the classical sense. However, the solution of environmental contamination problems is neither restricted to the application of any one scientific discipline nor the prerogative of any one jurisdiction. To the contrary, timely and cost-effective solutions must be based on the collaborative expertise of many scientific disciplines as well as on the cooperation of industry, academia and public agencies. In summary then, the results presented here help to set the stage for further research on environmental contamination, a problem which is faced by most nations.

Both the workshop and these proceedings would not have been possible without the acitve support of many colleagues, for which I thank deeply. In particular, I wish to acknowledge Dr. Rod J. Allan, Chief of the Environmental Contaminants Division, National Water Research Institute, for his encouragement and support, Dr. Peter V. Hodson, Great Lakes Fisheries Research Branch, Department of Fisheries and Oceans, and Dr. D. George Dixon, Department of Biology, University of Waterloo for their help in planning and organizing the workshop, Messrs. Howard Hunter and Brandon Hickie for their assistance in running the workshop, Mrs. Gaye Scott for the preparation of the discussion transcripts, and her valuable assistance in editing, proof-reading, and indexing of the manuscripts, and Mrs. Dianne Crabtree for her skill and patience in the typing and pagination of this volume. Finally, I thank all authors, speakers, session chairmen and reviewers whose participation and efforts are greatly appreciated. I hope this volume proves beneficial to all readers, freshman or specialist alike.

February 1984 Klaus L.E. Kaiser

LIST OF CONTRIBUTORS AND PARTICIPANTS

R.J. Allan, Environmental Contaminants Division, National Water Research Institute, Burlington, Ontario L7R 4A6, Canada.

R. Bisson, Environmental Protection Service, Environment Canada, Ottawa, Ontario K1A 1C8, Canada.

A. Bobra, Institute for Environmental Studies, University of Toronto, Toronto, Ontario M5S 1A4, Canada.

S.A. Boliska, Department of Biology, University of Waterloo, Waterloo, Ontario N2L 3G1, Canada.

N.C. Bols, Department of Psychiatry, University of California, La Jolla, California 92093, U.S.A.

I. Chu, Environmental and Occupational Toxicology Division, Health and Welfare Canada, Ottawa, Ontario K1A 0L2, Canada.

G. Craig, Toxicity Unit, Ontario Ministry of the Environment, Rexdale, Ontario M4V 1P5, Canada.

P. Craig, Health Designs, Inc., Rochester, New York 14604, U.S.A.

J.M. Cullen, Department of Chemistry, Harvard University, Cambridge, Massachusetts 02138, U.S.A.

D.G. Dixon, Department of Biology, University of Waterloo, Waterloo, Ontario N2L 3G1, Canada.

W.J. Dunn, III, Department of Medicinal Chemistry & Pharmacognosy, University of Illinois, Chicago, Illinois 60680, U.S.A.

K. Enslein, Health Designs, Inc., Rochester, New York 14604, U.S.A.

D. Freitag, GSF, Institute for Ecological Chemistry, D-8042 Neuherberg, Federal Republic of Germany.

D.F. Hagen, 3M Central Research Laboratories, St. Paul, Minnesota 55133, U.S.A.

E. Halfon, Aquatic Physics and System Division, National Water Research Institute, Burlington, Ontario L7R 4A6, Canada.

D.J. Hallett, Environment Canada, Toronto, Ontario M4T 1M2, Canada.

D.R. Hart, IEC Beak Consultants Ltd., Mississauga, Ontario L4V 1P1, Canada.

P.V. Hodson, Great Lakes Fisheries Research Branch, Department of Fisheries and Oceans, Burlington, Ontario L7R 4A6, Canada.

D.E. Johnson, Office of Toxic Substances, U.S. Environmental Protection Agency, Washington, DC 20460, U.S.A.

K.L.E. Kaiser, Environmental Contaminants Division, National Water Research Institute, Burlington, Ontario L7R 4A6, Canada.

R. Koch, Department of Toxicology, Institute of Hygiene, DDR-6500 Gera, German Democratic Republic.

F. Korte, Institute for Chemistry, Technical University Munich, D-8050 Freising-Weihenstephan, Federal Republic of Germany.

K. Kwasniewska, Environmental Contaminants Division, National Water Research Institute, Burlington, Ontario L7R 4A6, Canada.

S.S. Lande, 3M Environmental Laboratory, St. Paul, Minnesota 55133, U.S.A.

T.R. Lander, Health Designs Inc., Rochester, New York 14604, U.S.A.

J.P. Lay, GSF, Institute for Ecological Chemistry, D-8042 Neuherberg, Federal Republic of Germany.

G.A. LeBlanc, Biology Department, University of South Florida, Tampa, Florida 33620, U.S.A.

R.L. Lipnick, Office of Toxic Substances, U.S. Environmental Protection Agency, Washington, DC 20460, U.S.A.

M. Lupp, MacLaren Plansearch, Inc., Toronto, Ontario M5E 1E7, Canada.

D. Mackay, Department of Chemical Engineering and Applied Chemistry, University of Toronto, Toronto, Ontario M5S 1A4, Canada.

L.S. McCarty, MacLaren Plansearch, Inc., Toronto, Ontario M5E 1E7, Canada.

B.A. Moulton, Department of Animal Science, College of Veterinary Medicine, University of Tennessee, Knoxville, Tennessee 37901, U.S.A.

L.D. Newsome, Office of Toxic Substances, U.S. Environmental Protection Agency, Washington, DC 20460, U.S.A.

B.G. Oliver, Environmental Contaminants Division, National Water Research Institute, Burlington, Ontario L7R 4A6, Canada.

R.S. Pearlman, College of Pharmacy, University of Texas at Austin, Austin, Texas, U.S.A.

J.M. Ribo, Environmental Contaminants Division, National Water Research Institute, Burlington, Ontario L7R 4A6, Canada.

T.W. Schultz, Department of Animal Science, College of Veterinary Medicine, University of Tennessee, Knoxville, Tennessee 37901, U.S.A.

A.S. Seaver, 3M Central Research Laboratories, St. Paul, Minnesota 55133, U.S.A.

M. Shea, MacLaren Plansearch, Inc., Toronto, Ontario M5E 1E7, Canada.

W.Y. Shiu, Institute of Environmental Studies, University of Toronto, Toronto, Ontario M5S 1A4, Canada.

D.L. Stalling, U.S. Fish and Wildlife Service, Columbia National Fisheries Research Laboratory, Columbia, Missouri 62500, U.S.A.

M.E.Tomb, Health Designs Inc., Rochester, New York 14604, U.S.A.

K. Trudel, S.L. Ross Environmental Research Ltd., Ottawa, Ontario, Canada

V.E. Valli, Biopath Analysts Limited, Guelph, Ontario N1E 2X7, Canada.

D.C. Villeneuve, Environmental and Occupational Toxicology Division, Health and Welfare Canada, Ottawa, Ontario K1A 0L2, Canada.

P. Wells, Toxic Chemicals Management Program, Environment Canada, Ottawa, Ontario K1A 1C8, Canada.

S. Wold, Research Group for Chemometrics, Umea University, S-90187 Umea, Sweden.

V. Zitko, Fisheries and Environmental Sciences, Department of Fisheries and Oceans, Biological Station, St. Andrews, New Brunswick E0G 2X0, Canada.

FOREWORD

This volume of collected papers represents a valuable and timely extension of quantitative structure-activity relationships (QSAR) to aquatic ecosystems. QSAR is a fundamental concept in the development of pharmaceutical drugs and toxic organic chemicals (biocides) used in agriculture, in forestry, and for general pest control. The proliferation of organic biocides which have been produced and which continue to be developed at a rate of some 600 per annum, and of industrial chemicals which have been disposed of in waste dumps in the past and are now being detected in many waterways, predicates an urgent need for the ability to predict biological activity of organic chemicals by means other than classical toxicology. QSAR allows researchers to make such predictions of the biological acitivity of organic chemicals and their degradation products. Intermediate steps in environmental QSAR, such as relating chemical structure to degradation or bioaccumulation, are ways to more rapidly estimate expected toxicity, and to assist in the selection of compounds for which long-term testing may still be needed. Another potential use of QSAR is the prediction of synergistic and antagonistic effects, a major obstacle to real environmental risk assessment.

In 1978, the Great Lakes Water Quality Agreement between Canada and the United States called for an expansion of QSAR research. Monies assigned to execute the Agreement have supported some of the work presented here. QSAR research has also been recognized as an important tool and has been supported under Canada's Toxic Chemicals Management Program.

Papers in this volume show that QSAR developed with lower aquatic organisms can be rapid, and inexpensive, and can be extrapolated to higher organisms. It is a pleasure to acknowledge here the cooperation of the staff of Environment Canada and the Department of

Fisheries and Oceans at the Canada Centre for Inland Waters; along with the staff of Universities, industry, and other government agencies in Canada, the United States and Europe.

R.J. Allan
Chief, Environmental Contaminants Division
National Water Research Institute
Canada Centre for Inland Waters

STRUCTURE-ACTIVITY RELATIONSHIPS FOR TOXICITY OF HYDROCARBONS, CHLORINATED HYDROCARBONS AND OILS TO *DAPHNIA MAGNA*

Alice M. Bobra, Wan Ying Shiu and Donald Mackay

Institute for Environmental Studies
University of Toronto
Toronto, Ontario M5S 1A4
CANADA

ABSTRACT

Structure-activity relationships between hydrocarbon toxicity and physical-chemical properties are investigated. The acute toxicities of aqueous solutions of hydrocarbons, chlorinated hydrocarbons and crude oils to *Daphnia magna* in a closed bioassay system are reported and the toxicities correlated with solubilities.

INTRODUCTION

Since the aquatic environment is frequently subjected to contamination by hydrocarbons and chlorinated hydrocarbons (NAS 1975), it is desirable to be able to assess the toxic effects which these chemicals may have on organisms. Thus, many bioassays have been devised and conducted with organisms, varying from bacteria to mammals, using both pure substances and mixtures. However, due to the vast number of existing and newly introduced chemicals in the environment, the number of toxicity assessments required is much greater than the number of corresponding experimental studies. An attractive approach, which would facilitate prediction of the ecological impact of toxic substances, is to seek structure-activity relationships present in toxicological data and to identify chemical groups with similar effects.

K. L. E. Kaiser (ed.), QSAR in Environmental Toxicology, 3–16.

In many of the toxicity studies of volatile organic chemicals, a common experimental difficulty is the occurrence of vapour losses from the solution during the bioassay. This can result in the organism being exposed to a decreasing concentration of toxicant. Thus, truly closed or air-tight systems are preferred for toxicants with large Henry's law constants or air/water partition coefficients, as reported recently by Hutchinson *et al.* (1980), Rogerson *et al.* (1983) and Bobra *et al.* (1983a). Such systems appear to give more reproducible exposure conditions. In the latter study, Bobra *et al.* (1983a) described a method of exposing the water flea, *Daphnia magna* to hydrocarbon solutions in the absence of an air space. If an air space is present, it must be tightly sealed to prevent exchange with the atmosphere and the extent of air/water partitioning calculated. Also, in cases where this partitioning plays an important role, it is necessary to calculate the aqueous concentration drop attributable to partitioning into the test organism or onto sorbing surfaces.

Hutchinson *et al.* (1980) measured the photosynthetic toxicity thresholds in glass stoppered flasks of several alkanes, cycloalkanes, mononuclear and polynuclear aromatics and chlorinated hydrocarbons to two algal cultures (*Chlamydomonas* and *Chlorella*). They related the toxicity thresholds to the physical-chemical properties of several hydrocarbons and predicted the toxicity of a given hydrocarbon from its aqueous solubility. Rogerson *et al.* (1983) following this approach, measured the acute toxicity threshold concentrations of a similar range of hydrocarbons to a ciliated protozoa (*Tetrahymena ellioti*) in a closed system and observed, as in the previously mentioned study, a good correlation between toxicity and hydrocarbon aqueous solubility.

This report describes a research program at the University of Toronto in which the acute toxicities of aqueous solutions of liquid hydrocarbons, chlorinated hydrocarbons and oils to *Daphnia magna* in a closed system were measured, interpreted thermodynamically (Bobra *et al.* 1983a; 1983b; 1983c) and compared with the results of Rogerson *et al.* (1983) and Hutchinson *et al.* (1980).

MATERIALS AND METHODS

Saturated aqueous solutions of liquid compounds with a density greater than that of water (see TABLE 1) were prepared by adding an excess quantity of the compound to double distilled water in 250 mL glass stoppered flasks. The solutions with liquid compounds were stirred gently for 24 hours, to prevent emulsion formation. The solutions were subsequently allowed to settle at room temperature for a least 48 hours before the bioassays.

Saturated aqueous solutions of the liquid compounds and oils (fresh and weathered) with a density less than that of water (see TABLES 1 and 2) were prepared by adding an excess quantity of the compound to double distilled water in 250 mL cylindrical separatory funnels. The mixtures were then shaken gently on a wrist action shaker for 12 hours and allowed to settle at room temperature for at least 48 hours before the bioassays.

Weathered samples of crude oils were prepared by evaporating the oil in a pan in a wind tunnel at 20°C or by using an air stripping apparatus. The samples were continuously weighed on a top-loading balance and the degree of weathering expressed as the percent evaporated.

Daphnia magna obtained from laboratory cultures of parthenogenetic females were used for all bioassays. The stock cultures were fed trout chow and algae and kept in distilled water with a pH of approximately 6.5. *Daphnia* of approximately 1.5 mm in length (four to six days old) were selected visually from the stock culture for the bioassays.

The bioassays were performed in 24 mL glass vials with Teflon-lined screw-on caps, 10 animals per vial. The vials had no air spaces, were not aerated, nor were animals fed during the bioassays. *Daphnia* mortalities were observed after 48 hours. Death was defined as immobility. Mortality data from each of the toxicity tests were used to estimate lethal concentrations to 50% of the population.

The saturated aqueous solutions were added directly to the vials. The vials were immediately closed tightly and shaken gently for a few minutes. No analyses were done on the pure hydrocarbon solutions since it was assumed that there was negligible loss of the compound from the saturated solution by volatilization. The amounts of saturated solutions added were determined by the differences in mass before and after addition. Concentrations of liquid hydrocarbons were calculated from available literature solubility data of compounds (Mackay and Shiu 1981) and from the extent of dilution. The total concentrations of the aqueous phase in equilibrium with the oils (i.e., the "water soluble fractions") were determined using a Hewlett Packard GC Model 5840A equipped with a flame ionization detector and a 7675A Purge and Trap sampler. The column was a 50 m x 0.5 mm I.D. glass WCOT column coated with SE 30. The GC oven was temperature programmed from 50 to 220°C at a rate of 5°C/min.

RESULTS AND DISCUSSION

It should be noted that concentrations are expressed here in units of $g{\cdot}m^{-3}$. It is generally preferable to use $mol{\cdot}m^{-3}$, but in the case of crude oils the molecular weight is unknown thus only $g{\cdot}m^{-3}$ can be used.

A summary of the acute toxicity results of the water soluble fraction of the various pure hydrocarbons, chlorinated hydrocarbons and various oils to *Daphnia magna* in a closed system is presented in TABLES 1 and 2. *Daphnia* varied widely in their response to the hydrocarbons, chlorinated hydrocarbons and oils, the LC50's extending over several orders of magnitude. Since aqueous solubility is believed to influence partitioning into organisms with large surface area to volume ratios, such as the type studied here, the LC50's ($g{\cdot}m^{-3}$) of the water soluble fractions were regressed against this physical parameter. An example for hydrocarbons is presented in FIGURE 1. Near linear relationships were obtained which extended over several orders of magnitude. This is in good agreement with previous hydrocarbons studies using algae (Hutchinson *et al.* 1980) and ciliated protozoa (Rogerson *et al.* 1983) as shown in TABLE 3. Results in which the extrapolated lethal concentrations results were greater than aqueous solubility, were eliminated from the calculations, as shown in TABLES 1 and 3. Thus, one approach for correlating the toxicity of aqueous solutions of pure hydrocarbons, chlorinated hydrocarbons and fresh and weathered oils to *Daphnia*, algae and protozoa is to measure the aqueous solubility (C_L), and calculate the LC50 from the

TABLE 1: The toxicity of hydrocarbons and chlorinated hydrocarbons in closed systems ($g \cdot m^{-3}$); for references see TABLE 3

Compound	Solubility ($g \cdot m^{-3}$)(C_L)	Daphnia magna 48-hour LC50	Tetrahymena 48-hour C_{Wo}[b]	Chlamydomonas 3-hour EC50[a]	Chlorella 3-hour EC50
Alkanes					
Pentane	38.4	9.6			
Hexane	9.5	3.9	9.0	8.08	$>C_L$
Octane	0.66	0.38	0.44	$>C_L$	$>C_L$
Decane	0.050	0.029	$>C_L$	$>C_L$	0.042
Dodecane	0.0038			$>C_L$	$>C_L$
Tetradecane	0.0023			$>C_L$	$>C_L$
3-Methyl pentane	12.8			$>C_L$	12.7
Cycloalkanes					
Cyclopentane	155.7	10.9		119.9	168.8
Methyl cyclopentane	42			38.2	26.5
Cyclohexane	54.9	3.8	24.2	38.4	31.8
Methyl cyclohexane	14	1.5	8.7	10.5	10.5
Cyclooctane	7.9			3.0	2.8
Mononuclear Aromatics					
Benzene	1777	35.5	390.9	462	319.9
Toluene	514	10.3	143.9	133.6	205.6
Ethyl benzene	151.7	1.5		50.8	62.2
p-Ethyl toluene	94.85			55.0	48.8
o-Ethyl toluene	93.05			18.6	40.9
o-Xylene	174.7	3.5	17.5		
p-Xylene	184.7	9.2	16.6	46.2	105.3
m-Xylene	162	9.7	55.1		
Propylbenzene	55		18.2	18.2	16.0
Cumene	49.9	0.5	3.0	8.5	21.0
n-Butylbenzene	12.6			3.5	3.0
Isobutyl benzene	10.1			3.1	3.5

a Photosynthetic toxicity threshold concentrations.

b Acute toxicity threshold concentrations.

TABLE 1:(Cont'd)

Compound	Solubility $(g \cdot m^{-3})(C_L)$	Daphnia magna 48-hour LC50	Tetra-hymena 48-hour C_{Wo} [b]	Chlamy-domonas 3-hour EC50[a]	Chlorella 3-hour EC50
1,2,4-Trimethyl benzene	56.9	3.4	11.4		
1,3,5-Trimethyl benzene	96.8	5.8			
1,2,4,5-Tetramethyl benzene	3.5	0.49		3.5	$>C_L$
Decalin	6.21			2.0	1.0
Polynuclear Aromatic					
1-Methyl naphthalene	28.4	1.4	9.4	1.7	5.1
Chlorinated Compounds					
Dichloromethane	19400	135.8		1474.4	2328
Trichloromethane	7950	79		381.6	405.5
1,1,1-Trichloro-	720	57.6		280.8	153.4
ethane				24.5	30.2
1,1,2,2,2-Penta-chloroethane	480				
1,2,3-Trichloro-	1900			112.1	169.1
propane					98.9
Trichloroethylene	1100	7.7			98.9
Chlorobenzene	471.1			56.5	
o-Dichlorobenzene	145.2	<5.9			
1-Chloronaphthalene	22.4	<0.90			

a Photosynthetic toxicity threshold concentrations.

b Acute toxicity threshold concentrations.

equation relating LC50 to the solubility (C_L). Solubility (C_L) can be estimated from the molecular structure and melting point or total surface area (Yalkowsky and Valvani 1979). Thus, potentially the LC50 can be estimated from molecular structure, at least for a homologous series of compounds and a specified organism.

TABLE 2: The acute toxicity of the water soluble fractions of various oils to *Daphnia magna* in a closed system

Oil	Degree of Weathering (Percent Evaporated)	Solubility $C_L (g \cdot m^{-3})$	48-hour LC50 $(g \cdot m^{-3})$
Norman Wells	0	32.3	9.0
	6	27	10.8
	12	14.6	6.0
	20.4	7.3	4.4
	36.7	0.68	0.49
	43.2	0.14	0.12
Prudhoe Bay	0	29.3	9.4
	9.8	4.9	3.9
	18.2	0.15	0.15
	24.4	0.10	0.1
Lago Medio	0	25.5	7.7
	22.3	0.6	0.6
Murban	0	28.6	7.4
Natural Gas Condensate	0	74.7	9
	42	8	2
	71	0.06	0.03
Synthetic Oil Mixture	0	203.4	13.5

From TABLE 3, it also can be seen that for *Daphnia* the presence of chlorine as a substitute for hydrogen in hydrocarbon has no significant effect apart, of course, from its effect on solubility.

It is also interesting to note that in FIGURE 1 after the regression line crosses the 45° diagonal, at where LC50 equals C_L, the LC50 values exceed the C_L values. These points of intersection for the *Daphnia*, algae and ciliated protozoa are listed in TABLE 3. Beyond these points of intersection, it is impossible to make an aqueous solution which is toxic, at these conditions, to the organisms. This may apply to very weathered oils, to high molecular weight hydrocarbons and to many polychlorinated aromatic compounds which have low C_L values.

TABLE 3: Summary of correlations using data from TABLE 2

Organisms	Correlation	Correlation Coefficient (r)	Number of Compounds (n)	Intersection Concentration (log)[a] ($g \cdot m^{-3}$)
Hydrocarbons				
Daphnia[b]	48-hour log LC50 = 0.599 log C_L - 0.533	0.88	18	-1.33
Tetrahymena[c]	48-hour log C_{Wo} = 0.780 log C_L - 0.177	0.91	13	-0.81
Chlamydomonas[d]	3-hour log EC50 = 0.893 log C_L - 0.253	0.90	19	-2.36
Chorella[d]	3-hour log EC50 = 0.944 log C_L - 0.292	0.97	19	-5.21
Chlorinated Hydrocarbons				
Daphnia	48-hour log EC50 = 0.540 log C_L - 0.234	0.68	4	-0.51
Chlamydomonas	3-hour log EC50 = 0.809 log C_L - 0.421	0.87	6	-2.20
Chlorella	3-hour log EC50 = 0.873 log C_L - 0.567	0.92	7	-4.46
Oils				
Daphnia[e]	48-hour log EC50 = 0.765 log C_L - 0.240	0.97	15	-1.02

a Point of intersection with LC50 = C_L.

b Bobra *et al*. (1983a).

c Rogerson *et al*. (1983).

d Hutchinson *et al*. (1980).

e Bobra *et al*. (1983b).

These studies, we believe, demonstrate that the concentrations required within the organisms to induce the same extent of mortality are relatively constant for hydrocarbons, chlorinated hydrocarbons and oils, even though their physical-chemical properties and toxicity thresholds range over several orders of magnitude. These concentrations are controlled by organism/water partitioning. The concentrations in the organism are proportional to the water concentrations and to the organism/water partition coefficients, which are in turn controlled by, and inversely proportional to, solubility. Thus, for these substances, toxicity is primarily determined by solubility, with the chemical's molecular composition playing only a secondary role.

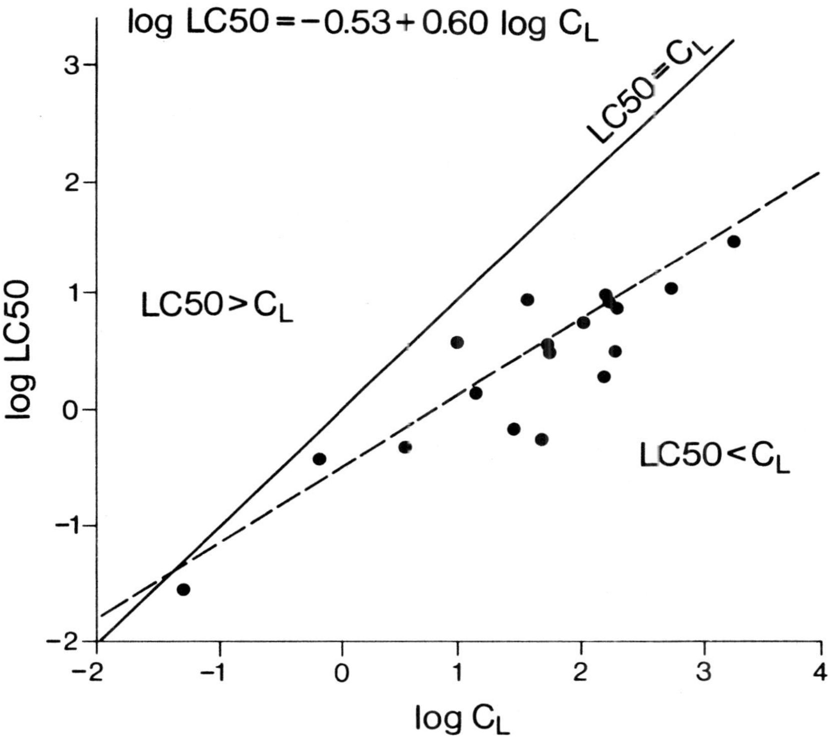

FIGURE 1: Acute median lethal concentrations (LC50) of hydrocarbons to *Daphnia magna* in a closed system versus their solubilities

It would be beneficial to examine other hydrocarbons, oils, halogenated hydrocarbons, and other more complex organisms to explore if similar relationships apply. If they do, the entire issue of hydrocarbon toxicology will become more predictable.

No solid chemicals are included in this analysis. It has been suggested that for solids the correct correlating solubility is that of the subcooled liquid (Rogerson *et al*. 1983) but this issue is not addressed here.

ACKNOWLEDGEMENT

The authors are grateful to the Inland Waters Directorate of Environment Canada for financial support.

REFERENCES

Bobra, A.M., Mackay, D. and Shiu, W.Y. 1983a. Toxicity of hydrocarbons and chlorinated hydrocarbons to the water flea (*Daphnia magna*). **Chemosphere 12:** 1121-1129.

Bobra, A.M., Shiu, W.Y. and Mackay, D. 1983b. Acute toxicity of fresh and weathered crude oils to *Daphnia magna*. **Chemosphere 12:** 1137-1149.

Bobra, A.M., Mackay, D. and Shiu, W.Y. 1983. Toxicity of hydrocarbons and crude oils. **Proceedings of the 6th Arctic Marine Oil Spill Program Technical Seminar,** Edmonton, Alberta, June 14-16.

Hutchinson, T.C., Hellebust, J.A., Tam, D., Mackay, D., Mascarenhas, R.A. and Shiu, W.Y. 1980. In **Hydrocarbons and Halogenated Hydrocarbons in the Aquatic Environment,** B.K. Afghan and D. Mackay (Eds.), Plenum Press, New York, pp. 577-586.

Mackay, D. and Shiu, W.Y. 1981. A critical review of Henry's law constants for chemicals of environmental interest. **J. Phys. Chem. Ref. Data 10:** 1175-1199.

NAS. 1975. National Academy of Sciences. **Petroleum in the Marine Environment.** National Academy of Sciences, Washington, DC, 107 p.

Rogerson, A., Shiu, W.Y., Huang, G.L., Mackay, D. and Berger, J. 1983. Determination and interpretation of hydrocarbon toxicity to ciliate protozoa. **J. Aquat. Toxicol. 3:** 215-228.

Yalkowsky, S.H. and Valvani, S.C. 1979. Solubilities and partitioning II. Relationships between aqueous solubilities, partition coefficients and molecular surface areas of rigid aromatic hydrocarbons. **J. Chem. Eng. Data 24:** 127-129.

WORKSHOP DISCUSSION

SCHULTZ:

You mentioned you used four to six-day old Daphnia, but isn't that getting to about the fifth instar?

BOBRA:

The reason I used four to six-day old Daphnia is because neonates, which are recommended by the EPA for bioassays, would be stressed by the lack of food during the 48-hour period and I did not want food to interfere with the concentration in the water column.

SCHULTZ:

What about exuviae? Did they molt?

BOBRA:

Yes, they did molt but I did not look at that; I was solely looking for death.

SCHULTZ:

I was wondering if the stress of molting may have affected the relationship?

BOBRA:

We did not examine that.

SCHULTZ:

You said you defined death just with the naked eye, what about immobility?

BOBRA:

You can see movements of the internal organs, not of the heart itself but of those in the abdomen. I used a magnifying glass in some cases where death was questionable. Usually I would look at the Daphnia for a few minutes and if they did no show any sign of movement after stimulation, then I assumed they were dead.

DUNN:

Did I hear you correctly when you said that solubility could be predicted from structure? How do you do that?

BOBRA:

Yes, from structure. Properties, in particular the total surface area and melting point as described by Yalkowsky and Valvani.

DUNN:

One other question; you have a considerable range in your toxicity of your crude oils, can you account for that in terms of components or is that known?

BOBRA:

Yes. We tested some of the major components in the water column, like benzene, toluene, xylenes and we attempted to add their separate effects. We were able to account for the difference in toxicity by the difference in concentration of chemicals in the water column.

DUNN:

Do you know what they are? Can you see variations in toxicity with variations in composition?

BOBRA:

Yes. Especially in the comparison of fresh and weathered crude oils, compounds like benzene and toluene evaporate very quickly. Both chemicals have a relatively high solubility so they usually have an immediate impact. They reach lethal concentrations in the organisms very quickly, as is found with fresh oil where these compounds are present in large amounts.

MACKAY:

The oil composition doesn't matter except for the influence of the composition in determining the solubility?

BOBRA:

Yes, the solubility is the determining factor.

DUNN:

It doesn't matter if there are chlorinated compounds in the crude oil.

HODSON:

I it is interesting that you have focused on very short-term bioassays. We are interested in long-term impact, so short-term results may be quite irrelevant. You are describing bioassays which, in essence are rather inefficient mechanisms for predicting solubility. If we pick materials that are not very soluble we get one answer, if we pick soluble ones, we get another.

The toxicity was a function of solubility because you used a very short-term test and limited the rate of uptake. This relates to your comment that it doesn't matter what was in the mixture, that it was the solubility that was important. It seems here that if you would allow the material to come to equilibrium in the animal, then you would measure the inherent toxicity of the material. I think we have to recognize that short-term tests really relate to what goes on inside an effluent plume or in the immediate vicinity of an oil spill, where the chemical and biological reactions are very fast. In terms of the ultimate environmental impact in, say fish, which might live ten to twelve years and accumulate a substantial dose, time to equilibrium and uptake rate are no longer important. The inherent toxicity governs and the compound's interaction with biological processes.

MACKAY:

We have an absolutely fundamental disagreement there because if you plot the concentration in the _Daphnia_ as a function of time, it will go up within an hour or so and so it is in equilibrium. Kinetics are completely irrelevant until you are up to a log K_{OW} of about 5 or 6 as Barry Oliver just recently showed and then you have to worry about it. So I think what Alice is looking at is an equilibrium constant concentration which, when sustained over a period of a couple of days, causes death and the kinetics are largely irrelevant.

KAISER:

That actually has been shown by some older studies on uptake of PCB's from water. It showed uptake into the fish body by autoradiography cross-section of the entire fish after twenty-four hours; there were significant radioactivity levels in the brain and various tissues, the abdomen and skin too.

COMPARATIVE TOXICITY AND METABOLISM OF TETRACHLOROBENZENE ISOMERS

Ih Chu[1], David C. Villeneuve[1] and Victor E. Valli[2]

[1] Environmental Health Directorate
Environmental and Occupational Toxicology Division
Health and Welfare Canada
Ottawa, Ontario K1A 0L2
CANADA

[2] Biopath Analysts Limited
Guelph, Ontario N1E 2X7
CANADA

ABSTRACT

LD50 values for 1,2,3,4-, 1,2,4,5- and 1,2,3,5-tetrachlorobenzene (TCB) were found to be 1470, 3105 and 2297 $mg \cdot kg^{-1}$, respectively, in male rats. In females the LD50 values were found to be 1167 and 1727 $mg \cdot kg^{-1}$ for 1,2,3,4- and 1,2,3,5-TCB, respectively. Clinical signs of toxicity included depression, flaccid muscle tone, prostration, piloerection, loose stool, hypothermia, dacryorrhea, coma and death. In a subacute study, rats were fed diets containing 0, 0.5, 5.0, 50 or 500 ppm 1,2,3,4-, 1,2,4,5- or 1,2,3,5-TCB for 28 days. At 500 ppm 1,2,4,5-, but not 1,2,3,4- or 1,2,3,5-TCB, caused a significant increase in the liver weight, and serum cholesterol of male and female rats. Hepatic microsomal aniline hydroxylase was induced by 500 ppm 1,2,4,5-TCB. Hepatic microsomal aminopyrine demethylase activity was increased by the administration of this compound at 50 ppm and higher in males and at 500 ppm in the females. Rats fed 1,2,3,4- and 1,2,3,5-TCB at 500 ppm also showed a significant increase in aminopyrine demethylase activity. Moderate to severe histological changes were found in the liver, thyroid, kidney and

K. L. E. Kaiser (ed.), QSAR in Environmental Toxicology, 17–37.

lungs of rats fed 500 ppm 1,2,4,5-TCB. Histological changes in the tissues produced by the administration of the 1,2,3,4- and 1,2,3,5-isomer were mild even at the highest dose levels. Similar results were observed in rats fed TCB isomers for 90 days. However, histopathological changes in the kidney of rats fed 1,2,4,5-TCB were more severe than those of the subacute study. Tissue residue data showed that 1,2,4,5-TCB accumulated in a dose-dependent manner and at much higher concentrations than the other two isomers. In a separate study adult male rats were given orally single doses of ^{14}C-1,2,3,4-, 1,2,3,5- or 1,2,4,5-TCB at 10 mg•kg^{-1} body weight, and were housed in individual metabolism cages to collect urine and feces for radioassay. The fat, skin and liver of rats dosed with 1,2,4,5-TCB contained the highest levels of radioactivity, whereas little ^{14}C was detected in 1,2,3,4- and 1,2,3,5-TCB treated animals. For 1,2,3,4- and 1,2,3,5-TCB, approximately 46% to 51% of the administered compound were excreted in urine and feces within 48 hours after administration. During the same period only 8% of the administered 1,2,4,5-Tetrachlorobenzene was excreted. Analysis of urine indicated that the tetrachlorobenzenes were biotransformed to a number of polar metabolites. The metabolites for each of the three TCB's in decreasing order of quantities were as follows; for 1,2,3,4-TCB: 2,3,4,5- and 2,3,4,6-tetrachlorophenol and a trace of tetrachlorothiophenol and 2,3,4-trichlorophenol; for 1,2,3,5-TCB: 2,3,4,6-tetrachlorophenol, isomeric mercaptotrichlorophenols and a trichlorophenol; for 1,2,4,5-TCB: 2,3,5,6-tetrachlorophenol, tetrachloroquinol and a trichlorophenol. Data indicate that 1,2,4,5-TCB was much more toxic than the other two TCB isomers, and that the differences in toxicity were associated with a higher degree of bioaccumulation and slower metabolic degradation of this compound.

INTRODUCTION

Tetrachlorobenzenes (TCB's) are industrial chemicals which have wide application in chemical synthesis and are being used for electrical insulation. TCB residues have been identified in fish in the Great Lakes (Great Lakes Water Quality Board 1978) and in the industrial pollutant coastal areas in Norway (Lunde and Ofstad 1976). Toxic changes have been demonstrated in laboratory animals treated with these chlorinated benzenes. Braun *et al.* (1978) observed an increase in serum alkaline phosphatase activity and bilirubin in beagle dogs fed 5 $mg \cdot kg^{-1} \cdot day^{-1}$ 1,2,4,5-TCB for two years. Fomenko (1965) found that the administration of 1,2,4,5-TCB resulted in neurotoxic changes in rats together with decreased liver function in rats and rabbits. 1,2,4,5-TCB was also shown to induce hepatic microsomal P-450 enzymes in rats (Ariyoshi et al. 1975a,b). At present, the toxic effects of 1,2,4,5-TCB have not been fully delineated. In addition, there is little toxicity information available on 1,2,3,4- and 1,2,3,5-TCB. In view of this and the potential for human exposure, the present study was designed to investigate the comparative toxicity and metabolism and pharmacokinetic behaviour of these three tetrachlorobenzene isomers in the rat.

MATERIALS AND METHODS

Materials

1,2,3,4-, 1,2,3,5- and 1,2,4,5-TCB's were purchased from Aldrich Chemical Co. (Milwaukee, Wis.). These compounds were purified by fractional crystallization from 95% ethanol to a purity of greater than 99.5%; [^{14}C(U)]-labelled tetrachlorobenzenes were custom sythesized by New England Nuclear (Boston, Mass.). The stated radiochemical purity and specific activity for each of the isomers were as follows: 1,2,3,4-TCB (99%, 1.33 $mCi \cdot mmol^{-1}$); 1,2,3,5-TCB (99%, 2.55 $mCi \cdot mmol^{-1}$); 1,2,4,5-TCB (99%, 5.8 $mCi \cdot mmol^{-1}$). The radiochemical purities of the compounds were confirmed by thin layer chromatography using a radiochromatography scanner. All other chemicals were of reagent grade and obtained commercially.

Acute Toxicity

Male and female Sprague-Dawley rats weighing 200 to 250 g were purchased from Charles River Breeding Laboratory (Wilmington, Mass.) and were acclimatized to laboratory conditions (temperature 20 ± 2°C, humidity 40% to 60%) for one week before treatment. Groups of 10 fasted animals were administered single oral doses of 1,2,3,4-, 1,2,4,5- or 1,2,3,5-TCB in corn oil (Mazola) at a constant volume of 1.0 $mL \cdot (100\ g)^{-1}$ body weight. At least five dose levels ranging from 200 to 4000 $mg \cdot kg^{-1}$ were administered for each compound in order to determine LD50 values*. The controls received an equal volume of corn oil. The animals, housed individually in stainless steel mesh cages, were given food (Master Fox, Purina Ralston) and water ad libitum. Clinical observations were made daily for 14 days and a postmortem examination was performed on all animals which died during the test period. All surviving animals were killed at the end of the observation period and subjected to gross pathological examination. The LD50 values for each compound were calculated using the Probit Analysis (Finney 1952).

Subacute Toxicity

Weanling male and female Sprague-Dawley rats were randomly divided into 13 groups of 10 animals each. The groups were fed powder diets (Purina Ralston) containing 0, 0.5, 5.0, 50 or 500 ppm of each TCB for 28 days. The test compounds were dissolved in corn oil and incorporated into the diets while the control group was given the diet containg 4% corn oil only. The animals were allowed free access to both food and water. Body weight and food consumption were measured weekly and clinical observations made daily. At the termination of the feeding period, all animals were lightly anesthetized with ether and exsanguinated via the abdominal aorta. All animals were examined grossly at the time of necropsy. Serum biochemical determinations were made using a Technicon microanalyzer, model 12/60 and included sodium, potassium, inorganic phosphate, total bilirubin, alkaline phosphatase, glutamic oxalacetic transaminase (GOT), total protein, calcium, cholesterol, glucose, uric acid and lactic dehydrogenase. Sorbitol dehydrogenase (SDH) activity was determined according to an automated method previously described (Yagminas and Villeneuve 1977). The brain, heart, liver, spleen and kidneys were excised and weighed. A sample of liver was taken for the determination of microsomal aniline hydroxylase (AH; Fouts 1963), aminopyrine demethylase (APDM; Cochin and Axelrod 1959) and ethoxyresorufin deethylase activities (ER; Burke and Mayer 1974). Liver porphyrin concentrations were also measured according to the method by Schwartz and Wikoff (1952). Liver

---

* Because of its limited solubility, 1,2,4,5-TCB was given to rats as a suspension in corn oil. The highest dose level that could be incorporated into the vehicle was 3000 $mg \cdot kg^{-1}$.

protein concentrations were determined by the biuret method (Gornall *et al.* 1948). The following tissues were fixed in buffered formalin and examined microscopically: brain, pituitary, liver, adrenal, thyroid, parathyroid, thymus, lungs, trachea, bronchi, thoracic aorta, esophagus, gastric cardia, fundus and pylorus, duodenum, pancreas, colon, kidney, spleen, bone marrow, mesenteric and mediastinal lymph nodes, testes, epididymis, skeletal muscle and heart. Fatty degeneration of liver was evaluated by the method described by Villeneuve *et al.* (1979).

Hematological parameters examined in this study included hemoglobin concentration, packed cell volume, erythrocyte count, total and differential count of leukocytes, mean corpuscular volume, and mean corpuscular hemoglobin concentration.

TCB residues in diets and in the fat and liver of rats were determined using a Hewlett Packard gas chromatograph model 5820A equipped with an electron capture detector. The procedure used for residue analysis was similar to the one described elsewhere (Villeneuve *et al.* 1979).

Subchronic Toxicity

Groups of 15 male and 15 female rats were fed diets containing each of the three TCB isomers at 0.5, 5.0, 50 or 500 ppm for 90 days. Body weight and food consumption were determined weekly and clinical observations made daily. At necropsy, organ weights, serum biochemical profiles, hematological and histopathological examination and residue analysis were carried out as described for the 28-day study.

Data were treated by one-way analysis of variance and when significant differences ($P < 0.05$) were indicated the data were subjected to the Student-Newman-Kuels procedure to identify the groups that differed.

Metabolism and Pharmacokinetic Studies

Groups of five male Sprague-Dawley rats (Charles River, Boston) weighing 300 to 350 g were administered orally single doses of each of three tetrachlorobenzene isomers dissolved in corn oil at 10 mg•kg^{-1} body weight (0.4 μCi•mg^{-1}). The animals were housed individually in metabolism cages to collect urine and feces for the determination of ^{14}C-content. A 1-mL aliquot of urine was dissolved in 15 mL Aquasol (New England Nuclear, Boston) and counted for radioactivity. The feces were dried and 0.2 g of the pulverized feces was dissolved in 1 mL of Soluene-350 (Packard Instruments, Des Moines, Illinois), and counted for radioactivity in Dimulume (Packard Instruments). The animals were killed seven days later and tissues excised and analyzed for ^{14}C-content. To characterize metabolites in the urine, rats were repeatedly administered with the test compounds in order to collect a sufficient quantity of urine for isolation and purification of metabolites.

General Procedures for Isolation, Purification and Identification of TCB Metabolites

Pooled urine (500 to 1000 mL) was acidified with 6 N HCl to pH 4.0 to 4.5, and centrifuged to remove suspended materials. The supernatant was decanted and extracted with three, 250-mL portions of diethyl ether. The suspended materials were similarly extracted. The combined extracts were evaporated to dryness using a rotary evaporator with a bath temperature of 30 to 35°C. The dark brown residue that remained after evaporation of ether was dissolved in chloroform (20 mL) and a few drops of methanol, applied to preparative TLC plates (2000 μ, silica gel GF, Analtech, Newark, Delaware), and developed in ether/petroleum ether b.p. 30 to 60° (1:3). The bands of radioactive materials as detected by the radiochromatography scanner were scraped from the plate and extracted with three, 100-mL portions of chloroform/methanol (3:1). After the solvents were evaporated to dryness the residues were dissolved in methanol, and analyzed by HPLC. The HPLC fractions that contained radioactivity, as determined by liquid scintillation counting, were collected and concentrated. The concentrates were methylated with diazomethane (Diazald, Aldrich Chemicals) and further analyzed by GC and GC/MS.

The urine that remained after extraction at pH 4.0 was acidified further with 6 N HCl to a pH 1.0 and extracted with three, 250-mL portions of diethyl ether. The combined ether extracts were evaporated to dryness and chromatographed on thin layer plates using chloroform/methanol/acetic acid (143:7:2). The extracts were analyzed by HPLC, GC and GC/MS as described above.

Purification of urinary metabolites of 1,2,4,5-TCB was slightly different from the other two isomers. Four liters of pooled urine were acidified with 6 N HCl to pH 1.0 and extracted with four, 400-mL portions of ether. The combined extracts were evaporated to dryness and chromatographed on thin layer plates using petroleum ether/ether (7:3), in a manner similar to those described for the other TCB isomers.

RESULTS

Acute Toxicity

After single oral administration the following, clinical signs common to three TCB isomers were observed: depression, flaccid muscle tone, prostration, piloerection, loose stool, hypothermia, dacryorrhea and coma. Deaths occurred at 48 to 72 hours after dosing. Postmortem examination of the animals that died during the study and on those killed after 14 days revealed distended and slightly hemorrhagic GI-tracts. The LD50 values of these compounds ranged from 1167 to 3105 $mg \cdot kg^{-1}$ for both male and female rats (TABLE 1). In male rats the relative acute toxicity is ranked as follows:

1,2,3,4-TCB > 1,2,3,5-TCB > 1,2,4,5-TCB.

In the female, 1,2,3,4-TCB was slightly more toxic than the 1,2,3,5-isomer. The LD50 of 1,2,4,5-TCB in females could not be determined as the dose levels selected in the experiment were not high enough to cause lethality.

Subacute Toxicity

All animals survived until the termination of the 28-day feeding period. No clinical signs of toxicity were observed. Growth rate and food consumption were not affected by treatment.

TABLE 1: Oral LD50 values of tetrachlorobenzenes in male and female rats

Chemical	LD50 ($mg \cdot kg^{-1}$)	95% Confidence Limits ($mg \cdot kg^{-1}$)	Estimated Slope
Male			
1,2,3,4-TCB	1470	1186 - 1819	8.19
1,2,4,5-TCB	3105	2498 - 3915	9.25
1,2,3,5-TCB	2297	1854 - 2828	11.2
Female			
1,2,3,4-TCB	1167	934 - 1448	8.19
1,2,4,5-TCB	> 2700	-	9.25
1,2,3,5-TCB	1727	1396 - 2143	11.2

A significant increase in liver weight was observed in rats fed 1,2,4,5-TCB at 500 ppm:

	Liver Weight[a]	
	Male	Female
Control	4.5 ± 0.54	4.4 ± 0.30
500 ppm 1,2,4,5-TCB	5.9 ± 0.42	5.3 ± 0.51

a Values represent mean ± SD (in percentage of body weight) obtained from 10 rats.

Of the three TCB isomers only 1,2,4,5-TCB at 500 ppm resulted in elevated serum cholesterol levels in both male and female rats. TABLE 2 shows the hepatic mixed function oxidase activities of rats treated with these chlorinated benzenes. Among the TCB's, 500 ppm of the 1,2,4,5-isomer produced significant increases in aniline hydroxylase (AH) and aminopyrine demethylase (APDM) activities. All three TCB's at 500 ppm induced hepatic APDM activity of male rats, but the 1,2,4,5-isomer was more active than 1,2,3,5-TCB by one order of magnitude. In general, male animals were more sensitive than females to enzyme induction by these compounds. Liver porphyrin concentrations were not affected by treatment. No treatment-related hematological changes were detected in rats fed TCB isomers.

TABLE 2: Hepatic mixed function oxidase activities of rats fed tetrachlorobenzenes for 28 days[a]

Treatment	AH $\frac{\text{nmole PAP}}{\text{hr}\cdot\text{mg protein}}$	APDM $\frac{\text{nmole HCHO}}{\text{hr}\cdot\text{mg protein}}$
Male		
Control	12 ± 5.1(10)	20 ± 7.1(10)
1,2,4,5-TCB (ppm)		
50	19 ± 4.0(10)	35 ± 7.6(10)[b]
500	33 ± 6.5(10)[b]	64 ± 18(10)[b]
1,2,3,5-TCB		
500	18 ± 5.4(9)	35 ± 9.0(9)[b]
Female		
Control	20 ± 6.9(9)	16 ± 3.0(0)
1,2,4,5-TCB (ppm)		
50	23 ± 7.2(10)	17 ± 2.7(10)
500	34 ± 1.5(9)	36 ± 8.8(9)[b]
1,2,3,5-TCB		
500	27 ± 6.8(8)	24 ± 4.7(8)

a Results represent mean ± SD obtained from the number of animals indicated in parentheses.

b Significant difference from the control at $P \leq 0.05$ level.

Histological changes were found in the liver, thyroid, kidney and lung of rats fed chlorinated benzenes. The prevalence of these changes is dose-dependent and is shown in TABLE 3. The severity of the changes in the liver was also dose-dependent and consisted of cytoplasmic vacuolation of hepatocytes and anisokaryosis accompanied by redution in aggregated basophilia in the perivenous area. These changes were qualitatively similar for all three chlorinated benzenes tested. Severity was more pronounced with 1,2,4,5-TCB than with 1,2,3,4- and 1,2,3,5-TCB. The changes caused by 1,2,4,5-TCB were considered moderate to severe at the highest dose level, whereas with the other two isomers the changes were mild. Histological changes associated with the thyroid consisted of an increase in epithelial height and angular collapse of thyroid follicles with a reduction in colloid density. As with the liver, the prevalence of lesions in the

TABLE 3: Prevalence of lesions in the liver, thyroid, kidney and lungs of rats fed chlorinated benzenes for 28 days[a]

Treatment	Liver	Thyroid	Kidney	Lungs[b]
Male				
Control	0/10	0/10	0/10	0/10
1,2,3,4-TCB (ppm)				
0.5	0/10	3/10	1/10	3/10
5.0	2/10	4/10	1/10	4/10
50	1/10	4/10	1/10	4/10
500	0/10	5/10	1/10	4/10
1,2,4,5-TCB				
0.5	2/10	5/10	0/10	4/10
5.0	0/10	3/10	0/10	5/10
50	4/10	4/10	5/10	2/10
500	10/10	9/10	8/10	6/10
1,2,3,5-TCB				
0.5	1/10	4/10	1/10	2/10
5.0	2/10	2/10	1/10	2/10
50	1/10	3/10	2/10	6/10
500	3/10	5/10	3/10	4/10
Female				
Control	0/10	0/10	4/10	3/10
1,2,3,4-TCB (ppm)				
0.5	0/10	2/10	3/10	-
5.0	2/10	1/10	2/10	-
50	1/10	3/10	1/10	-
500	7/10	3/10	1/10	0/10
1,2,4,5-TCB				
0.5	1/10	2/10	1/10	-
5.0	1/10	3/10	2/10	-
50	3/10	4/10	3/10	-
500	10/10	6/10	3/10	2/10
1,2,3,5-TCB				
0.5	0/10	2/10	2/10	-
5.0	2/10	3/10	3/10	-
50	5/10	3/10	2/10	-
500	5/10	2/10	3/10	2/10

a Figures denote animals showing lesions/animals examined.

b Only control animals and those on the highest dose groups were examined.

thyroid was dose-dependent. Comparing the relative effect of these compounds on the thyroid, it was apparent that 1,2,4,5-TCB was the more active than the other two isomers. Even at the highest dose level of these chemicals the changes were considered mild. Changes elicited by the administration of 1,2,3,4- and 1,2,3,5-TCB were very mild in nature. Morphological changes in the kidney were characterized by the presence of eosinophilic inclusions in the proximal convoluted tubule of the renal cortex. As the dose increased, the inclusions became more prominent and bulged into the tubular lumina. Mild treatment-related changes occured in the lung which consisted of focal alveolar emphysema and a variable amount of alveolar mixed inflammatory cell reaction. In general, histological changes in the target organs were less severe in the female rat.

Subchronic Toxicity

Growth rate and food consumption were not affected by treatment. Four animals (two male and two female) died spontaneously. The cause of death could not be determined but did not appear to be treatment-related. In general, the biochemical, hematological and histological changes observed in the subchronic experiment were similar to those described for the subacute study. 1,2,4,5-TCB was the only isomer producing enlarged liver and kidney, elevated serum cholesterol levels, and increased hepatic microsomal AH and APDM activities (TABLE 4).

Histopathological changes in the liver, kidney and thyroid were similar to those observed in the subacute study except that the renal injury was more severe in nature for the animals fed 500 ppm 1,2,4,5-TCB. The changes consisted of swollen kidneys, extensive epithelial necrosis and formation of large intratubular cellular casts throughout the cortex. As with the 28-day study, histological changes were less severe in the female rat.

Analysis of fat and liver for TCB residues showed that 1,2,4,5-TCB accumulated in fat and liver in a dose-dependent manner (TABLE 5). Very little 1,2,3,4- and 1,2,3,5-isomers were found in these tissues. 1,2,4,5-TCB reached its steady-state concentration in these tissues in the 28-day exposure because prolonged exposure, such as in the 90-day subchronic study, did not result in any significantly higher concentrations of the TCB levels.

TABLE 4: Organ weights and biochemical changes of rats fed 1,2,4,5-tetrachlorobenzene for 90 days[a]

Treatment	Liver Weight (g)	Kidney Weight (g)	Serum Cholesterol (mg/100 mL)	Hepatic AH nmole PAP / hr•mg protein	Hepatic APSM nmole HCHO / hr•mg protein
Male					
Control	17±3.1	1.6±2.2	75±13	27±5.5	44±7.3
1,2,4,5-TCB					
50 ppm	-	-	80±17	40±7.8[b]	55±14[b]
500 ppm	20±2.0[b]	2.3±0.45[b]	96±12[b]	72±11[b]	98±15[b]
Female					
Control	10.2±1.2	0.99±0.18	62±20	19±5.5	22±2.9
1,2,4,5-TCB					
50 ppm	-	-	65±15	24±6.8	25±2.9
500 ppm	12.8±1.4[b]	1.07±0.11[b]	88±18[b]	48±24[b]	54±2.3[b]

a Results denote mean ± SD obtained from groups of more than 11 animals.

b Significantly different from the control group at $P \leq 0.05$ level.

TABLE 5: Tetrachlorobenzene residues in rat tissues[a]

Treatment	Liver	Fat
Male		
Control	<0.1	1.4 ± 0.29
1,2,3,4-TCB (ppm)		
0.5	<0.1	1.1 ± 0.28
5.0	<0.1	1.1 ± 0.28
50	0.16 ± 0.14	1.1 ± 0.31
500	0.71 ± 0.17	9.6 ± 1.54
	(1.1 ± 0.40)[b]	(15 ± 14)[b]

a Results denote mean ± SD obtained from groups of five animals.

b Values in parentheses are 28-day exposure data. They are included here for the purpose of comparison.

TABLE 5: (Cont'd)

Treatment	Liver	Fat
Male (Cont'd)		
1,2,4,5-TCB		
0.5	0.25 ± 0.03	2.5 ± 0.4
5.0	1.7 ± 0.38	19 ± 1.1
50	25 ± 13	669 ± 720
500	113 ± 19	1540 ± 728
	(63 ± 13)[b]	(1590 ± 818)[b]
1,2,3,5-TCB		
0.5	1.7 ± 3.0	15 ± 28
5.0	0.19 ± 0.01	2.6 ± 0.13
50	1.5 ± 0.32	19 ± 4.94
500	11 ± 0.93	203 ± 23
	(6.5 ± 5.8)[b]	(16 ± 16)[b]
Female		
Control	<0.1	<0.1
1,2,3,4-TCB (ppm)		
0.5	<0.1	<0.1
5.0	<0.1	0.1 ± 0.01
50	<0.1 ± 0.1	0.63 ± 0.13
500	0.26 ± 0.1	6.0 ± 0.46
	(0.61 ± 0.15)[b]	(15 ± 4.9)
1,2,4,5-TCB		
0.5	0.12 ± 0.06	2.8 ± 0.56
5.0	0.77 ± 0.15	18 ± 1.4
50	9.0 ± 2.3	205 ± 29
500	79 ± 9.0	1890 ± 178
	(68 ± 0.92)[b]	(1641 ± 193)[b]
1,2,3,5-TCB		
0.5	0.1 ± 0.02	1.5 ± 0.3
5.0	0.1 ± 0.02	1.4 ± 0.16
50	0.46 ± 0.16	12 ± 2.3
500	5.3 ± 1.2	135 ± 15
	(2.9 ± 0.55)[b]	(89 ± 48)[b]

a Results denote mean ± SD obtained from groups of five animals.

b Values in parentheses are 28-day exposure data. They are included here for the purpose of comparison.

Metabolism and Pharmacokinetic Studies

Tissue distribution data using ^{14}C-labelled compounds are shown in FIGURE 1. Amongst the three TCB isomers, 1,2,4,5-TCB showed the highest concentrations of ^{14}C in all tissues examined. Little residual radioactivity due to 1,2,3,4- and 1,2,3,5-TCB was found. The levels of 1,2,4,5-TCB in the fat and liver were two orders of magnitude higher than those derived from the other two isomers (FIGURE 1).

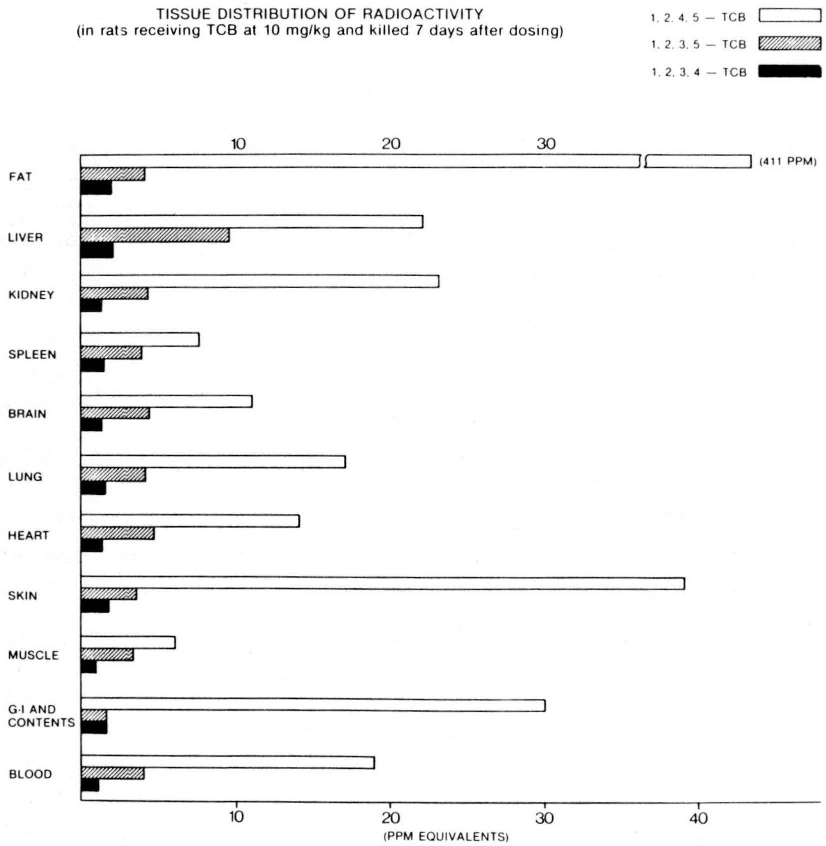

FIGURE 1: Tissue distribution of radioactivity

Consistent with the tissue distribution data are the excretion profiles of TCB isomers (TABLE 6). 1,2,3,4- and 1,2,3,5-TCB were readily excreted and approximately 46% to 51% of the dosed compound were disposed of within 48 hrs after dosing. Elimination after this time was slow with only small quantities of radioactivity detected in days two to seven excreta. In contrast, only 8% of the administered 1,2,4,5-isomer was excreted in 48 hours, however, the excretion process continued at a steady rate and by day seven a total of 21% was eliminated (TABLE 6).

TABLE 6: Cumulative excretion of radioactivity[a]

Compound	Urine		Feces	
	48 hours	7 days	48 hours	7 days
Tetrachlorobenzene				
1,2,3,4-	15 ± 6.9	16 ± 7.1	36 ± 9.6	39 ± 109
1,2,3,5-	25 ± 3.1	27 ± 3.3	21 ± 2.4	24 ± 0.82
1,2,4,5-	6.3 ± 0.95	17 ± 3.1	1.8 ± 0.23	4.2 ± 0.42

a The rats received single oral doses of ^{14}C-Tetrachlorobenzene at 10 mg•kg^{-1}. Results represent mean ± SD obtained from groups of five rats.

Analysis of urine indicated that the tetrachlorobenzenes were biotransformed to a number of polar compounds. The metabolites for each of the three TCB's, in decreasing order of quantitites, were as follows: 1,2,3,4-TCB: 2,3,4,5- and 2,3,4,6-tetrachlorophenol and traces of tetrachlorothiophenol and 2,3,4-trichlorophenol; 1,2,3,5-TCB: 2,3,4,6-tetrachlorophenol, isomeric mercaptotrichlorophenols and trichlorophenol; 1,2,4,5-TCB: 2,3,5,6-tetrachlorophenol, tetrachloroquinol and trichlorophenol (SCHEME 1).

DISCUSSION

While acute toxicity data revealed that 1,2,3,4- and 1,2,4,5-TCB's were the most and least toxic compounds, respectively, (TABLE 1), this was not the case with the subacute studies. In the subacute and subchronic studies the toxic effects of these organochlorines were directly related to their tissue accumulation. Of the three TCB isomers, 1,2,4,5-TCB was found to possess the highest concentrations in fat and liver. In conjunction with these findings, only this isomer at 500 ppm dietary level caused hepatomegaly, hepatic microsomal enzyme induction and serum biochemical changes.

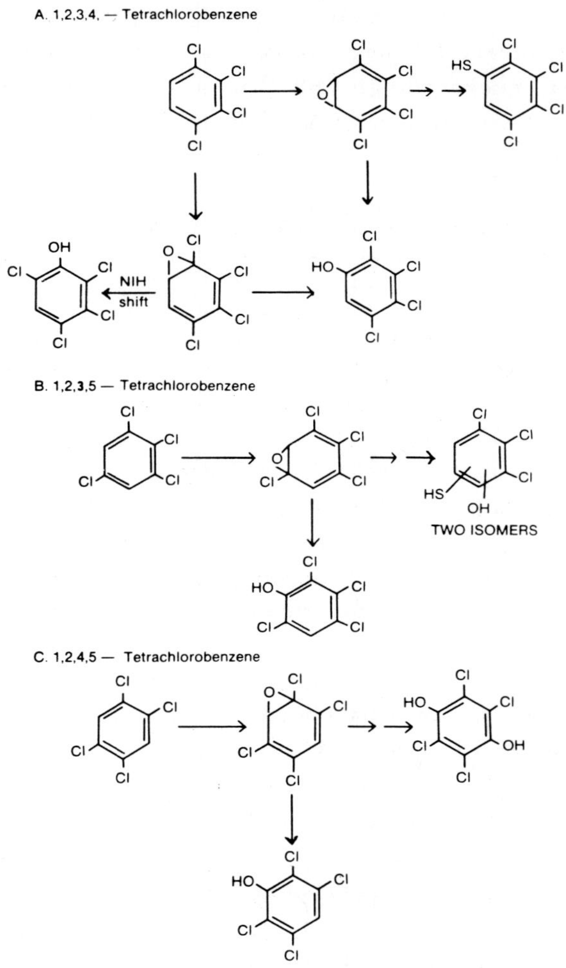

SCHEME 1: Metabolism of tetrachlorobenzene

All three TCB's were determined to be P-450 type liver microsomal enzyme inducers because they induced hepatic APDM but not ER. Of the three, 1,2,4,5-TCB was the most potent APDM inducer.

Histologically, the liver was the prime target organ in this study. Of the compounds tested, 1,2,4,5-TCB produced the greatest change followed by 1,2,3,4- and 1,2,3,5-TCB. With these latter two chemicals the changes were very mild. The types of changes were characteristic of organohalogen exposure in general and consisted of cytoplasmic ballooning and anisokaryosis of hepatocytes. Changes in the thyroid were considered very mild in all treatment groups. Changes in the kidney were only considered significant in males fed 50 or 500 ppm 1,2,4,5-TCB and these renal injuries were more severe with prolonged exposure.

1,2,3,4-TCB was metabolized to 2,3,4,5- and 2,3,4,6-tetrachlorophenol, 2,3,4-trichlorophenol and a tetrachlorothiophenol (SCHEME 1). Based on these metabolites two arene oxide intermediates may be involved: 2,3,4,5- and 3,4,5,6-tetrachlorobenzene-oxide (TCBO). The formation of 2,3,4,6-tetrachlorophenol suggests that 2,3,4,5-TCBO is the only intermediate which rearranges to yield the product through the NIH shift (Jerina and Daly 1974). 2,3,4,5-tetrachlorophenol, on the other hand, may be formed by a spontaneous rearrangement of 2,3,4,5- and/or 3,4,5,6-TCBO. An isotope labelling technique would be required in order to prove whether or not the NIH shift is involved. Tetrachlorothiophenol could be formed from either 3,4,5,6- and/or 2,3,4,5-TCBO, the intermediate reacted with endogenous sulfhydryl compound, followed by aromatization. Hydrolysis of the side chain resulted in the thiophenol. The formation of 2,3,4-trichlorophenol probably represents an intestinal bacterial action rather than an enzymatic metabolism.

The metabolism of 1,2,3,4- and 1,2,3,5-TCB may involve the epoxide intermediate, however, there is not sufficient evidence to indicate the presence of epoxide intermediates (SCHEME 1). The three TCB isomers have similar octanol/water partition coefficients and therefore the lipid solubility could not account for the differences in toxicity observed. It was found that the 1,2,4,5-TCB was more resistant to metabolism such as the formation of tetrachlorobenzene-epoxide, and perhaps this feature of TCB made it more toxic than the other isomers. In a subsequent study with three tetrachlorophenols, we noted that 2,3,5,6-tetrachlorophenol was the most toxic isomer (Shen et al. 1983). This isomer was the metabolite of 1,2,4,5-TCB.

ACKNOWLEDGEMENT

The authors wish to thank Drs. L. Whitehouse and E. Nestmann for useful discussion, Dr. F.M. Benoit for interpretation of GC/MS,

N. Beament, A. Yagminas, B. Reed, V. Secours and A. Viau for technical assistance and J. Ireland for typing the manuscript.

REFERENCES

Ariyoshi, T., Ideguchi, K., Ishizuka, Y., Iwasaki, K. and Arakaki, M. 1975a. Relationship between chemical structure and activity I. Effects of number of chlorinated atoms in chlorinated benzenes on the components of drug-metabolizing system and the hepatic constituents. **Chem. Pharm. Bull. 23:** 817-823.

Ariyoshi, T., Ideguchi, K., Iwasaki, K. and Arakaki, M. 1975b. Relationship between chemical structure and activity II. Influences of isomers in dichlorobenzene, trichlorobenzene and tetrachlorobenzene on the activities of drug-metabolizing enzymes. **Chem. Pharm. Bull. 23:** 834-830.

Braun, W.H., Sung, L.Y., Keyes, D.G. and Kociba, R.J. 1978. Pharmacokinetic and toxicological evaluation of dogs fed 1,2,4,5-tetrachlorobenzene in the diet for 2 years. **J. Toxicol. Environ. Health 4:** 727-734.

Burke, M.D. and Mayer, R.T. 1974. Ethoxyresorufin: Direct fluorometric assay of microsomal O-delakylation which is preferentially inducible by 3-methylcholanthrene. **Drug Metab. Dispos. 2:** 583-588.

Cochin, J. and Axelrod, J. 1959. Biochemical and pharmacological changes in the rat following chronic administration of morphine, nalorphine and normophine. **J. Pharmacol. Exp. Therap. 125:** 105-110.

Finney, D.J. 1952. **Probit Analysis.** 2nd ed., Cambridge University Press, Cambridge.

Fomenko, V.N. 1965. Determination of the maximum permissible concentration of tetrachlorobenzene in water basins. **Hyg. Sanit. (U.S.S.R.) 30:** 8-15.

Fouts, J.R. 1963. Fators influencing the metabolism of drugs in liver microsomes. **Ann. N.Y. Acad. Sci. 104:** 875-880.

Gornall, A.G., Bardawill, C.J. and David, M.M. 1948. Determination of serum proteins by means of biuret reaction. **J. Biol. Chem. 177:** 751-755.

Great Lakes Water Quality Board. 1978. Great Lakes Water Quality. Status report on organic and heavy metal contaminants in the Lakes Erie, Michigan, Huron and Superior Basins. **International Joint Commission,** Windsor, Ontario, pp. 373.

Jerina, D.M. and Daly, J.W. 1974. Arene oxides: A new aspect of drug metabolism. **Science 185:** 573-582.

Lunde, G. and Ofstad, E.B. 1976. Determination of fat soluble chlorinated compounds in fish. **J. Anal. Chem. 282:** 395-399.

Schwartz, S. and Wikoff, H.H. 1952. The relation of erythrocyte coproporphyrin and protoporphyrin to erythropoiesis. **J. Biol. Chem. 194:** 563-573.

Shen, S.Y., Villeneuve, D.C. Chu, I., Kelly, J. and Gilman, A.P. 1983. The acute dermal toxicity of tetrachlorophenols in the rat. **Bull. Environ. Contam. Toxicol.,** in press.

Villeneuve, D.C., Valli, V.E. Chu, I., Secours, V.E., Ritter, L. and Becking, G.C. 1979. Ninety-day toxicity of photomirex in the male rat. **Toxicology 12:** 235-250.

Yagminas, A.P. and Villeneuve, D.C. 1977. An automated continuous flow assay for serum sorbitol dehydrogenase activity and its use in experimental liver damage. **Biochem. Med. 18:** 117-125.

WORKSHOP DISCUSSION

OLIVER:

Why do you think the toxicity of the 1,2,3,4 is higher than that of the 1,2,4,5 isomer?

CHU:

I think in the acute studies we are looking at a central nervous system effect. What I mean is that you get a very high concentration in the blood stream and it reaches the nervous system to exert its toxic effect. In the sub-acute tests you are dealing with the concentration of the chemical in the organs such as the liver, kidney or thyroid.

OLIVER:

Then you say that chronic testing is much more useful than acute toxicity testing?

CHU:

We measure the different endpoints in the acute and chronic studies.

KWASNIEWSKA:

To obtain reliable and fast results on acute and chronic toxicities, it is advantageous to use yeast strains as they are also eukaryots with typical mammalian secretion system and mitochondria and yet you can get results in about 24 hours.

CHU:

Then the question arises as to the general public concern, namely, is this fish safe to eat? Then we have to answer what are the levels of the toxic chemicals in fish and what are the safe levels.

KAISER:

Have you tried to compare the chlorobenzene results with those of other compounds? I am sure you have done some work on other chlorinated compounds.

CHU:

Yes, we have completed studies on dichlorobenzenes, trichlorobenzenes and we have also used hexachlorobenzene as standards to compare it with these results. For instance you will notice that 1,2,4,5-tetrachlorobenzene, in fact, is slightly more toxic than pentachlorobenzene which has a higher number of chlorine atoms and frequently higher toxicity which is not the case here.

Question:

The rest of the chlorobenzenes, did that go down in toxicity again?

CHU:

Dichlorobenzenes are less toxic than the higher chlorinated benzene congeners.

KAISER:

Did you try to do any quantitative correlations?

CHU:

No, not really. It is very difficult. In order to do this you have to plot a dose/response curve, you have to do a separate dose level and again, you have to see if the dose/response are parallel curves before you can consider comparisons, which is very difficult here.

ZITKO:

After your presentation, I see we have a long way to go in aquatic toxicology.

CHU:

We need the information from aquatic toxicology in order to know as to where we stand.

ZITKO:

I know. I only wish that we could have all the data the way you have it for mammalian toxicology.

AN EXAMINATION OF THE ROLE OF ROTATIONAL BARRIERS IN THE TOXICOLOGY OF PCB'S

John M. Cullen[1] and Klaus L.E. Kaiser

Environmental Contaminants Division
National Water Research Institute
PO Box 5050, Burlington, Ontario, L7R 4A6
CANADA

ABSTRACT

The internal rotational barriers of PCB isomers were computed using the quantum mechanical INDO method and corrected for the known deficiencies of this semi-empirical model. Based on the hypothesis that the active molecule must be planar, a correlation coefficient (r^2) equal to 0.98 was found between the arylhydrocarbon hydroxylase activation data and the fraction of planar conformers predicted from a Boltzmann like distribution.

INTRODUCTION

Polychlorinated biphenyls (PCB's) are the most widespread class of halogenated aromatic pollutants found in the environment (TFPCB 1976). This is a result of their past ubiquitous use in industry, their considerable resistance to chemical, thermal and photo-chemical degradation and their high lipophilicity. Those isomers which prove to be the most toxic are found to be structurally

[1] Present Address: Department of Chemistry, Harvard University, Cambridge, Massachusetts 02138, U.S.A.

K. L. E. Kaiser (ed.), QSAR in Environmental Toxicology, 39–66.

similar to the highly toxic 2,3,7,8-tetrachlorodibenzo-p-dioxin (TCDD). For example, the 3,4,3',4'-tetrachlorobiphenyl when placed in the planar trans conformation has chlorines which almost directly superimpose onto the TCDD chlorines (FIGURE 1). Although PCB's, unlike dioxins, are inherently nonplanar with opposing aromatic rings twisted to each other, the meta- and para-substituted compounds have sufficiently low rotational barriers to enable a small fraction to assume a planar conformation. Work by Goldstein *et al.* (1977), Poland and Glover (1973; 1977) have identified the 3,4,3',4'-tetrachloro-, 3,4,5,3',4'-pentachloro- and the 3,4,5,3',4',5'-hexachlorobiphenyls to act toxicologically in the same manner as TCDD. It was further postulated that substitutions of chlorines into ortho positions would abolish all TCDD-like activity due to steric hindrance encountered when these compounds were rotated into planar conformations. Any activity observed was attributed to trace contamination, as for example, the 2,3,7,8-tetrachlorodibenzofuran (TCDF) found in the Ullman preparation of the 2,4,5,2',4',5-hexachlorobiphenyl (Goldstein *et al.* 1978; Goldstein 1979).

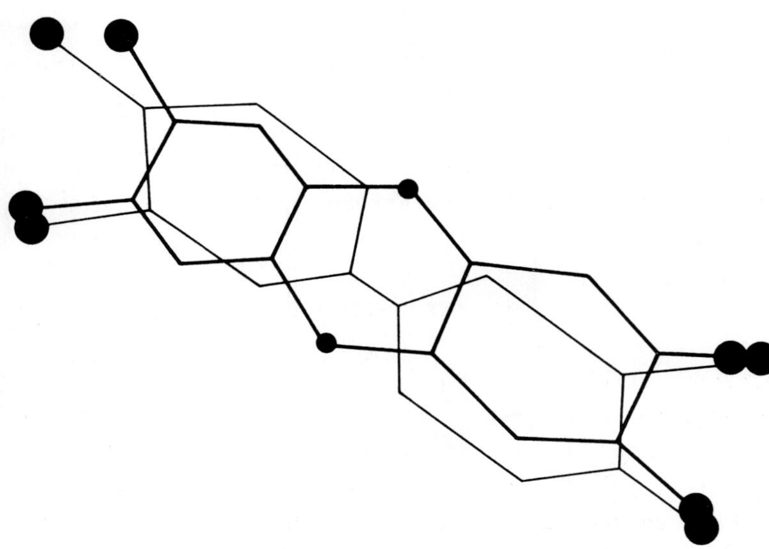

FIGURE 1: Superposition of the 2,3,7,8-tetrachlorodibenzo-p-dioxin and trans-3,4,3,4'-tetrachlorobiphenyl molecular skeletons

However careful studies by Parkinson et al. (1980a; 1980b; 1981) have now revealed that not only do some mono-ortho substituted PCB's behave in part like TCDD but also some of the di-ortho substituted isomers as well. This raises the question as to whether a compound needs to be planar in order to be active in the same manner as TCDD. In continuation of earlier work on this subject (Kaiser 1974), a semi-empirical quantum mechanical study of the internal rotational barriers of the polychlorinated biphenyls has therefore been carried out to determine quantitatively the degree of steric hindrance encountered in these compounds.

QUANTUM CHEMICAL THEORY

In principle, all chemical and physical properties of a molecule can be found from the solution of its Schrödinger equation:

$$H\Psi = E\Psi \tag{1}$$

where H is the Hamiltonian or energy operator of the system and Ψ the corresponding wave or state function. Since the nuclei move much slower than the electrons, a common practice (Born and Oppenheimer 1927) is to approximate Ψ as a product of a fast varying electronic function, Ψ_e, and a slow varying nuclear function, Ψ_N.

$$\Psi \simeq \psi_e \cdot \Psi_N \tag{2}$$

This results in the decomposition of the Schrödinger equation into an electronic component:

$$H_e(\vec{R},\vec{r})\ \Psi_e = E_{ELECT}(\vec{R})\Psi_e \tag{3a}$$

which is solved for fixed positions, $\vec{R}$ of the nuclei, and a nuclear component.

$$[H_N(\vec{R}) + E_{ELECT}(\vec{R})]\ \Psi_N = E\Psi_N \tag{3b}$$

Exact solutions to equation 3a are generally not possible and further approximations must be made. In the intermediate neglect of diatomic overlap (INDO) method (Pople et al. 1967; 1970) the exact operator H_e is replaced by a much simpler model Hamiltonian

$$H_e \simeq H_{INDO} \tag{4}$$

H_{INDO} ignores many of the smaller interactions occurring in the system incorporating their effects into atomic parameters found or fitted from experiment. Furthermore, if one is only interested in the potential energy surface on which the nuclei lie and not in the corresponding vibrational or rotational states that result, then all nuclear motion may also be neglected. The total energy then becomes simply the sum of the electronic energy and the electrostatic repulsions of the positively charged nuclei.

$$E_{TOTAL}(\vec{R}) = E_{ELECT}(\vec{R}) + E_{REPUL}(\vec{R}) \qquad (5)$$

Approximate INDO solutions to equation 5 yielding useful geometry predictions can be economically computed at the Hartree Fock molecular orbital level.

In the case of a polychlorinated biphenyl, the electronic energy, E_{ELECT}, is composed of an internal energy contribution, E_{INT}, from each ring, a nonbonded component, E_{NB}, from inter-ring steric interactions and finally a conjugation energy term, E_{π}, due to resonance stabilization over the central carbon-carbon bond. The total energy is then given by

$$E_{TOTAL}(\vec{R}) = E_{INT}(\vec{R}) + E_{NB}(\vec{R}) + E_{\pi}(\vec{R}) + E_{REPUL}(\vec{R}) \qquad (6)$$

An informative first approximation is to consider the biphenyl as a rigid molecular skeleton with the only degree of freedom being the torsional angle θ between the rings. The internal energy, E_{INT}, which is not a function of θ but remains constant can then be eliminated by redefining the energy relative to the planar conformation ($\theta = 0°$):

$$E_{TOTAL}(\theta) = E_{NB}(\theta) + E_{\pi}(\theta) + E_{REPUL}(\theta) \qquad (7a)$$

with

$$E_{TOTAL}(0°) = 0 \qquad (7b)$$

The resulting INDO calculated energy (Birner and Hofmann 1982) as a function of Θ is depicted in FIGURE 2. A monotonic decreasing curve with a minimum at 90° and barrier height of 5.0 $kcal{\cdot}mol^{-1}$ is obtained. In contrast, more accurate ab-initio calculations by Almlöf (1974) predict a barrier height of 1.2 $kcal{\cdot}mol^{-1}$ with a minima at an angle $\theta = 32°$. This is in reasonable agreement with experimental NMR studies by Casalone _et al._ (1968) who found a barrier height of 2.5 $kcal{\cdot}mol^{-1}$ at an equilibrium angle of 35°. The erroneous predictions of the INDO method have also been observed by Gropen and Seip (1971) in CNDO/2 calculations. An analysis by Tyrrell _et al._ (1981) has attributed the fault of these methods to the spherically symmetric integral approximations made on the p orbitals. However, INDO calculations do yield useful results for sigma bonded systems. For example, the rotational barrier in ethane is predicted to be 2.4 $kcal \cdot mol^{-1}$ compared with the experimental 2.9 $kcal \cdot mol^{-1}$ value (Weiss and Leroi 1968). It might therefore be expected that errors resulting from the use of the INDO method for biphenyl calculations will occur mostly in the pi conjugation energy term, E_π. This is verified when comparisons are made to previous studies by Fischer-Hjalmars (1963).

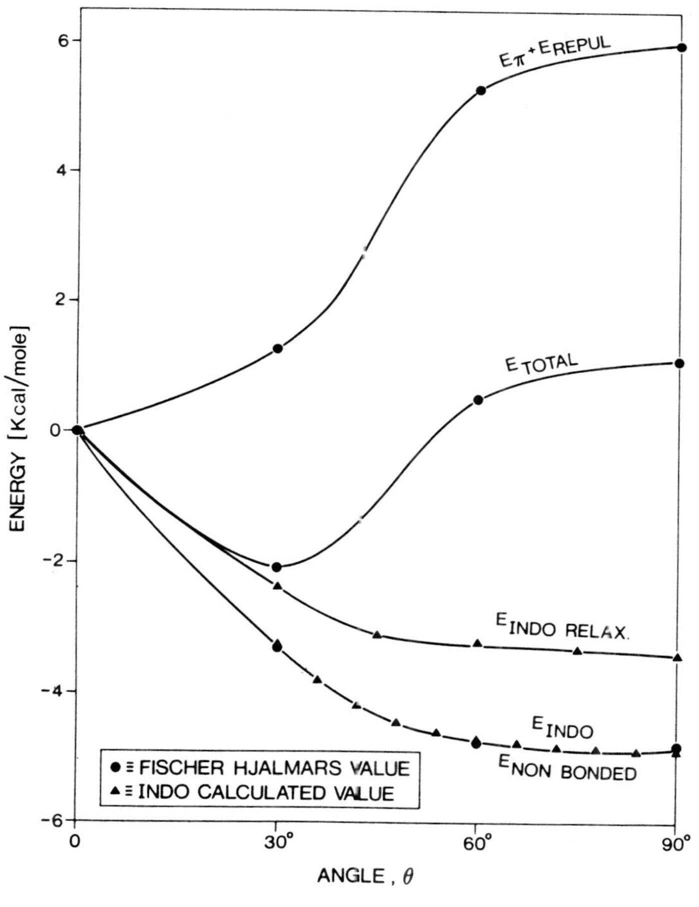

FIGURE 2: Molecular energies of the biphenyl skeleton versus the rotational angle θ

In that work the rotational barrier of biphenyl was computed with nonbonded interactions estimated from empirical atom-atom Lennard Jones or Buckingham potentials (Bartell 1960) and pi conjugation energy determined quantum mechanically (Pariser and Parr 1953; Pariser and Parr 1953; Pople 1953). As in the INDO case, all calculations presumed a rigid molecular skeleton with the only degree of freedom being the torsional angle θ. Calculated results from the nonbonded energy contributions are found to a good approximation to lie directly on the computed INDO energy curve (FIGURE 2).

$$E_{TOTAL}^{INDO}(\theta) = E_{NB}(\theta) \tag{8}$$

In comparison, the INDO results are well fitted by the equation

$$E_{TOTAL}^{INDO}(\theta) = -5.0\,(\sin\theta)^{0.52428} \tag{9a}$$

$$(r^2 = 0.95)$$

while the nonbonded energy results of Fischer-Hjalmars obey a similar approximate square root sinθ dependence:

$$E_{NB}(\theta) = -4.8\,(\sin\theta)^{0.52695} \tag{9b}$$

From equations 8 and 7 it is seen that the INDO model predicts the pi conjugation energy should cancel against the electrostatic nuclear repulsions:

$$E_{\pi}^{INDO} + E_{REPUL}^{INDO} \simeq 0 \tag{10}$$

However, the Fischer-Hjalmars work shows (FIGURE 2) that both E_{π} and E_{REPUL} have an angular dependence approximately proportional to the square of sin θ.

$$E_{\pi} + E_{REPUL} \simeq 6.22\,(\sin\theta)^{2.034} \tag{11}$$

Addition of the nonbonding energy contribution (equation 9b) yields a final total energy curve in reasonable agreement with experiment. Hence the INDO method underestimates the pi conjugation term, E_{π}. Furthermore, the nonbonded energy, E_{NB}, is approximately a slowly varying $(\sin\theta)^{\frac{1}{2}}$ function while in contrast E_{π} and E_{REPUL} change rapidly having a $(\sin\theta)^2$ dependence. This suggests that the

INDO results can be corrected by simply adding a correction factor, C with a $(\sin\theta)^2$ dependence

$$E_{TOTAL}(\theta) = E^{INDO}_{TOTAL}(\theta) + C(\sin\theta)^2 \tag{12}$$

Removal of the constraint of a rigid molecular skeleton and allowing all atoms to relax to the conformation of lowest energy for fixed angle θ does not significantly alter the computed INDO energy curve (FIGURE 2). The curve does however become shallower due to the strong repulsive nonbonded interactions which occur when the biphenyl is twisted into a planar conformation being reduced in favour of distortions to the molecular skeleton. Since there is no *a priori* reason now to expect a total cancelation of the pi conjugation term, E_π against nuclear repulsions, the angular dependence of the computed INDO energy is given by the general expression:

$$E^{INDO}_{TOTAL}(\theta) \simeq A(\sin\theta)^{\frac{1}{2}} + B(\sin\theta)^2 \tag{13a}$$

where:

$$A = E^{INDO}_{NB}(\theta) + \Delta E^{INDO}_{A} \tag{13b}$$

$$B = E^{INDO}_{\pi}(\theta) + E^{INDO}_{REPUL} + \Delta E^{INDO}_{B} \tag{13c}$$

Changes to the internal energy, E_{INT} are assumed to be small and approximated by the fit

$$E_{INT}(\theta) \simeq \Delta E_A(\sin\theta)^{\frac{1}{2}} + \Delta E_B(\sin\theta)^2 \tag{13d}$$

Combining equations 12 and 13a yields the final result:

$$E_{TOTAL}(\theta) = A(\sin\theta)^{\frac{1}{2}} + B(\sin\theta)^2 + C(\sin\theta)^2 \tag{14}$$

The constants A,B are therefore first fitted to the computed INDO energy curve and the remaining factor C is then used to correct the underestimation of E_π. A value for C of 1.766 $kcal \cdot mol^{-1}$ was found to give good agreement with the experimental work of Casalone *et al.* (1968) and the theoretical results of Almlöf (1974).

Fortunately there is strong evidence that E_π remains relatively constant for the substituted biphenyls. This can be seen developed by Field and coworkers (Field and Sternhell 1981; Bott *et al.* 1980; Field *et al.* 1977) from their dynamic NMR data to predict free energies and torsional angles. Also theoretically, the calculated bond orders over the central pair of carbon p orbitals for substituted biphenyls in a planar conformation should give a direct measure of the pi conjugation term E_π (Allinger and Sprague 1973). The INDO computed bond orders for over 110 PCB isomers show a change of less than three tenths of a percent. Therefore, the value found for the correction factor C in the biphenyl case may also be used along with equation 14 to estimate rotational barriers of other substituted biphenyl isomers.

ROTATIONAL BARRIER RESULTS AND DISCUSSION

INDO calculations were carried out for over 110 PCB isomers for both cis ($\theta = 0°$) and trans ($\theta = 180°$) planar geometries as well as perpendicular ($\theta = 90°$) conformations. The nuclei were allowed to relax in these calculations to the molecular geometry of lowest energy for fixed angle θ. In order to estimate the parameters A,B of equation 14 computations were also performed at the angles $\theta = 45°$, 135° for small representative sets of isomers containing structurally the same ortho substituents. B constants (equation 13c) which have an E_π dependence were then solved from this information. A substantial amount of computing time was thus saved by performing the 45° and 135° calculations for only a small number of the PCB's. Moreover the error introduced by using an average value of B for isomers which are structurally identical in their ortho positions was generally less than a few percent for mean values of B ranging from 1 to 3 $kcal{\cdot}mol^{-1}$. The A constants (equation 13b) which measure the nonbonded interactions ranged in value from 4 to 60 $kcal{\cdot}mol^{-1}$ and were computed on an individual compound by compound basis. The final results are presented in TABLES 1 through 8.

No Ortho Substituents

In the case of zero orthochloro substitutions (TABLE 1) an average energy barrier to the planar conformation of 1.94 $kcal{\cdot}mol^{-1}$ with a root mean square error of 0.4 $kcal{\cdot}mol^{-1}$ was predicted. This agrees well with previous estimates (Almlöf 1974; Casalone et al. 1962) of 1.2 to 2.5 kcal•mol^{-1} for the barrier height. The corresponding minima at an angle of approximately θ = 28½° lies slightly below the 32° to 35° found by other groups (Almlöf 1974; Gropen and Seip 1971; Casalone et al. 1968). The alternate barrier to the perpendicular conformation calculated at 1.0 kcal•mol^{-1} however does not agree with the established 3 to 4 kcal•mol^{-1} estimate by Almlöf (1974) and Casalone et al. (1968). Fortunately the property of biological interest is the more accurately predicted rotational barrier to the planar conformation.

TABLE 1: Computed energies (E) and angles of lowest energy conformation of various chlorobiphenyl congeners without orthochloro substituents

Ring 1	Ring 2	E(INDO)	E(0.0 Deg.)	E(90.0 Deg.)	Angle
0	0	3.346	2.402	0.822	32.329
3	0	2.414	1.746	1.098	27.125
4	0	2.223	1.618	1.161	26.035
34	0	2.032	1.493	1.227	24.933
35	0	2.749	1.976	0.993	29.010
3	3	2.557	1.844	1.052	27.936
3	4	2.486	1.795	1.075	27.531
4	4	2.971	2.132	0.927	30.251
345	0	1.912	1.416	1.270	24.238
34	3	1.936	1.431	1.261	24.378
35	3	2.940	2.110	0.936	30.078
34	4	2.008	1.477	1.236	24.795
35	4	2.916	2.093	0.943	29.945
345	3	4.087	2.959	0.638	36.406
345	4	2.390	1.730	1.106	26.989
34	34	2.390	1.730	1.106	26.989
35	34	3.418	2.455	0.803	32.725
35	35	3.370	2.420	0.816	32.461
Average value		2.67	1.935	1.026	28.564
Root mean square		0.59	0.411	0.175	3.282

Note: Energies in kcal•mol^{-1}.

One Ortho Substituent

Similarly the mono-ortho-chlorinated PCB's (TABLE 2) were found to have an average barrier to planar rotation of 6.9 $kcal{\cdot}mol^{-1}$ with a root mean square error of 0.8 $kcal{\cdot}mol^{-1}$. The corresponding minima occurred at an average angle of 56°. Variations observed in rotational barriers of individual isomers are probably the result of theoretical errors rather than actual physical effects. For example, the 2,4-dichlorobiphenyl is predicted to have a barrier almost 2 $kcal{\cdot}mol^{-1}$ higher than the 2,3-dichlorobiphenyl. Rotational barriers which are computed as small energy differences on the order of 10^{-3} to 10^{-2} atomic units* from total energies of the order 10^2 atomic units are very sensitive to small changes in molecular conformation. Hence any incompleteness in the convergence of the molecular goemetries of these isomers to their lowest energy for fixed angle θ will have a strong effect. These random errors however should approximately cancel when an average value is taken over a large set of structurally similar compounds.

TABLE 2: Computed energies (E) and angles of lowest energy conformation of various chlorobiphenyl congeners (cis conformation) with one orthochloro substituent

Ring 1	Ring 2	E(INDO)	E(0.0 Deg.)	E(90.0 Deg.)	Angle
2	0	8.566	6.934	0.143	55.990
23	0	8.078	6.502	0.189	53.390
24	0	9.967	8.246	0.046	64.690
2	4	7.935	6.374	0.205	52.633
234	0	8.580	6.955	0.141	56.123
235	0	9.536	7.840	0.069	61.817
245	0	6.907	5.474	0.333	47.439
2	34	8.174	6.588	0.180	53.900
2	35	7.290	5.805	0.281	49.330
23	3	8.294	6.695	0.168	54.544

Note: Energies in $kcal{\cdot}mol^{-1}$.

* 1 Atomic Unit = 627.50 $kcal{\cdot}mole^{-1}$.

TABLE 2: Cont'd

Ring 1	Ring 2	E(INDO)	E(0.0 Deg.)	E(90.0 Deg.)	Angle
24	3	8.748	7.108	0.127	57.067
25	3	7.098	5.639	0.307	48.379
23	4	8.341	6.738	0.163	54.804
24	4	8.533	6.912	0.145	55.857
25	4	7.290	5.805	0.281	49.330
2345	0	9.011	7.350	0.106	58.591
2	345	9.034	7.372	0.104	58.732
234	3	9.680	7.975	0.061	62.748
235	3	10.229	8.497	0.034	66.586
245	3	7.385	5.888	0.269	49.810
234	4	9.369	7.683	0.080	60.761
235	4	7.505	5.993	0.254	50.414
23	34	8.652	7.021	0.135	56.526
23	35	7.983	6.416	0.199	52.884
24	34	8.843	7.196	0.119	57.616
24	35	7.576	6.056	0.246	50.779
25	34	8.772	7.130	0.125	57.204
25	35	9.919	8.201	0.048	64.358
2345	3	9.560	7.862	0.068	61.971
2345	4	8.389	6.782	0.159	55.065
24	345	8.652	7.021	0.135	56.526
25	345	8.413	6.803	0.156	55.196
234	34	9.202	7.527	0.091	59.734
235	34	9.728	8.020	0.058	63.063
245	34	7.146	5.680	0.300	48.616
234	35	9.465	7.773	0.074	61.361
235	35	9.728	8.020	0.058	63.063
245	35	7.672	6.140	0.234	51.268
2345	35	8.078	6.502	0.189	53.390
234	345	9.034	7.372	0.104	58.732
235	345	9.321	7.638	0.083	60.465
245	345	7.768	6.225	0.223	51.761
Average value		8.56	6.947	0.155	56.250
Root mean square		0.88	0.798	0.080	4.973

Note: Energies in kcal·mol^{-1}.

Two Ortho Substituents

In the case of di-orthochloro substitutions, there are three structural possiblities. These are 2,2'-cis; 2,2'-trans and the 2,6 conformers. The sterically hindered 2,2'-cis conformation, as one might expect, is totally repulsive (TABLE 3) with a predicted torsional angle of 90° and an energy barrier of 22.4 kcal•mol^{-1}. In contrast, the 2,2'-trans conformers (TABLE 4) are predicted to have an angle of 70° and a barrier of 11.3 kcal•mol^{-1}. Buttressing effects, that is having one large steric group beside another to give an overall increase in the nonbonded interaction were not evident.

For comparison, an independent estimate of the barrier height was also derived from the studies of Bott _et al_. (1980). In that work, rotational barriers of 33 biphenyl derivatives with methyl groups substituted at the 2 position and various groups in the 2' position were determined from dynamic NMR experiments. A plot of the barrier height against the van der Waals radii of the 2' groups yielded a smooth monotonic increasing curve. Hence 2' groups with similar van der Waals radii are linearly related by the approximate expression

$$\frac{E_1}{E_2} \simeq \frac{R_1}{R_2} \tag{15}$$

where the E's are rotational energy barriers and R's the corresponding van der Waals radii of the attached 2' groups. Therefore from the experimentally determined barrier of 11.8 kcal•mol^{-1} for 2-CH_3, 2'-Cl biphenyl derivative and van der Waals radii of 1.80Å and 1.73Å for the CH_3 and Cl groups respectively, the barrier for the 2,2'-dichloro isomer is estimated to be 11.3 kcal•mol^{-1} in exact agreement with the INDO corrected calculations.

The class of 2,6-di-orthochloro compounds (TABLE 5) are found to have an average barrier to planar rotation of 12.3 kcal•mol^{-1} with a root mean square error of 0.7 kcal•mol^{-1}. The predicted torsional angle of 68.2° is in excellent agreement with the experimental 68.1° value of the 2,4,6-trichloro-4'-bromobiphenyl (Field _et al_. 1977).

TABLE 3: Computed energies (E) and angles of lowest energy conformation of various chlorobiphenyl congeners (cis conformation) with two orthochloro substituents

Ring 1	Ring 2	E(INDO)	E(0.0 Deg.)	E(90.0 Deg.)	Angle
2	2	22.586	20.820	0.000	90.000
23	2	21.511	19.745	0.000	90.000
24	2	25.072	23.306	0.000	90.000
25	2	23.972	22.206	0.000	90.000
234	2	20.005	18.239	0.000	90.000
235	2	24.450	22.648	0.000	90.000
245	2	22.084	20.318	0.000	90.000
23	23	26.506	24.740	0.000	90.000
23	24	21.678	19.912	0.000	90.000
23	25	32.099	30.333	0.000	90.000
24	25	22.777	21.011	0.000	90.000
25	25	20.626	18.860	0.000	90.000
2345	2	23.423	21.657	0.000	90.000
234	23	23.590	21.824	0.000	90.000
235	23	22.299	20.533	0.000	90.000
245	23	21.009	19.243	0.000	90.000
234	24	22.490	20.724	0.000	90.000
235	24	22.323	20.557	0.000	90.000
245	24	22.108	20.342	0.000	90.000
234	25	23.279	21.513	0.000	90.000
235	25	23.064	21.298	0.000	90.000
245	25	22.849	21.083	0.000	90.000
2345	23	31.119	29.353	0.000	90.000
2345	24	24.140	22.374	0.000	90.000
2345	25	21.917	20.151	0.000	90.000
234	234	32.122	30.356	0.000	90.000
234	235	27.533	25.767	0.000	90.000
245	234	36.353	34.587	0.000	90.000
235	235	22.180	20.414	0.000	90.000
235	245	22.873	21.107	0.000	90.000
245	245	21.869	20.103	0.000	90.000
Average value		24.19	22.424	0.000	90.000
Root mean square		3.76	3.755	0.000	0.000

Note: Energies in $kcal \cdot mol^{-1}$.

TABLE 4: Computed energies (E) and angles of lowest energy conformation of various chlorobiphenyl congeners (trans conformation) with two orthochloro substituents

Ring 1	Ring 2	E(INDO)	E(0.0 Deg.)	E(90.0 Deg.)	Angle
23	2	11.544	9.878	0.100	60.978
24	2	12.859	11.126	0.033	68.130
25	2	11.544	9.878	0.100	60.978
234	2	12.739	11.011	0.038	67.402
235	2	12.548	10.827	0.045	66.278
245	2	12.548	10.827	0.045	66.278
23	24	12.381	10.667	0.053	65.330
23	25	13.623	11.868	0.011	73.438
24	24	13.504	11.752	0.014	72.513
24	25	12.620	10.896	0.042	66.694
25	25	11.950	10.259	0.075	63.021
2345	2	12.428	10.713	0.051	65.598
234	23	14.508	12.742	0.000	83.296
235	23	13.528	11.775	0.013	72.694
245	23	13.074	11.333	0.026	69.497
234	24	14.269	12.504	0.002	79.692
245	24	12.094	10.395	0.067	63.772
234	25	13.522	11.798	0.013	72.877
2353	25	13.121	11.380	0.024	69.812
245	25	13.241	11.495	0.021	70.619
2345	23	13.767	12.009	0.008	74.617
2345	24	12.835	11.103	0.034	67.983
2345	25	13.791	12.032	0.008	74.822
234	234	14.269	12.504	0.002	79.692
235	234	14.460	12.694	0.000	82.439
234	245	13.074	11.333	0.026	69.497
235	235	13.432	11.682	0.015	71.978
235	245	12.596	10.873	0.043	66.555
245	245	12.357	10.645	0.054	65.197
Average value		13.04	11.310	0.033	70.058
Root mean square		0.80	0.771	0.027	5.818

Note: Energies in $kcal \cdot mol^{-1}$.

TABLE 5: Computed energies (E) and angles of lowest energy conformation of various chlorobiphenyl congeners (cis conformation) with two orthochloro substituents on the same ring

Ring 1	Ring 2	E(INDO)	E(0.0 Deg.)	E(90.0 Deg.)	Angle
26	0	14.293	12.554	0.027	69.656
236	0	13.360	11.660	0.065	64.636
246	0	13.647	11.933	0.052	66.084
26	3	13.265	11.569	0.070	64.169
26	4	12.142	10.516	0.140	59.107
2346	0	14.006	12.277	0.037	68.006
2356	0	13.599	11.887	0.054	65.838
236	3	12.787	11.118	0.097	61.927
246	3	13.982	12.254	0.038	67.873
236	4	14.173	12.438	0.031	68.955
246	4	15.272	13.512	0.005	76.588
26	34	13.838	12.116	0.044	67.092
26	35	12.811	11.140	0.095	62.036
23456	0	14.747	12.996	0.015	72.549
2346	3	14.101	12.369	0.034	68.543
2356	3	14.412	12.670	0.024	70.380
2346	4	14.054	12.323	0.036	68.273
2356	4	14.197	12.462	0.031	69.093
26	345	13.337	11.637	0.066	64.519
236	34	13.862	12.139	0.043	67.220
246	34	13.982	12.254	0.038	67.873
236	35	13.599	11.887	0.054	65.838
246	35	13.958	12.231	0.039	67.741
23456	4	14.771	13.019	0.014	72.713
2346	34	15.607	13.843	0.002	79.923
2356	34	14.460	12.716	0.022	70.676
2346	35	14.986	13.230	0.010	74.264
2356	35	14.245	12.508	0.029	69.373
Average value		13.98	12.259	0.043	68.248
Root mean square		0.74	0.707	0.030	4.311

Note: Energies in kcal·mol^{-1}.

Three and Four Ortho Substituents

As in the case of 2,2'-cis-dichlorobiphenyl, the tri- and tetra-orthochloro substituted PCB's are totally repulsive with the rings twisted 90°. The 2,6,2',6'-tetrachloro derivatives (TABLE 8) are found to have very large rotation barriers of 57.5 kcal•mol^{-1}. The cis and trans conformers of the 2,6,2'-trichloro derivatives (TABLES 6 and 7) provide a further check on the magnitude of buttressing effects. In principle, the influence of buttressing should be more strongly felt in cis isomers, as for example, the 2,3,6,2',3',4' than in the corresponding trans compounds. Therefore, if buttressing is significant, the average value for the cis rotational barrier should be higher than found in the trans case. However, our results show virtually identical values of 31.589 and 31.583 kcal•mol^{-1}, respectively which indicate little or no buttressing occurring.

TABLE 6: Computed energies (E) and angles of lowest energy conformation of various chlorobiphenyl congeners (cis conformation) with three orthochloro substituents

Ring 1	Ring 2	E(INDO)	E(0.0 Deg.)	E(90.0 Deg.)	Angle
236	2	32.433	30.667	0.000	90.000
246	2	33.891	32.125	0.000	90.000
26	23	31.381	29.615	0.000	90.000
26	24	32.027	30.261	0.000	90.000
26	25	35.851	34.085	0.000	90.000
2346	2	31.501	29.735	0.000	90.000
2356	2	32.553	30.787	0.000	90.000
236	23	32.816	31.050	0.000	90.000
246	23	33.485	31.719	0.000	90.000
236	24	32.433	30.667	0.000	90.000
246	24	33.198	31.432	0.000	90.000
236	25	31.788	30.022	0.000	90.000
246	25	32.959	31.193	0.000	90.000
26	245	31.692	29.926	0.000	90.000
23456	2	33.509	31.743	0.000	90.000
2346	23	33.580	31.814	0.000	90.000
2356	23	34.393	32.627	0.000	90.000

Note: Energies in kcal•mol^{-1}.

TABLE 6: Cont'd

Ring 1	Ring 2	E(INDO)	E(0.0 Deg.)	E(90.0 Deg.)	Angle
2346	24	33.819	32.053	0.000	90.000
2356	24	39.269	37.503	0.000	90.000
2356	25	31.931	30.165	0.000	90.000
236	234	32.457	30.691	0.000	90.000
246	234	33.485	31.719	0.000	90.000
236	235	32.744	30.978	0.000	90.000
246	235	32.266	30.500	0.000	90.000
236	245	33.772	32.006	0.000	90.000
246	245	38.002	36.236	0.000	90.000
Average value		33.36	31.589	0.000	90.000
Root mean square		1.82	1.821	0.000	0.000

Note: Energies in kcal•mol^{-1}.

TABLE 7: Computed energies (E) and angles of lowest energy conformation of various chlorobiphenyl congeners (trans conformation) with three orthochloro substituents

Ring 1	Ring 2	E(INDO)	E(0.0 Deg.)	E(90.0 Deg.)	Angle
2346	2	33.819	33.053	0.000	90.000
236	23	32.122	30.356	0.000	90.000
236	24	37.643	35.877	0.000	90.000
236	25	32.194	30.428	0.000	90.000
2346	23	33.102	31.336	0.000	90.000
2346	24	32.911	31.145	0.000	90.000
2346	25	32.744	30.978	0.000	90.000
236	234	32.481	30.715	0.000	90.000
236	235	34.011	32.245	0.000	90.000
236	245	32.457	30.691	0.000	90.000
Average value		33.35	31.583	0.000	90.000
Root mean square		1.55	1.553	0.000	0.000

Note: Energies in kcal•mol^{-1}.

TABLE 8: Computed energies (E) and angles of lowest energy conformation of various chlorobiphenyl congeners with four ortho-chloro substituents

Ring 1	Ring 2	E(INDO)	E(0.0 Deg.)	E(90.0 Deg.)	Angle
26	26	56.238	54.472	0.000	90.000
236	26	57.600	55.834	0.000	90.000
246	26	58.580	56.814	0.000	90.000
2346	26	59.489	57.723	0.000	90.000
2356	26	59.011	57.245	0.000	90.000
236	236	57.242	55.476	0.000	90.000
236	246	66.396	64.630	0.000	90.000
246	246	60.158	58.392	0.000	90.000
Average value		59.34	57.573	0.000	90.000
Root mean square		2.92	2.920	0.000	0.000

Note: Energies in $kcal \cdot mol^{-1}$.

BIOLOGICAL CORRELATION RESULTS AND DISCUSSION

The biological activity of TCDD-like compounds is expressed in many nonspecific ways. These include a wasting syndrome characterized by weight loss, lymph gland atrophy, decreased food consumption, skin endocrine and reproductive disorders, and chemical carcinogenesis (Parkinson and Safe 1981). Initial studies by Poland and Glover (1977; 1973) have, however, established a direct link between whole animal toxicity data by dibenzo-p-dioxins and their ability to induce the liver enzymes arylhydrocarbon hydroxylase (AHH) and 5-aminolevulinic acid synthetase. The correlation found in that work, between ED50's or estimated doses for 50% of maximal AHH induction and corresponding toxicity has been extended to the PCDF and PCB series of congeners by Poland _et al._ (1976) and Yoshimura _et al._ (1976). Recent research has focused on measuring only the maximal amounts of induced AHH rather than total log-dose response curves needed for calculation of ED50 values. Compounds with no ortho-chlorines such as 3,4,5,3',4',5'-hexachlorobiphenyl are found to cause maximum AHH induction at dose levels of around 10 $mg \cdot kg^{-1}$ or approximately 30 $\mu mole \cdot kg^{-1}$ in the rat (Goldstein _et al._ 1977). In the case of ortho-chlorinated derivatives such as the 2,4,5,2',4',5'-hexachlorobiphenyl, larger doses of 50 $mg \cdot kg^{-1}$ or approximately 150 $\mu mol \cdot kg^{-1}$ are required. For the purpose of this paper, we define the AHH specific activity as the maximum amount of AHH induced divided by the corresponding PCB dose.

Working Hypothesis

In order to reduce the complexity of the problem, the PCB molecule was partitioned into two functional areas: (i) ortho positions whose main effects are steric and (ii) meta, para positions whose main effects are electronic. As a working hypothesis the specific acivity, A, was taken to be proportional to that fraction of conformers found from a Boltzmann distribution to have enough energy to become planar;

$$A = A_o \cdot e^{-G/RT} \qquad (16a)$$

The parameter A_o is defined as the specific activity of a congener in the absence of rotational barriers and is dependent on the electronic effects of the meta and para substituents. G is defined as the free energy of activation required to rotate a given compound into its planar conformation and is a function of the rotational barrier energy, E, and the entropy of activation, S. Increasing E causes the system to become more rigid and ordered. This results in an increase in entropy of activation relative to the state of the freely rotating conformer. If, as a first approximation, S is taken to be linearly related to the rotational barrier energy, E, then, equation (16a) may be reexpressed as:

$$A = A_o' \cdot e^{-\alpha E} \qquad (16b)$$

where A_o' and α are constants determined from the experimental specific activities.

According to the above hypothesis, any PCB will undergo changes in its ability to induce AHH due to strictly steric effects when substitutions or deletions of groups are made at ortho positions. Therefore, as a test, the four PCB congeners known to be pure AHH inducers (Safe _et al_. 1982), namely 3,4,3',4'-tetrachloro, 3,4,5,3',4'-pentachloro, 3,4,5,3',4',5'-hexachloro and the 3,4,5,4'-tetrachlorobiphenyls were selected as reference compounds. The AHH specific activities of these compounds and their corresponding ortho-chlorinated derivatives were determined from the experimental literature (Yoshimura _et al_. 1978; Parkinson _et al_. 1981; 1980a; 1980b; Goldstein _et al_. 1978), and a structure-activity correlation was then calculated based on equation 16. The results of this work are presented in TABLE 9.

Since the hydroxylation rate of benzo-a-pyrene is used to indirectly measure the AHH present, the amount of induced AHH is generally expressed as a percent increase over that induced solely by the corn oil solvent in which the PCB's are dissolved. Hence in TABLE 9, the specific activites are in units of percent increase over the control corn oil, divided by the dose in mmol$\cdot$kg^{-1}. With the exception of the 3,4,5,3',4'-penta- and 3,4,5,3',4',3'-hexachloro-biphenyl, all specific activities were determined from the experimental studies of Parkinson, Safe and coworkers (Parkinson _et al_. 1980a; 1980b; 1981). In the case of the 3,4,5,3',4' congener, no satisfactory quantitative value for the AHH specific activity could be found. As an approximation the specific activity was taken to be the same as the 3,4,3',4'-tetrachlorobiphenyl since both these compounds have almost identical enzyme induction profiles (Yoshimura _et al_. 1978). In the case of the 3,4,5,3',4',5'-hexachlorobiphenyl the relative activity in comparison to the 3,4,3',4' congener was first determined from the work of Goldstein _et al_. (1977). The specific activity was then estimated using the 3,4,3',4' value found from the study by Parkinson _et al_. (1980a).

RESULTS

Inspection of TABLE 9 reveals that with the exception of the 3,4,5,3',4',5' and 3,4,5,4' congeners good agreement is found between the experimental and predicted mean specific activities. Mean values were used in order to average out any differences in activity of the individual isomers due to unaccounted electronic effects of ortho-chlorines. The root mean square errors between the mean and individual isomer specific activities range from 14% to 50%.

TABLE 9. Measured and predicted arylhydrocarbon hydroxylase activation activities of chlorobiphenyl isomers with one or two orthocloro substituents. Predicted values from equation 18

Isomer	Specific Activity	Isomer	Specific Activity	Isomer	Specific Activity	Isomer	Specific Activity
Reference Compounds							
343'4'	474[a]	3453'4'	~474[b]	3453'4'3'	~297[c]	3454'	81[d]
Predicted average	466	Predicted average	431	Predicted average	81	Predicted average	319
Mono-ortho Chlorinated Derivatives							
2343'4'	105[a]	2343'4'5'	95[a]	23453'4'5'	16[a]	23454'	73[d]
2453'4'	79[a]	2453'4'5'	18[a]			3452'4'	53[d]
		23453'4'	75[a]				
Average	92	Average	85			Average	63
RMS error**	13	RMS error***	10			RMS error	10
Predicted average	98	Predicted average	93	Predicted average	17	Predicted average	67
Di-ortho Chlorinated Derivatives							
2342'3'4'	22[e]	23452'3'4'	36[e]	23451'3'4'5'	0[e]	23452'4'	1[e]
2342'4'5'	46[e]	23452'4'5'	13[e]	234563'4'5'	10[e]	234564'	28[e]
2452'4'5'	14[e]	234563'4'	13[e]			2463'4'5'	0[e]
23463'4'	32[e]	23463'4'5'	17[e]				
Average	28	Average	20	Average	5	Average	10
RMS error	6	RMS error	5	RMS error	5	RMS error	7
Predicted average	23	Predicted average	20	Predicted average	4	Predicted average	14

* In % increase over control [kg·mmol^{-1}].

** RMS = root mean square.

*** The novel inducer 2453'4'5' was excluded from the average (see Safe *et al.* 1982).

a Parkinson *et al.* (1980a).

b Estimated from Yoshimura *et al.* (1978).

c Estimated from Goldstein *et al.* (1977).

d Parkinson *et al.* (1980b).

e Parkinson *et al.* (1981).

However, the experimental error is also large, for example, the specific activity of the 3,4,3',4'-tetrachlorobiphenyl has a determined value of 474 with an error of ± 122. Therefore the magnitude of the electronic effects of ortho-chlorines on AHH induction cannot be generally estimated from these data.

Electronic effects of ortho-chlorines do however influence AHH induction in the 3,4,5,3',4',5'hexa- and 3,4,5,4'-tetrachlorobiphenyls. Particularly intriguing is the excellent agreement to experiment that can be obtained if predicted specific activities are simply interchanged. If rotational barriers are momentarily neglected, then the addition of an ortho-chlorine to the 3,4,5,3',4',5 congener produces an activity equivalent of the removal of two meta chlorines from this compound to yield the 3,4,5,4' congener. In contrast, the addition of an ortho-chlorine to the 3,4,5,4'-tetrachlorobiphenyl produces an activity in the absence of rotational barriers equivalent to the substitution of two meta chlorines yielding the 3,4,5,3',4',5'-hexachlorobiphenyl. These strong electronic effects can be averaged out by taking mean values over the two congeners. Good agreement is then obtained between the averaged predicted and experimental activities.

Structure-Activity Correlation

The predicted mean specific activites found in TABLE 9 were determined from the structure-activity relationship:

$$A/A_{REF} = Qe^{-\alpha(E-E_{REF})} \tag{17}$$

where A_{REF}, E_{REF} are reference specific activities and activation energies. Due to the experimental uncertainty found for the meta, para congeners without ortho-chlorines, the mono-ortho-chlorinated compounds were chosen as references for these calculations. A least squares fit to equation (17) excluding the 3,4,5,3',4',5' and 3,4,5',4' congeners yielded:

$$A/A_{REF} = 1.064\ e^{-0.3115(E-E_{REF})} \tag{18}$$

$$n = 10;\ r^2 = 0.98$$

with a correlation coefficient of $r^2 = 0.98$ over the 10 points. The 1.064 value for Q is in good agreement with the expected value of 1.0

predicted from equation 16 of our working hypothesis. From the α value of 0.3115 mol•kcal^{-1}, equation 18, the increase in entropy of activation, S, with the rotational barrier energy, E, may be examined. For example, at T = 300 °K, the increase in S, going from the ortho mono-orthochloro congener to the di-ortho-2,2'-trans-dichlorobiphenyl is estimated to be 8.7 cal•mol^{-1}. The corresponding increase in E is 4.3 kcal•mol^{-1}. In comparison, the dynamic NMR results of Bott et al. (1980), show a measured increase in entropy of activation going from the 2-CH_3-2'-F-biphenyl to the 2-CH_3-2'-Cl-biphenyl of 8.1 cal•mol^{-1} as the rotational barrier increases by 6.3 kcal•mol^{-1}. Hence, the value of α fitted from the biological data is in reasonable agreement with what might be expected of the chemical entropy data of PCB's were known. Finally, the high value found for the correlation coefficient, r^2 is somewhat misleading since the experimental values of A/A_{REF} cluster around only three points. Structure-activity data for ortho substituted groups different from chlorine would prove valuable in clarifying this tentative correlation.

CONCLUSIONS

In summary, from the quantum chemical studies, the rotational energy barriers for the planar conformation of the biologically active PCB's have been calculated to an accuracy of approximately 1 kcal•mol^{-1}. Furthermore these barriers appear to be additive. For example, the rotational barrier of 1.94 kcal•mol^{-1} in biphenyl can be considered as the sum of two hydrogen-hydrogen ½ barriers with values of 0.97 kcal•mol^{-1} each. From the rotational energy barrier of 6.97 kcal•mol^{-1} for the monochlorinated PCB's, a Cl-H ½ barrier of 6.00 kcal•mol^{-1} is then estimated. The value of 12.0 kcal•mol^{-1} predicted from the sum of two Cl-H ½ barriers is in good agreement with actual calculated barriers of 11.3 kcal•mol^{-1} and 12.3 kcal•mol^{-1} of the 2,2' trans and 2,6-dichloro congeners, respectively. This linearity however breaks down for the higher ortho-chlorinated compounds such as the 2,6,2'- and 2,6,2',6'-PCB's.

The corresponding biological studies resulting in the structure-activity relationship of equation 18 support the principle that a chlorobiphenyl molecule must be approximately planar in order to be active in the same manner as TCDD.

ACKNOWLEDGEMENTS

We thank the Great Lakes Water Quality Program for support of this study and Dr. M.C. Zerner for a copy of the INDO program. One of us (J.M.C.) gratefully acknowledges a Visiting Fellowship of the Natural Sciences and Engineering Research Council of Canada.

REFERENCES

Allinger, N.L and Sprague, J.T. 1973. Calculation of the structures of hydrocarbons containing delocalized electronic systems by the molecular mechanics method. **J. Am. Chem. Soc. 95:** 3893-3907.

Almlöf, J. 1974. Ab initio calculations on the equilibrium geometry and rotation barriers in biphenyl. **Chem. Phys. 6:** 135-139.

Bartell, L.S. 1960. The effects of intramolecular van der Waals forces. **J. Chem. Phys. 32:** 827-831.

Birner, P. and Hofmann, H.J. 1982. Quantum chemical calculations for the determination of the molecular structure of conjugated compounds. Part XVII. **Int. J. Quantum Chem. 21:** 833-843.

Born, M. and Oppenheimer, J.R. 1927. Quantum theory of the molecules. **Ann. Physik. 84:** 457-484.

Bott, G., Field, L.D. and Sternhell, S. 1980. Steric effects. A study of a rationally designed system. **J. Am. Chem. Soc. 102:** 5618-5626.

Casalone, G.L., Mariani, C., Mugnoli, A. and Simonetta, M. 1968. Molecular structure of biphenyl in the gas and solid phases. **Mol. Phys. 15:** 399-348.

Field, L.D. and Sternhell, S. 1981. The conformation of biphenyls in nematic liquid crystalline solution. The relationship between the torsional angles and the size of the halogens in 2,6-dihalogenated derivatives. **J. Am. Chem. Soc. 103:** 738.

Field, L.D. Sternhell, S. and Tracey, A.S. 1977. The conformation of biphenyls in nematic liquid crystalline solution. An investigation of the torsional angles in 2,6-dihalogenated derivatives. **J. Am. Chem. Soc. 99:** 5249-5253.

Fischer-Hjalmars, I. 1963. Conjugation and non-bonded interaction. **Tetrahedron 19:** 1805-1815.

Goldstein, J.A. 1979. The structure-activity relationahips of halogenated biphenyls as enzyme inducers. **Ann. N.Y. Acad. Sci. 320:** 164.

Goldstein, J.A. Hickman, P., Bergman, H., McKinney, J.D. and Walker, M.P. 1977. Separation of pure polychlorinated biphenyl isomers into two types of inducers on the basis of induction of cytochrome P-450 or P-448. **Chem.-Biol. Interact 17:** 69-87.

Goldstein, J.A., Hass, J.R., Linko, P. and Harvan, D.J. 1978. 2,3,7,8-Tetrachlorodibenzofuran in a commercially available 99% pure polychlorinated biphenyl isomer identified as the inducer of hepatic cytochrome P-448 and arylhydrocarbon hydroxylase in the rat. **Drug Metab. Dispos. 6:** 258-264.

Gropen, O. and Seip, H.M. 1971. Failure of the CNDO/2 method to predict the barriers and conformations in some conjugated systems. **Chem. Phys. Lett. 11:** 445-449.

Kaiser, K.L.E. 1974. On the optical activity of polychorinated biphenyls. **Environ. Pollut. 7:** 93-101.

Pariser, R. and Parr, R.G. 1953. A semiempirical theory of the electronic spectra and electronic structure of complex unsaturated molecules. II. **J. Chem Phys. 21:** 767-776.

Parkinson, A., Cockerline, R. and Safe, S. 1980a. Polychlorinated biphenyl isomers and congeners as inducers of both 3-methylcholanthrene- and phenobarbitone-type microsomal enzyme activity. **Chem. Biol. Interact 29:** 277-289.

Parkinson, A., Robertson, L.W., Safe, L. and Safe, S. 1980b. Polychlorinated biphenyls as inducers of hepatic microsomal enzymes: Structure-activity rules. **Chem.-Biol. Interact 30:** 271-285.

Parkinson, A. and Safe, S. 1981. Arylhydrocarbon hydroxylase induction and its relationship to the toxicity of halogenated arylhydrocarbons. **Toxicol. Environ. Chem. 4:** 1-45.

Parkinson, A., Robertson, L.W., Safe, L. and Safe, S. 1981. Polychlorinated biphenyls as inducers of hepatic microsomal enzymes: Effects of di-ortho substitution. **Chem.-Biol. Interact 35:** 1-12.

Poland, A. and Glover, E. 1973. Chlorinated dibenzo-p-dioxins. Potent inducers of δ-aminolevulinic acid synthetase and arylhydrocarbon hydroxylase. Structure-activity relation. **Mol. Pharmacol. 9:** 736-747.

Poland, A. and Glover, E. 1977. Chlorinated biphenyl induction of arylhydrocarbon hydroxylase activity: A study of the structure-activity relationship. **Mol. Pharmacol. 13:** 924-938.

Poland, A., Glover, E. and Kende, A.S. 1976. Stereospecific, high affinity binding of 2,3,7,8-Tetrachlorodibenzo-p-dioxin by hepatic cytosol. **J. Biol. Chem. 251:** 4936-4946.

Pople, J.A. and Beveridge, D.L. 1970. **Approximate Molecule Orbital Theory,** McGraw Hill, New York.

Pople, J.A., Beveridge, D.L. and Dobosh, P.A. 1967. Approximate self-consistent molecular-orbital theory. V. Intermediate neglect of differential overlap. **J. Chem. Phys. 47:** 2026-2033.

Pople, J.A. 1953. Electron interaction in unsaturated hydrocarbons. **Trans. Faraday Soc.** 49: 1375-1385.

TFPCB. 1976. Background to the regulation of polychlorinated biphenyls (PCB) in Canada. Task Force on PCB. **Environment Canada** and **Health & Welfare Canada,** Technical report 76-1, pp. 169.

Tyrrell, J., Weinstock, R.B. and Weinhold, F. 1981. Bond-antibond analysis of internal rotation barriers in glyoxal and related molecules: Where INDO fails. **Int. J. Quantum Chem. 19:** 781-791.

Weiss, S. and Leroi, G.E. 1968. Direct observation of the infrared torsional spectrum of C_2H_6, CH_3CD_3, and C_2D_6. **J. Chem. Phys. 48:** 962-967.

Yoshimura, H. Ozawa, N. and Saeki, S. 1978. Induction effect of polychlorinated biphenyls mixture and individual isomers on the hepatic microsomal enzymes. **Chem. Pharm. Bull. 26:** 1215-1221.

Yoshimura, H., Yoshihara, S., Ozawa, N. and Miki, M. 1979. Possible correlation between induction modes of hepatic enzymes by PCBs and their toxicity in rats. **Ann. N.Y. Acad. Sci. 320:** 179-192.

WORKSHOP DISCUSSION

McCARTY:

Just to summarize that, you say that the closer the PCB's approximate dioxin, the more similar is the effect?

CULLEN:

If you can rotate it into the planar form and the positions of the chlorines in the meta and para positions are similar to TCDD chlorine positions, yes.

McCARTY:

You can make a model here to predict the various activities of various isomers of PCB compounds, substituted compounds. If you change that around, can you predict how close you are going to come to a toxicity of a known compound?

CULLEN:

It would be interesting to substitute in the ortho position other groups besides chlorines to see how their toxicities compare with those predicted from this model.

McCARTY:

So you think if you substitute something else, you might approximate the toxicity of the compound even closer?

CULLEN:

Yes, if it is a smaller group; for example, a fluorine there would be more probability.

KAISER:

We have two enzyme systems being activated by various PCB isomers, namely methylcholanthrene and phenobarbitol type inductions. Some isomers activate one, the other isomers the other and some both enzyme systems. Any PCB isomer which is a methylcholanthrene type inducer is in the planar form. In fact, that is what our results seem to bear out. If you add two ortho-chlorines or so which give a high energy barrier in getting both rings into a planar conformation then they are no longer methylcholanthrene type inducers.

CHU:

This is precisely what is observed for Aroclors which are mixtures of some planar molecules and some nonplanar molecules, they are mixed type inducers. You can use both P-448 and P-450, but what is pointed out here is very interesting, the more planar molecules are type P-448 inducers for example 3,4,3',4'-tetrachlorobiphenyl. On the other hand, molecules with the ortho substituents have a greater angle, and are phenobarbitol type inducers, am I right?

CULLEN:

Yes, also it has been shown with the biphenyl blocked into a planar conformation in the compound bi-phenylene. This was done by Poland and Grover in the early days and its toxicity was very comparable to TCDD.

THE USE OF RAPID BIOCHEMICAL INDICATORS OF TOXICANT STRESS TO GENERATE BIOLOGICAL DATA BASES FOR QSAR

D. George Dixon[1], Peter V. Hodson[2] and Klaus L.E. Kaiser[3]

[1] Department of Biology
University of Waterloo
Waterloo, Ontario N2L 3G1
CANADA

[2] Great Lakes Fisheries Research Branch
Department of Fisheries and Oceans
PO Box 5050, Burlington, Ontario L7R 4A6
CANADA

[3] Environmental Contaminants Division
National Water Research Institute
PO Box 5050 Burlington, Ontario L7R 4A6
CANADA

ABSTRACT

The assay of serum sorbitol dehydrogenase and plasma leucine aminonaphthylamidase activities to indicate liver damage of toxicant to rainbow trout (Salmo gairdneri) was asssessed using para-methylphenol (para-cresol) as a model toxicant. Both tests appear to be useful indicators for the rapid assessment of toxicant effects on fish.

INTRODUCTION

One of the major difficulties in developing sound Quantitative Structure-Activity Relationships (QSARs) is obtaining uniform and consistent biological data sets on large numbers of chemicals. Tests directed at obtaining this relative toxicity data

K. L. E. Kaiser (ed.), QSAR in Environmental Toxicology, 67–73.

should be rapid, cost effective and relatively easy to perform. A method which lends itself to these requirements is the determination of blood enzymes in fish following intraperitoneal toxicant dosing. Changes in blood levels of two enzymes, sorbitol dehydrogenase and leucine aminonaphthylamidase, were determined in rainbow trout following dosing with para-methylphenol to determine the relative merit of these two enzymes as toxicity screening tests.

Leucine aminonaphthylamidase (LAN) is a proteolytic enzyme normally restricted to lysosomes (Kaulen *et al.* 1970). On cell death, LAN is released from lysosomes as a function of autolysis, and eventually appears in the blood. Plasma LAN (PLAN) is therefore considered as a nonspecific indicator of cell death (Hess 1963). In mammals, elevated PLAN levels have been used as indicators of liver damage (Smith *et al.* 1965), vitamin deficiency (Cheeseman *et al.* 1967), wasting disorders (Hess 1963) and allergic response (Rabb 1966). Bouck *et al.* (1975, 1978) have used PLAN as a nonspecific indicator of environmental stress during field studies with rainbow trout. They found that it responds to the stresses of capture, increased salinity, starvation, repetitive blood sampling, reproductive activity and dietary borate toxicity.

Sorbitol dehydrogenase catalyzes the reversible interconversion of fructose and the polyhydric alcohol sorbitol, an oxidation-reduction reaction which occurs predominantly in the liver (Wolf *et al.* 1973). The use of serum sorbitol dehydrogenase as an indicator of hepatic injury in mammals was first proposed by Gerlach (1957). SSDH is now used extensively to diagnose mammalian hepatic insult (Alemu *et al.* 1977; Yagminas and Villeneuve 1977). Elevated SSDH has not been reported in organ disease other than liver insult, and is considered in mammals to be liver specific (Asada and Galamos 1963).

EXPERIMENTAL

For each dose response relationship, 60 rainbow trout (*Salmo gairdneri*) (10 fish x 5 dose levels + control) weighing from 80 to 100 g were injected intraperitoneally with para methylphenol (para cresol) (PMP) at doses of from 10% to 75% of the 96-hour intraperitoneal LD50 (Hodson *et al.* 1984). Following a 96 hour post-injection holding period, a blood sample was taken from each fish by caudal peduncle puncture. Serum or plasma was isolated depending on the enzyme to be determined. SSDH levels were assayed using an ultraviolet kinetic assay (Anon. 1980). PLAN levels were determined with an ultraviolet endpoint assay (Bouck _et al._ 1975). In both cases serum or protein levels were also assayed for each sample (Anon. 1974). Enzyme activities were expressed as milli international units (mU) per mg serum or plasma protein. Dose/response curves for the two enzymes are given in FIGURES 1 and 2.

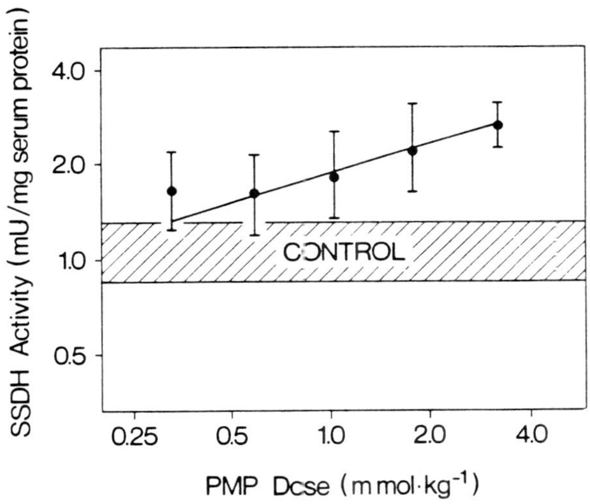

FIGURE 1: Serum sorbitol dehydrogenase (SSDH) activity versus para methylphenol (PMP) dose for rainbow trout 96-hour post injection. Data points indicate the means and 95% confidence intervals for ten fish at each dose and for control fish

RESULTS AND DISCUSSION

Additional studies were undertaken to determine the impact of various modifying factors on the activity of the two enzymes. The duration of post-injection holding was found to have a significant impact on PLAN activity in control and PMP/dosed rainbow trout with the largest increase after 96 hours. A waterborne concentration of 0.028 mM PMP (0.41 96-hour LC50) resulted in statistically significan increases in the PLAN activity shown by rainbow trout after 48, 96 and 192 hours of exposure. PLAN activity increased 38% to 87% relative to controls.

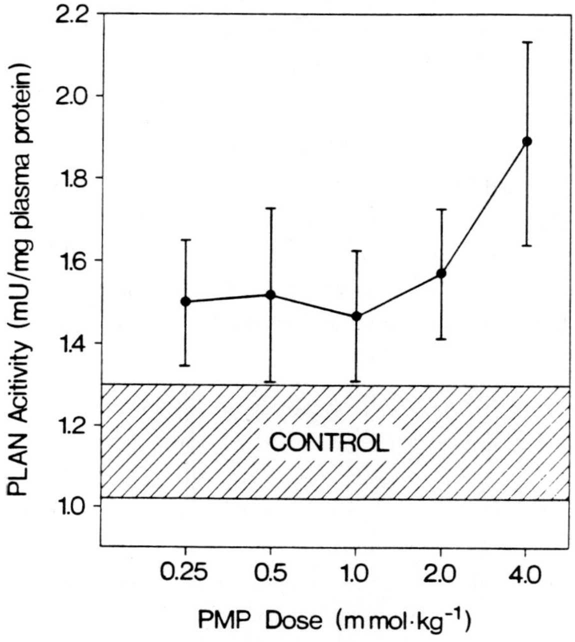

FIGURE 2: Plasma leucine aminonaphthylamidase (PLAN) activity versus para methylphenol (PMP) dose for rainbow trout 96-hour post injection. Data points indicate the means and 95% confidence intervals for ten fish at each dose and for control fish

The temperature and duration of plasma storage were found to have a significant effect on PLAN activity. No suitable long-term storage method has been found, and it is suggested that activity be determined within six hours of bleeding. Elevated dietary carbohydrate levels were found to result in significantly increased PLAN activity. No significant difference in PLAN response was apparent between male and female fish.

A post-injection holding period of 96 hours was also found to be optimal for maximizing SSDH activity. Waterborne exposure to either 0.028 mM PMP or $2.1 \cdot 10^{-6}$ mM copper (0.35 x 96-hour LC50) resulted in statistically significant increases in SSDH activity after 96 hours of exposure. Storage of serum in liquid nitrogen for up to three months resulted in no change in SSDH activity. Neither dietary carbohydrate level nor gender were found to be significant modifiers of SSDH activity.

Of the two enzymes, SSDH appears to be more appropriate as a screening test. It responds at lower doses of toxicant than does PLAN and is less variable. This has subsequently been confirmed by tests with a number of other toxicants. In addition, SSDH would appear to be less susceptible to the impact of modifying factors and is more convenient since the samples can be stored without loss of activity.

In summary, the plasma leucine aminonaphthylamidase and, in particular, the serum sorbitol dehydrogenase assays appear to be useful indicators for the rapid determination of toxicant stress on fish. For this reason, they are of interest as screening methods by regulatory agencies and for the development of data bases on the biological response of known and potential aquatic contaminants. The latter are necessary for the development of quantitative structure-activity correlations which will allow the prediction of effects of specific compounds from those of other compounds already tested.

REFERENCES

Alemu, P., Forsyth, G.W. and Searcy, G.P. 1977. A comparison of parameters used to assess liver damage in sheep treated with carbon tetrachloride. **Can. J. Comp. Med. 41:** 420-427.

Anon. 1980. The ultraviolet quantitative determination of sorbitol dehydrogenase (SDH) in serum at 340 nm. **Sigma Tech. Bull. 50-uv:** 1-11.

Anon. 1974. The colorimetric determination of total protein at 540 to 545 nm. **Sigma Tech. Bull. 540:** 1-10.

Asada, M. and Galamos, J.T. 1963. Sorbitol dehydrogenase and hepatocellular injury: an experimental and clinical study. **Gastroenterology 44:** 518-526.

Bouck, G.R., Cairns, M.A. and Christian, A.R. 1978. Effect of capture stress on plasma enzyme activities in rainbow trout (*Salmo gairdneri*). **J. Fish. Res. Board Can. 35:** 1485-1488.

Bouck, G.R., Schneider, P.W. Jr., Jacobson, J. and Ball, R.C. 1975. Characterization and subcellular localization of leucine aminonaphthylamidase (LAN) in rainbow trout (*Salmo gairdneri*). **J. Fish. Res. Board Can. 32:** 1289-1295.

Cheeseman, E.M., Copping, A.M. and Prebble, P.M. 1967. Early changes in enzyme activity during the development of and recovery from vitamin deficiency in the rat. **Br. J. Nutr. 20:** 413-420.

Gerlach, U. 1957. Pathologischer Übertritt von Sorbitdehydrogenase ins Blut bei Lebererkrankungen. **Klin. Wochenschr. 35:** 1144-1151.

Hess, E. 1963. **Enzymes in Blood Plasma.** Academic Press, New York, 167 p.

Hodson, P.V., Dixon, D.G. and Kaiser, K.L.E. 1984. The measurement of median lethal doses (LD50's) as a rapid indication of contaminant toxicity to fish. **Environ. Toxicol. Chem.,** in press.

Kaulen, H.D., Henning, R. and Stoffel, W. 1970. Biochemical analyses of the pinocytotic process: I. Comparison of some enzymes of the lysosomal and the plasma membrane of the rat lower cell. **Hoppe Seyler's Physiol. Chem. 351:** 1551-1563.

Rabb, W.P. 1966. Renal enzyme excretation following anaphylactic shock. **Nature (Lond.) 212:** 953-956.

Smith, E.L., Kaufman, J.T. and Rutenburg, A.M. 1965. The partial purification of an amino acid naphthylamidase from human liver. **J. Biol. Chem. 240:** 1718-1721.

Wolf, P.L., Williams, D. and Von der Muehll, E. 1973. **Practical Clinical Enzymology.** John Wiley and Sons, Toronto, 580 p.

Yagminas, A.P. and Villeneuve, D.C. 1977. An automated continuous-flow assay for serum sorbitol dehydrogenase activity and its use in experimental liver damage. **Biochem. Med. 18:** 117-125.

THE USE OF SIMCA PATTERN RECOGNITION IN THE ANALYSIS OF COMPLEX CHROMATOGRAPHIC DATA

William J. Dunn, III[1], David L. Stalling[2] and Svante Wold[3]

[1] Department of Medicinal Chemistry and Pharmacognosy
College of Pharmacy
University of Illinois at Chicago
833 South Wood Street
Chicago, Illinois 60680
U.S.A.

[2] U.S. Fish and Wildlife Service
Columbia National Fisheries Research Laboratory
RR#1, Columbia, Missouri 62500
U.S.A.

[3] Research Group for Chemometrics
Umea University
S-90187 Umea
SWEDEN

ABSTRACT

A technique is proposed for the quantitative determination of constituents in complex mixtures characterized by gas chromatography data. The technique is the SIMCA pattern recognition method and the data to which it is applied are gas chromatograms of the Aroclors 1242, 1248, 1252, and 1260. The problem is formulated as one of classification and a sample of used transformer oil from a waste dump is classified as to its composition.

INTRODUCTION

The nature and properties of organic contaminants that have entered the environment range widely and vary in complexity and

K. L. E. Kaiser (ed.), QSAR in Environmental Toxicology, 75–88.

chemical composition with each source. Some commercial products are extremely complex formulations. An example of such a mixture are the polychlorinated biphenyls (PCB's), commonly known as Aroclors. TABLE 1 is a summary of the possible number of PCB isomers that may be in a given commercial formulation. A typical capillary gas chromatogram of a 1:1:1:1 mixture of the Aroclors 1242, 1248, 1254 and 1260 shows approximately 100 isomers. At this time, 69 PCB isomers have been characterized but soon twice as many will be characterized. Other groups of environmental contaminants, such as the dioxins are equally complex mixtures.

TABLE 1. Number of possible PCB isomers with number of chlorine atoms per congener

Number of Chlorines	1	2	3	4	5	6	7	8	9	10
Number of Isomers	3	12	24	42	46	42	24	12	3	1

With the recent advances in instrumentation it is possible to collect large amounts of data regarding the identity, total number and the concentration of each component of these mixtures in environmental samples. Such data contain information which can be extracted and manipulated using appropriate methods of data analysis. For example, it is possible to obtain information about the origin, bioaccumulation, distribution, environmental alteration, and so forth, about such contaminants.

Due to the complexity of the mixtures and sources, and the multitude of factors which affect their alteration, it is necessary to apply multivariable methods of statistics to the data. In this laboratory we have used the SIMCA method of pattern recognition (Wold 1976) since it was developed for analysis of multivariate chemical data.

A large number of environmental problems can be formulated as classification problems based on the analysis of multivariate chemical data. In this report, the characterization of environmental samples by capillary gas chromatograms is first discussed and this is followed by the application of SIMCA pattern recognition to such data.

MATERIALS AND METHODS

Characterization of Samples by Gas Chromatograms

FIGURE 1 is a capillary gas chromatogram obtained from a sample of used transformer oil from a waste dump in New Jersey. The analytical protocol used to obtain this chromatogram was selective for PCB's so the sample appears to be similar to an Aroclor. Thus the chromatogram is a "fingerprint" or pattern which uniquely describes that sample. By systematically comparing it to the gas chromatographic patterns of the Aroclors 1242, 1248, 1254 and 1260 it is possible to classify it as to its origin.

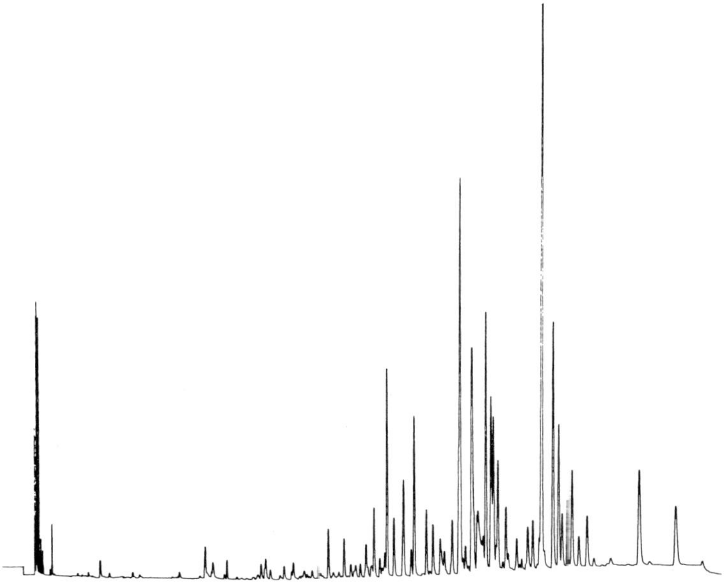

FIGURE 1: Capillary column gas chromatogram of a used transformer oil sample

The patterns of the commercial Aroclors are given in FIGURE 2. A visual comparison of FIGURES 1 and 2 indicates that the transformer oil is similar to that of Aroclor 1260. Analysis by pattern recognition methods provide a quantitative confirmation of this assignment.

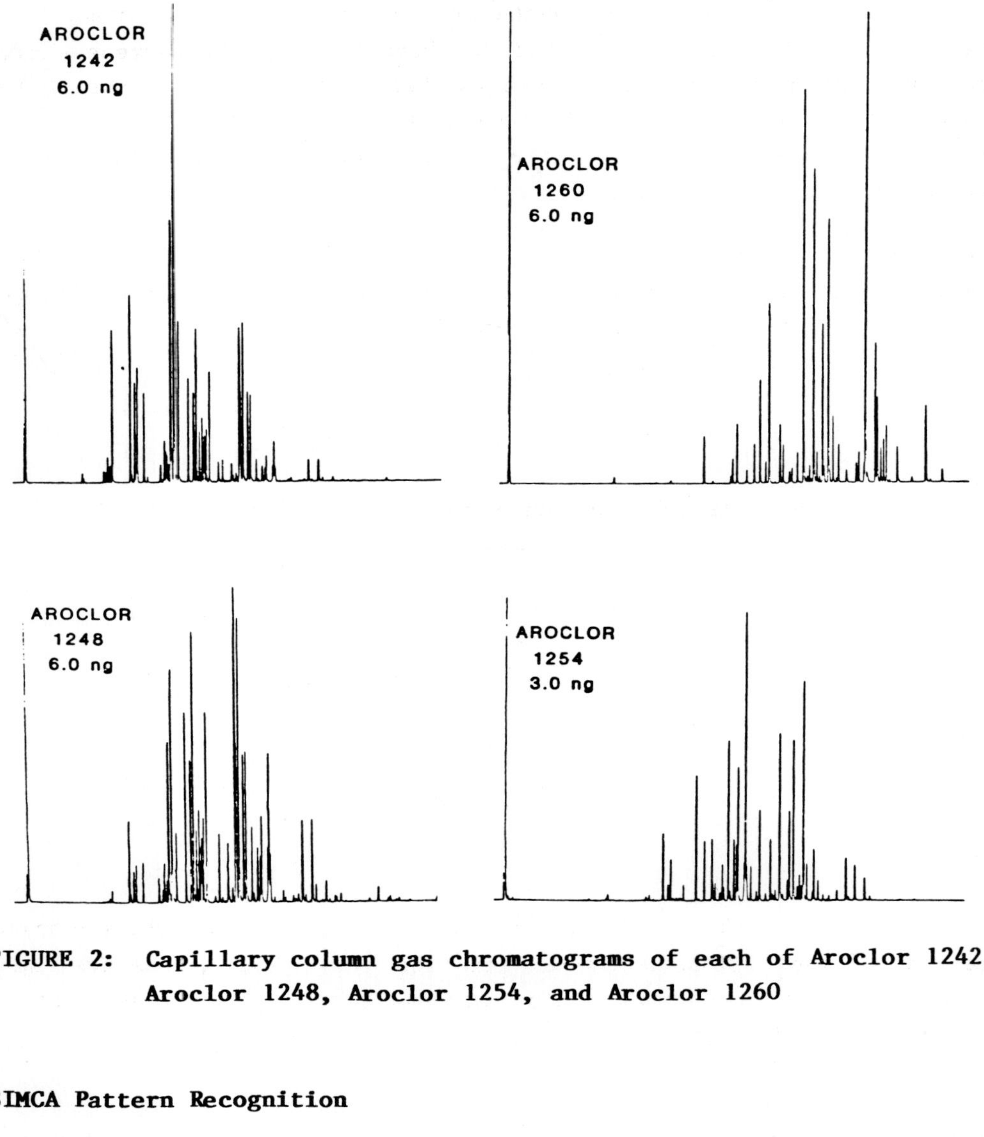

FIGURE 2: Capillary column gas chromatograms of each of Aroclor 1242, Aroclor 1248, Aroclor 1254, and Aroclor 1260

SIMCA Pattern Recognition

The objective of a pattern recognition study is to establish a set of rules which can be used for recognition of the patterns of samples of unknown class assignment. The SIMCA method derives and uses mathematical models; in this example, the models are derived from gas chromatograms of the samples of known classification, namely the Aroclors. These known samples are generally referred to as the <u>training set(s)</u> whereas the unknowns, in this example the transformer oil samples, are referred to as the <u>test set</u>.

These data are converted to numerical form by creating a table or matrix as shown in TABLE 2. Here each row is representative of a gas chromatogram with the entries being the concentrations of the respective isomers. In all, 69 isomers were considered in the Aroclor example. In matrix notation, the concentration of peak $\underline{i}$ in sample $\underline{k}$ is Y_{ki}.

TABLE 2: Generalized data matrix for the classification of PCB isomers for SIMCA pattern recognition

	Isomer Number								
	1	2	3	•	•	i	•	•	M
Sample	Isomer Concentration						Training Set		
1	Y_{11}	Y_{12}	Y_{13}	•	•	Y_{1i}	•	•	M
2						•	Aroclor 1242		
•						•			
•						•	Aroclor 1248		
k	•	•	•	•	•	Y_{ki}		•	
•								•	
N									

The matrix is divided by putting into separate classes those samples of known class assignment and the unknowns into the test set. The training sets correspond to the Aroclors and the test sets to environmental transformer oil samples.

In the next phase of the analysis the training sets are modelled by principal components models which have the form of equation 1 where q is a number of classes modelled; mi is the mean of peak i in the data set, A is the number of product terms selected in the class model and t_{ki} is the difference in observed Y_{ki} and its estimated value.

$$Y^q_{ki} = m^q_i + \sum_{a=1}^{A} b \qquad (1)$$

The product term consists of the loadings, b_i, and the principal components, t_k, the former being a peak specific term and the latter a sample specific term.

This approach to pattern recognition has a geometric interpretation. An m-dimensional (m=69) coordinate system is constructed with the axes being the concentration of the isomers. The data for each sample are projected into the coordinate system with the samples appearing as points in space. This is illustrated in FIGURE 3 in three-dimensions for convenience. In this figure the training sets have different stuctures.

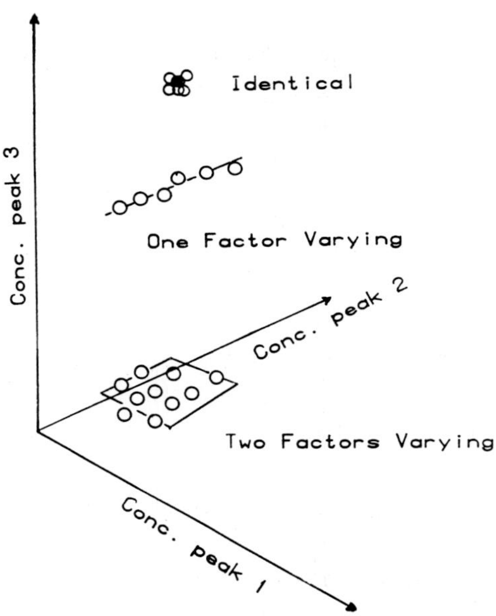

FIGURE 3: Three-dimensional projections of the SIMCA pattern recognition method for classification problems

A class containing objects which are identical (i.e., replicates) appear as one point in space. If there is systematic variation in the isomer composition within a class, this variation can be modelled by a line (equation 1 with A=1) or plane (equation 1 with A=z) in this space. Confidence intervals can be established around these class structures to define and determine space regions in descriptors in which the various classes of samples can be expected to occur.

To obtain a view of the data structure, the data for the classes can be projected onto a two-dimensional surface or into a three-dimensional projection. Such plots are called principal components plots and each sample will appear as a cluster of points with coordinates of the points being the principal components or t_{ka} terms in equation 1.

As in most classification problems based on higher dimensional data, some variables or isomers may not be important to the classification problem or may add "noise" to the analysis. With SIMCA, it is possible to determine which variables contribute to class structure. This is done by using the parameter modelling power, MPOW, defined in equation 2, where $S_{i,y}$ is the standard deviation of a variable before and S_i that after the class is modelled.

$$MPOW = 1 - \frac{S_i}{S_{i,y}} \qquad (2)$$

The range of MPOW is $0 \leq MPOW \leq 1$ with MPOW approaching 1 ideal and MPOW approaching 0 indicating noise.

RESULTS AND DISCUSSION

Application of SIMCA to Aroclor Data

To illustrate the utility of this method in classification studies SIMCA has been applied to data on the Aroclors and transformer oil (Schwartz *et al.* 1983; Dunn *et al.* 1983). In the Aroclor analysis, four training sets were established from capillary gas chromatography data on four to five replicate analyses of each commercial product and triplicate analyses of the transformer oil. Since the data have been published (Dunn *et al.* 1982) they will not be given here. A principal component projection (two-dimensional) is given in FIGURE 4 from which it is apparent that the transformer oil is similar to Aroclor 1260.

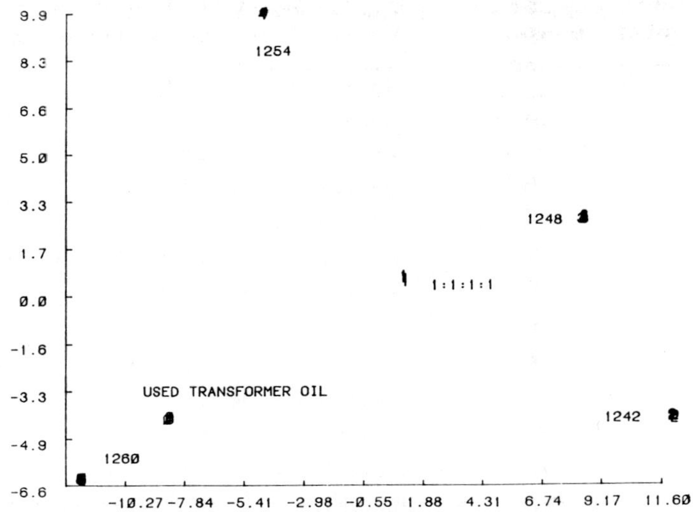

FIGURE 4: Principal components plot of the Aroclor samples and the used transformer oil sample, showing that the latter is similar to Aroclor 1260

To obtain quantitative classification of the transformer oils, a principal component model with two component terms was derived for each class and the transformer oils samples were fit to each class model. The results are given in TABLE 3.

TABLE 3. Standard deviation of fit of each transformer oil sample to each Aroclor class model[a]

Transformer Oil Sample	Aroclor			
	1242	1248	1254	1260
1	1.26	1.12	0.87	0.27
2	1.28	1.13	0.87	0.26
3	1.27	1.12	0.87	0.27

a Compare with the standard deviations of the Aroclors fit to their respective class of 1242 (0.014); 1248 (0.014); 1254 (0.012); 1260 (0.012).

It can be seen that the best fit of the transformer oil data is to Aroclor 1260 with a standard deviation of about 0.27. This is still very far from the class, however, since the fit of 1260 data to the class model has a standard deviation of 0.012, a difference of about 25 times. Thus the transformer oils is most similar to 1260 and apparently a mixture.

This example illustrates the power of SIMCA pattern recognition as applied to environmental problems. It also suggests other directions which can be taken after a classification analysis has been completed. For example, it would be good to know what the Aroclor compositon of the used transformer oil is. This is a calibration problem.

In the analysis of PCB residues in biological systems, alteration of the Aroclor pattern is obvious. Certain isomers appear to accumualte in preference to others. Also there are observable differences in the accumulation and contamination which is related to source, species, etc. Such questions require correlation methodology and we are presently exploring the use of these methods for application to environmental problems.

CONCLUSIONS

The SIMCA pattern recognition analysis of multivariate data is demonstrated on the example of the identificaiton of the compositon of PCB-like waste oils and their similarity to existing commercial mixtures. This method can equally be applied to the analysis of other problems with multivariate data, such as commonly found in quantitative structure-activity relationships.

REFERENCES

Dunn, W.J. III, Stalling, D.L., Schwartz, T.R., Hogan, J.W., Petty, J.D., Johansson, E. and Wold, S. 1983. Classification and quantification of PCB's in environmental samples using SIMCA pattern recognition. **Anal. Chem.**, submitted.

Schwartz, T.R., Campbell, R.D., Stalling, D.L., Little, R.L., Petty, J.D., Hogan, J.W. and Kaines, E.M. 1983. An isomer specific method for the analysis of PCB's in environmental samples. **Anal. Chem.**, submitted.

Stalling, D.L., Dunn, W.J. III, Schwartz, T.R., Hogan, J.W., Petty, J.D. and Johansson, E.J. 1983. Application of SIMCA, a principal components method, in isomer specific analysis of PCB's. ACS Symposium Series, D. Kurtz (Ed.), **American Chemical Society,** Washington, D.C.

Wold, S. 1976. Pattern recognition by disjoint principal components models. **Pattern Recognition 8:** 127-134.

WORKSHOP DISCUSSION

MACKAY:

Did I interpret correctly that the distribution of PCB's in fish changes markedly depending on the absolute concentration of PCB's in the fish?

DUNN:

We saw differences in absolute PCB concentrations and the variations we saw were due to differences in some of the isomer concentrations. Now, I did say further that this was not generally due to the total isomer accumulation but due to about eight or ten specific isomers which appear to accumulate preferentially. I am not familiar enough with the data to identify each of those peaks in terms of the specific isomers that accumulate in the fish samples. This could be due to species differences. All I know is that they were fish, mostly carp.

MACKAY:

What happens if some of these congeners degrade or react faster than others, will this cause your point to move around the diagram?

DUNN:

You will see that in the analysis. You will see a lot of variation of either one or several peaks in the samples. You can determine which isomers vary and then, if you want to go further, you can find out why and if you want to go even further than that, you can see if there are any effects that this variation may be inducing, but that was beyond the scope of this analysis.

HODSON:

Do you think that it would be possible to classify your fish according to rivers of origin or predatory fish according to their PCB content?

DUNN:

You very well might be able to but these are all variables which are external to the analysis and hopefully with a complete analysis you will be able to correlate this variation with these external variables which can be absorbed in the system. The only thing I can discuss at this point is variation in the data. I can't say anything other than that. We do know that the isomers are accumulating specifically in these plants, but we also see the same things in fish.

ZITKO:

The standards, are they all injected at the same concentration or did you have standards that were injected at different concentrations as well?

DUNN:

The same amount of sample was injected every time so that the concentration levels of the isomers and the total concentration that was there was the same allowing for experimental error.

ZITKO:

I was wondering what was happening if one injects different concentrations so as to simulate different concentrations in fish or plants?

DUNN:

I think that was in the analysis. We took the concentration differences away by normalizing the data. We left it in one part of the analysis.

ZITKO:

I know, but what I wonder about are differences in specific electron capture detector sensitivities?

DUNN:

The detector was an electron capture detector but the values were converted to flame ionization detector concentrations. I am not sure how they did this, but there is a proportionality factor and the concentrations were determined from a three to four point calibration curve which was done simultaneously with each peak or with each analysis.

ZITKO:

Yes, well it probably eliminated any effect.

DUNN:

I can also say that there is at most, a 5% replicate variation in sample analysis using this technique.

KAISER:

What kind of computing requirements did you have?

DUNN:

The Aroclor analysis was done on a desk top microcomputer. We are running this program specifically on a Zenith 8085 micro processor machine with 64K memory; that is all that is required. The program is written in BASIC and we run it as a compiled BASIC program. It is rather slow, this particular analysis probably took four hours to do. We also compiled the program so that it would run on the 16 bit machine so that it would be equivalent to an IBM-PC, and it runs in about fifteen minutes so you get quite an increased speed using that. Now the fish and vegetation data were run on a Cyber machine. That program is now obsolete and we haven't used it for about six months. We are trying to get a BASIC version of the program running on the mainframe but we haven't had any success at that. We are trying to dump the micro version onto the big machine and we don't have any communications software that works at this point.

ZITKO:

Are you going to publish the program somewhere, could we get hold of it?

DUNN:

The program is provided to me at no cost by Svante Wold due to a collaboration we have. Dave Stalling is distributing the micro processor based version of the program at a cost of about six or seven hundred dollars. With that, you can run a data set, you can change the dimensions of the program up to what the machine memory will accommodate and it is a very user friendly, interactive program and it is very easy to run.

KAISER:

Even for simple linear regressions, we are always suffering from lack of data, whether physical or biological. In order to be able to look at things, you obviously have many, many, data points there, many samples. What sort of minimum data requirements would you have?

DUNN:

For the principal components analysis, the more data you have, the better and the more stable the analysis becomes that means in terms of what you would call independent variables as opposed to a number of samples. You don't need as many samples as you do with regression analysis. You can get by with a lot fewer because the analysis does not actually involve the dependent variable as such, it is external to the analysis. You only do your analysis on the descriptor data. The more of that you have, the more stable the analysis becomes but we have done principal components analysis on two variables. It is kind of iffy but you can still do it, we have about thirty options.

ZITKO:

Were all those fish and plants collected in one year?

DUNN:

No, over a period of three or four years. The project began in 1978 and it continued, I think, through 1982. That is when I got involved.

ZITKO:

Do you see any trend over the years?

DUNN:

That is something we haven't looked at and I think it is something that should be done.

ZITKO:

I looked at many of our east coast samples, they contain mostly the Aroclor 1264 and two of the peaks generally decrease with time.

DUNN:

Those do decrease with time? There is considerable variation in the environmental points that we see and we haven't compared these with the standard Aroclors and that should be done. We can deconvolute a sample into its Aroclor components and then get residuals and see what variations there are.

STRUCTURE-ACTIVITY MODELS OF BIOLOGICAL OXYGEN DEMAND[1]

Kurt Enslein, Michael E. Tomb, Tom R. Lander

Health Designs, Inc.
Rochester, New York 14610
U.S.A.

ABSTRACT

A structure-activity model of biological oxygen demand (BOD) has been developed. The independent parameters consist mostly of molecular connectivity indices (MCI), pioneered by Kier and Hall. A discriminant analysis model based on approximately 250 chemicals accurately classified 88% of the classifiable chemicals, producing approximately 11% false-positive and 1% false negatives, in the presence of approximately 42% of chemicals for which no estimate could be made.

The long-term outlook for structure-activity models of BOD is favorable, particularly with the inclusion of larger numbers of compounds and additional independent parameters.

INTRODUCTION AND PURPOSES

There has been for some time a need to be able to estimate biodegradability of chemicals in waste streams. One of the measures of biodegradability is biological oxygen demand (BOD). BOD is measured under laboratory conditions, and as such, should be reasonably reproducible across experiments.

[1] Work performed under an EPA-Duluth subcontract from Montana State University.

K. L. E. Kaiser (ed.), QSAR in Environmental Toxicology, 89–109.

A series of BOD's were collected from the literature under another project (Quinlivan et al. 1981). The question then arose as to whether these could be explained (modelled) by means of recently emerging structure-activity relationship techniques (SAR). While a variety of techniques are currently being used, in this project the techniques were limited to those embodied in molecular connectivity indices (MCI). The MCI techniques have been pioneered by Kier and Hall (1976), and embodied in a variety of computer programs.

In this report, then, we describe the results of a modelling effort given a set of chemicals for which BOD's had been measured or collected from the literature, and for which MCI's were calculated. Given robust enough equations, it would be possible to estimate the biodegradability of chemicals for which the measurements have not been performed, from their structures and the computed MCI's.

DATA BASE

Data Base Assembly

The data base was assembled from two sources: a collection of BOD's obtained for variable numbers of days of experimental exposure on the one hand, and MCI's calculated at EPA Duluth on the other hand (Weininger 1982). It was decided that rather than using data from experiments having different endpoint duration, the effort would be limited to those values obtained at day five. A total of 379 compounds with the required data thus became available.

These compounds were then coupled to the relevant MCI's. the net result was a data base of 252 to 254 compounds for which up to sixth order MCI's were available, and for which at least one BOD value at day five had been measured. For some of the chemicals MCI's up to tenth order were calculated. However, since these represented an inadequate number of chemicals, the modelling effort was limited to those with sixth order MCI's.

Below then, a description of the various parameters available on each of the compounds in the data base:

- BOD: For compounds with a single observation of BOD at day five, that value was used. For compounds with multiple observations of BOD's at day five, the highest was used when the model required the peak BOD, and the average was used when mean BOD was required.

- SEED CODE: Seeding information was available for a number of compounds. A seed code = 1 meant that seeding had taken place, and code = 2 indicated that the seeding had been acclimmated. A code of zero indicated that no information as to seeding was available; however, it was inferred from the BOD distribution as a function of seeding that most of the chemicals with code zero in fact were not seeded. The number of chemicals with various levels of seeding are shown below (when multiple observations with different seed codes were available, only the highest seed code is shown).

Seed Code	Number of Chemicals
0	171
1	22
2	62

- ATOM COUNTS: These were the simple counts of number of oxygen, carbon or hydrogen atoms as well as the sum of these for each compound.
- MCI'S: For each molecule a set of variables were calculated consisting of the following MCI types:

  Simple
  Valence-adjusted
  Framework

For each of these types, path, cluster, chain and path-cluster indices at the different levels were computed. A variety of totals were then also calculated based on the number of subgraphs, valence-adjusted, graphs, etc., as well as the total indices for path-cluster, etc., independently across all levels.

A total of 108 variables were thus generated. For many of these, due to the structure of the molecules, non-zero values were available for only a relatively small number. Consequently, those variables were removed from the data set which had distributions so skewed that it would be unlikely that they could contribute materially to the statistics, or that would bias the statistics severely. This left a total of 88 variabl~s with which the various regressions and discriminant analyses were intially performed. As will be seen further on, a number of additional variables were subsequently removed at different stages for a variety of reasons.

It was decided from the start that no transformation of MCI's would be performed; we felt that sufficient numbers of variables could be retained by eliminating those that might require transformation, without reducing the amount of information in the data base materially, as many of of the variables were expected to be highly correlated. This did, in fact, turn out to be the case. The simple statistics of the data in the complete data base are shown in TABLE 1[2].

The next step was to obtain distributions of the mean and peak BOD's and these are shown in FIGURES 1 and 2. Further discussion of these distributions will be deferred to the modelling section.

STATISTICAL METHODOLOGY

The problem in most modelling tasks in which a variety of correlated independent variables are available is to select those parameters which explain most of the variability in the endpoint, without introducing near-singularities which might cause the equation to be sensitive to small fluctuations in the data. Over the past several years, the practice has evolved to use stepwise aggregation and deletion techniques to this effect. In this instance, one first allows those candidate variables to enter the equation which have any chance at all contributing to the explanation of the endpoint. After all these have been entered into the equation, one then removes variables one at a time, starting with that variable which contributes

---

[2] Copies of TABLES 1 to 9 and of supplementary data on compounds, variables and discriminant analyses are available, on request, from the authors.

least to the explanation of the endpoint. This process terminates when the lowest variable still in the equation meets a given variance criterion (F). It has been shown that the information in the equation is a maximum when F = 1.7.

While one typically follows this step with a ridge regression procedure to remove variables which introduce near-singularity in the correlation matrix of those variables included in the equation, in this instance the number of such variables was low enough and their tolerance sufficiently high that it was concluded that the use of ridge regression was not indicated. The computer programs used in this project were those from the BMDP collection of the University of California (Dixon and Brown 1981).

MODELLING

In this section we will recite the step-by-step efforts embodied in various computer runs which led to the final discriminant model. We will not dwell on those runs which did not result in some degree of progress towards the final goal. We will describe the various computations in terms of run numbers so that these can, if desired, be traced back to computer outputs for future reference.

Run 1 Even though it was clear from the distributions of FIGURES 1 and 2 that one did not deal with a simple unimodal distribution, the first regression was nonetheless performed to get some idea of the power of the various variables in explaining peak BOD. The results are shown in TABLE 2. They indicate a rather poor fit with a large standard error of estimate (26.4%). An examination of the residuals shows that the BOD values near the extremes are particularly poorly fitted as are also several subpopulations in the midrange. From this, we concluded that the next step should be a piecewise linear regression to see if the assumption of multiple distributions was in fact justified.

Run 2 Before this run was undertaken, one compound (EPA #575) was removed because it had been identified as an obvious outlier from the previous run. In Run 2 we assumed that the data consisted of three subdistributions: those BOD's in the

FIGURE 1: Frequency distribution of mean biological oxygen demand (BOD)

All Compounds in Database

Midpoint Mean BOD		FREQ	CUM. FREQ	PERCENT	CUM. PER-CENT
0	***********************************	35	35	13.73	13.73
2	****************	16	51	6.27	20.00
4	*******	7	58	2.75	22.75
6	**********	10	68	3.92	26.67
8	**	2	70	0.78	27.45
10	*******	7	77	2.75	30.20
12	*****	5	82	1.96	32.16
14	****	4	86	1.57	33.73
16	******	6	92	2.35	36.08
18	******	6	98	2.35	38.43
20	*******	7	105	2.75	41.18
22	***	3	108	1.18	42.35
24	*	1	109	0.39	42.75
26	*	1	110	0.39	43.14
28	******	6	116	2.35	45.59
30	**	2	118	0.78	46.27
32	***	3	121	1.18	47.45
34	*****	5	126	1.96	49.41
36	****	4	130	1.57	50.98
38	******	6	136	2.35	53.33
40	*****	5	141	1.96	55.29
42	*****	5	146	1.96	57.25
44	****	4	150	1.57	58.82
46	*****	5	155	1.96	60.78
48	**	2	157	0.78	61.57
50	****	4	161	1.57	63.14
52	********	8	169	3.14	66.27
54	*******	7	176	2.75	69.02
56	********	8	184	3.14	72.16
58	*****	5	189	1.96	74.12
60	*******	7	196	2.75	76.86
62	********	8	204	3.14	80.00
64	*******	7	211	2.75	82.75
66	*****	5	216	1.96	84.71
68	**	2	218	0.78	85.49
70	******	6	224	2.35	87.84
72	****	4	228	1.57	89.41
74	*****	5	233	1.96	91.37
76	******	6	239	2.35	93.73
78	****	4	243	1.57	95.29
80	*	1	244	0.39	95.69
82	*	1	245	0.39	96.08
84	***	3	248	1.18	97.25
86	*	1	249	0.39	97.65
88		0	249	0.00	97.65
90		0	249	0.00	97.65
92		0	249	0.00	97.65
94		0	249	0.00	97.65
96		0	249	0.00	97.65
98	****	4	253	1.57	99.22
100	**	2	255	0.78	100.00

```
-----+----+----+----+----+----+----+
     5   10   15   20   25   30   35
```

FIGURE 2: Frequency distribution of peak biological oxygen demand (BOD)

All Compounds in Database

Midpoint Peak BOD		FREQ	CUM. FREQ	PERCENT	CUM. PER-CENT
0	***********************************	35	35	13.73	13.73
2	***************	15	50	5.88	19.61
4	****	4	54	1.57	21.18
6	**********	10	64	3.92	25.10
8	****	4	68	1.57	26.67
10	***	3	71	1.18	27.84
12	****	4	75	1.57	29.41
14	**	2	77	0.78	30.20
16	***	3	80	1.18	31.37
18	*	1	81	0.39	31.76
20	***	3	84	1.18	32.94
22	**	2	86	0.78	33.73
24	****	4	90	1.57	35.29
26	****	4	94	1.57	36.86
28	*	1	95	0.39	37.25
30	********	8	103	3.14	40.39
32	***	3	106	1.18	47.57
34	***	3	109	1.18	42.75
36	*****	5	114	1.96	44.71
38	*****	5	119	1.96	46.67
40	****	4	123	1.57	48.24
42	*	1	124	0.39	48.63
44	***	4	127	1.18	59.80
46	***	3	130	1.18	50.98
48	***	3	133	1.18	52.16
50	***	3	136	1.18	53.33
52	**	2	138	0.78	54.12
54	*******	7	145	2.75	56.86
56	***	3	148	1.18	58.04
58	******	6	154	2.35	60.39
60	******	6	160	2.45	62.75
62	********	8	168	3.14	65.88
64	******	6	174	2.35	68.24
66	*****	5	179	1.96	70.20
68	**********	10	189	3.92	74.12
70	******	6	195	2.35	76.47
72	*********	9	204	4.54	80.00
74	**	2	206	0.78	80.78
76	****	4	210	1.57	82.35
78	******	6	216	2.35	84.71
80	*******	7	223	2.75	87.45
82	*	1	224	0.39	87.84
84	*****	5	229	1.96	89.80
86	**	2	231	0.78	90.59
88	*	1	232	0.39	90.98
90		0	232	0.00	90.98
92	*	1	233	0.39	91.37
94	****	4	237	1.57	92.94
96	***	3	240	1.18	94.12
98	*************	13	253	5.10	99.22
100	**	2	255	0.78	100.00

```
-----+----+----+----+----+----+----+
     5   10   15   20   25   30   35
```

region of 24% and below, those between 24% and 56%, and those above 56%. Appropriate indicator variables were then inserted for these three ranges and a stepwise regression again performed. Part of the results are shown in TABLE 3.

While the coefficient of determination is of course very large due to the insertion of the indicator variables, it is interesting to note that the standard error of estimate has been reduced to approximately 8.8%. In this regression, mean BOD rather than peak BOD was used on the theory that the mean would be a more stable value than the peak.

An examination of the residual plots indeed confirmed the hypothesis that the data consisted of more than one population of compounds. One also found that a number of compounds near 100% BOD were very poorly fitted. It is not clear whether these represent unusual chemicals or poor experimentation.

In view of the fact that the distributional hypothesis was confirmed, the next step consisted of a discriminant analysis shown below.

Run 3 For this run the data were separated into two groups: those with BOD $\leq 24\%$ (n = 105) And those with BOD $>24\%$ (n = 151). The resulting classification functions are shown in TABLE 4. While approximately 75% of the compounds are correctly classified, an examination of the posterior probabilities of classification shows that many of the misclassifications are near p = 0.5, i.e., in the indeterminate region (see definition below). This is further confirmed by the histogram of the canonical variable shown in FIGURE 3. In other words, if the results from this discriminant run were to be used, one would undoubtedly find that the degree of misclassification (i.e., false-positives and false-negatives) would be relatively small but that a rather large fraction of compounds would fall in the indeterminate range around P = 0.5.

We were now concerned as to whether those compounds that had been evaluated in seeded experiments contained more reliable information than the others. We were also concerned about compounds with BOD values >85. The latter were redistributed more or less randomly to values near 85. As to compounds with seed codes, a variety of regressions were run, usually witn peak BOD as the independent variable. The distribution of the compounds with seed code resembled the distribtuion of the entire population remarkably well. Perhaps because of this, the regressions based on seeded compounds alone did not provide information beyond what had already been learned.

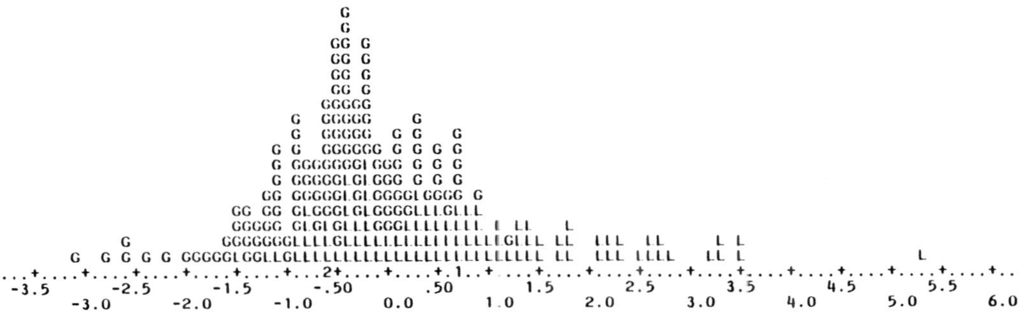

FIGURE 3: Stepwise discriminant analysis on mean BOD histogram of canonical variable, L = BOD ≤ 24%, G = BOD >24%

The next question that arose was whether the "total" parameters introduced any near-singularities in the correlation matrix, and might possibly have a deleterious effect. As a result, the next run reported below was performed.

Run 8 In this run, most of the "total" variables were removed. A few of the "total" variables were left in the data set because they represented totals for variables that had been removed due to their containing too many zeros. For example, Valtot-4 (the sum of all fourth-order valence-adjusted indices) has been kept because it totals the fourth-order valence path and path-cluster indices as well as the valence fourth-order chain and cluster indices, which had too few non-zero values to be used. In addition, a parameter for seed code was included, i.e., zero if no seed code and one if any seed code was associated with a particular BOD.

The results are shown in TABLE 5. It turned out that seed code was not important enough to be included in the equation. The removal of the "total" variables also did not materially influence the results. The resulting equation may be somewhat more robust than that derived from Run 1.

In view of the fact that the distributions shown in FIGURES 1 and 2 showed a peak near zero BOD we wanted to see next if one could fit the remainder of the distribution beyond the "low" BOD region more adequately by removing that part of the distribution.

Run 9 In this run, a regression was done on peak BOD for the 156 chemicals with BOD values greater than 28.8%. The results are shown in TABLE 6. The standard error of estimate (14.9%) is lower than when all compounds are included, but still of course higher than for Run 2, the piecewise linear regression experiment. The equation shows that two parameters dominate the expression, i.e., seed code and simple fifth-order clusters. Note that the coefficient for seed code is negative. While this would seem to be a paradoxical result, it is due to the fact that compounds near the positive end of the range tend to be the result of experiments performed without seeding, whereas this is not the case for the lower end of the range. In fact, the simple correlation coefficient between seed code and BOD for the compounds in this regression run is -0.31. One would imagine this to be the natural result of experimental design in that one would first perform an experiment without seeding, and if that experiment produced a large enough BOD one would never proceed to a seeding experiment.

Thus, it would seem that a single parameter, simple fifth-order clusters does most of the explaining for the compounds with BOD >28.8% since the seeding code would simply act as a covariate. However, the correlation of this term with BOD is only 0.1654, thus explaining only approximately 2.7% of the variance in BOD.

This relatively poor performance led us to decide that the only appropriate approach to modelling the BOD data in hand was a discriminant analysis procedure. This is shown in the next section.

Run 10 In this discriminant run, two groups were defined, one with BOD $\leq 28.8\%$, the other with BOD $> 60\%$, i.e., a group with "low" BOD and a group with "high" BOD. This selection resulted in 184 chemicals, 90 chemicals in the low BOD group and 94 in the high BOD group. The classification table is shown in TABLE 7, and the functions in TABLE 8.

While one would be concerned about the relatively low accuracy of classification for the lower BOD group, one should bear in mind that a compound is called misclassified if its posterior probability of classification is not at least 0.500. Thus, even marginally misclassified compounds are included in the 25 shown in TABLE 7. In our previous work, we have usually assumed that a region near chance, (i.e., p = 0.5 should be considered an indeterminate region), i.e., a region in which no decision can be made. Our custom has been to use 0.5 ± 0.2 as a definition for this region. With this assumption, the classifications shown in TABLE 9 result. From this TABLE it is seen that a large proportion (almost 42%) of the chemicals fall in the indeterminate range, that the false-positives represent about 11% and the false-negatives less than 1% of the compounds which can be classified. Thus, one can conclude that for those compounds to which it is applicable, the equation can distinguish non-biodegradable compounds from biodegradable compounds with a high level of accuracy, i.e., it is very unlikely that a compound will be identified as non-biodegradable when it is in fact biodegradable. As such within the limited applicability of the equation it can be used for predictive purposes.

DISCUSSION

Misclassifications

In the analysis of the results from Run 10 (the last discriminant analysis), it was noted that most of the misclassifications (TABLE 7) occurred in the first set of compounds received from EPA. Virtually no errors of classification occurred in the second set. An exhaustive investigation was made to determine if the two sets differed in the types of compounds, their complexity, the distribution of the BOD values, etc. No reasonable relationship could be detected. It is therefore most likely that some of the BOD values themselves in the first set of compounds are in error. This is particularly so since the probability of misclassification is only mild, i.e., not much larger than 0.5 for most of the misclassified chemicals. The net result is that an unusually large fraction of chemicals are classified into the indeterminate range (TABLE 9). More accurate BOD values thus would be likely to move many of these chemicals into the correctly classified or severely misclassified region.

Usefulness

It is clear that even with these limitations (and several others to be detailed below) the discriminant structure-activity model is useful in that it will call a compound non-biodegradable when in fact it is degradable only about 1% of the time, and will call a compound degradable when in fact it is non-biodegradable about 11% of the time. It should be pointed out that, as in all discriminant analysis procedures, these error rates can be manipulated between the false-positive and the false-negative groups, so as to, for example, emphasize the false-negatives as contrasted to the false-positives. The results quoted above have been achieved without such adjustments.

Limitations

Two major limitations to the applicability of this model exist: It is limited to compounds of relatively low complexity since these were the compounds that formed the basis of the current model; and some 40% of the compound of that complexity which would be submitted to the model in its current state would probably not be estimable since they would be likely to fall in the indeterminate range.

Possible Improvements

From the various computer runs it would appear that some parameters besides molecular connectivity indices might be useful to further improve this model. These parameters would include substructural keys, area and volume, and vapor pressure, among others. All these parameters are currently in existence and only limitation of funds has prevented us from experimenting with them. It would also of course be helpful to include a larger number of compounds, together with compounds of greater complexity, in order to make the model more generally useful.

Asymptotic Outcome

In the long run, given further parameters and compounds, it is still relatively unlikely that one will be able to model BOD by means of regression equations due to the likelihood that the observations will be concentrated at the extremes of the distribution. Of course, this is an ideal situation for the discriminant analysis approach. It is likely that the misclassification percentages can be improved through the application of the steps outlined above.

REFERENCES

Dixon, W.J. and Brown, M.B., Eds. 1981. **BMDP-81 Biomedical Computer Programs, P-Series.** University of California Press, Berkely, California.

Kier, L.B. and Hall, L.H. 1976. **Molecular Connectivity in Chemistry and Drug Research.** Academic Press, New York.

Quinlivan, S., Powers, M. and Painter, P. 1981. **Compilation of data on biochemical oxygen demand of organic chemicals.** TRW, Redondo Beach, California. EPA Contract #68023174, November 2, 1981.

Weininger, D. 1982. **INDEX: A Computer Program for the Computation of Molecular Connectivities.** Environmental Protection Agency - Duluth Laboratory, Version 7.

WORKSHOP DISCUSSION

Question:

Could you explain your variables?

LANDER:

Yes. These indicate molecular connectivity parameters, the number of hydrogens, the number of oxygens and so forth. There are three types of connectivity parameters: there is simple, valence-adjusted and framework. Valence-adjusted is calculated as: (number of building electrons minus number of hydrogens) over (number of core electrons minus one); for example, formaldehyde °Xpath = 1.11. Simple connectivity is calculated by reducing all valences to account only for the number of non-hydrogen bonds, thus the °Xpath for formaldehyde is 1.41. Framework connectivity is calculated by reducing the entire skeleton to that of a saturated hydrocarbon, hence formaldehyde is depicted as carbon-carbon, and its °Xpath then is 2.00. Under each of the three types you have got path, which is simply a linear path of atoms, and zero order which is simply the number of atoms. First order if from one atom to the next, second order if three atoms in a row and so on. You will notice that we never get anything over sixth order because these compounds are so simple. Chain is a ring compound and clusters are branch points.

Question:

If you look at your BOD numbers that you based all this on, did you go back and check the classification and find out whether or not there was some discrepancy in procedure or something in the literature that might be wrong?

LANDER:

We did not go back to the original literature. We assumed that they had been screened properly. That is one thing we would like to have done. We did look at the outliers during modelling to see if there was any similarity in their structures; there was not.

Comment:

Sometimes that happens when the data you are working with have some numbers that aren't exactly right and that might affect the correlation. You may want to check that, it might improve this.

Question:

In classifying your compounds initially into two groups, was this an arbitrary group definition?

LANDER:

This was done based on the BOD values. We based it on those distributions which I showed at the beginning and we just tried moving the break points several times.

Question:

Was there a natural grouping that you could discern?

LANDER:

It was pretty obvious that quite a few were down around zero and 1% so that was an obvious grouping.

SCHULTZ:

You said that these were, for the most part, simple compounds and in the end you basically regrouped them into groups that would not be quantitative. Did you look at the substituents associated with these different compounds? For example, those groups you could not define? Did they contain groups that tend to interact with water such as carboxyl or hydroxyl groups?

LANDER:

No, we were not able to do that.

WELLS:

The BOD data that you originally used came, I assume, from one of the standardized BOD methods. Is there any correlation between the BOD method that generated that data and BOD or degradation rates that you would expect under field conditions? In other words, is there any field verification for the BOD data that you used?

LANDER:

We were not involved in the actual data collection. They were screened at the University of Wisconsin.

KAISER:

I have noted a few gross errors in the Registry of Toxic Substances like milligrams instead of nanograms or vice versa. Normally, it is quite obvious if you look at the whole series of particular data. The compilation isn't very accurate and I have noticed wrong formula and molecular weights too. What I would like to know is whether it is the perception here that we can really use this or should use models like that, for example, to find out whether there are any new contaminants which may be of concern in the Great Lakes? Do you find it satisfying to say with 70% to 80% certainty that this may or may not be of concern? Is it better than having no idea?

Comment:

It is probably better than having no idea. I think it has a utility. It is probably good, as someone suggested, to use QSAR in general as a screen, a preliminary. I'm not sure that we will ever get to the point where we can have ultimate confidence in the predictions we make with QSAR but as a first step, I think it is very important. In many instances it will give us the trigger to know what we need to measure or perhaps to forget about that compound. I don't know whether we can ever become confident enough to say that we are past the preliminary screening.

ZITKO:

I think the important thing is to have fuzziness built in so that we can say, 'Okay, we don't know about these compounds, we can't tell'.

Comment:

Really, the problem is now that we don't have a rule to go by to measure exceptions, if we had a rule then we could start looking at exceptions to the rule but without a rule, we have nothing to work with at all.

HALLETT:

As I'm sitting here, I am interested in the last few comments, essentially you have got one test and you have got a model. If you took a series of correlations, it is like taking a battery of tests such as the genetic batteries and other batteries, and organize them mathematically into a seris of correlations, perhaps you may develop an even greater level of error but maybe you won't. Perhaps you will miss some in one battery, in one QSAR, and pick them up in another so that you will be prioritizing as well. If you take what you get from one test or QSAR, and put it through another QSAR you would see what was common in those and you would have a pretty good idea what was important and you would still have your outliers that you could question.

KAISER:

Were you thinking along the lines of a hierarchy of different tests?

HALLETT:

Developing QSAR on one series of structures and one series of tests and then developing another one and another series, that is the state of the art now, you've not built them all into one huge body.

KAISER:

Do you think that different tests have different usefulness? We can fit them the same way mathematically and you get 70%, 80%, or 90%. Do you feel that there are some tests which are better or let's say more useful in describing whether or not the compound has a potential to become a problem or do you figure they are all equally good or equally bad or whatever?

Question:

What do you mean by tests? Are you talking about bioassays?

KAISER:

Bioassays, yes. I include BOD or anything.

SCHULTZ:

Aren't we limited to that, one, sensitivity of organisms, two, the data base that we have available and at this point you can't really ignore Daphnia. You can't really ignore a fish of some sort just because there is such a wealth of data already available on those tests.

KAISER:

That is the question, what is more important, Daphnia or the fish, or are both equally important?

SCHULTZ:

I think it is important to look at those higher fishes that have the arylhydrocarbon hydroxylase activity and are able to activate chemicals that are carcinogens and you are able to see tumors in a fish; you certainly don't see tumors in Daphnia.

HALLETT:

I think you are looking at a very useful tool here in organizing data and expressing such data through models - one model type or another type and in that sense I would think that you could basically take them on a hierachy or models applied to data sets.

LANDER:

It all depends on what you are looking for. It depends on what test is important to you. It's not up to us to say what tests you want. Our models simply model the data that exist. The models are only as good as the data. It also depends on whether you want to regulate a chemical based on mutagenesis, based on carcinogenesis, etc.

HALLETT:

It is not just to test the compounds, but to get them crunched into these models and then you are using the data that you've got on the toxicity.

Comment:

So that when you get something new, you have some parameter, some chemical measure, the log P, or whatnot, you can just put that in each one of these tests and see where it fits and something may not occur in one of the models but it may show up in one of the other models and that may resolve the outliers.

LANDER:

I still want to point out that it is not meant to replace testing and you should not ignore anything that is predicted negative but maybe first concentrate on those for which positive values are predicted.

Comment:

False negatives and false positives would mean different things to different people and if you were in industry, false positive could be very expensive if someone suggested the chemicals could be controlled. If you are protecting the environment, a false negative could mean a real problem down the road. Could your models be set up in such a way that you could minimize either false positive or false negative?

LANDER:

Yes, you can set it so that it will minimize the false negatives at the expense of the false positives or vice versa.

CULLEN:

In the BOD model, is it classified as degradable or not degradable?

LANDER:

That is just the term we have to apply. It is really a biochemical oxidation. Anything less than 28% we called non-degradable. That doesn't mean that its not going to degrade by other methods.

Question:

Given that, your model predicts with an accuracy of approximately 50%, wouldn't a good organic chemist be able to just guess and get almost the same accuracy as your model?

LANDER:

I can't answer that.

KAISER:

I think the question in part is, what types of compounds were included or excluded in the set, where one could make a good estimate such as PCB's or anything like that?

HALLETT:

One thing that is going to come out of this is basically a really good list of data that is incorrect. I think with the systematic use, you are going to come up with a great deal of data and it may be actually useful to organize that and basically publish it some way.

LANDER:

We found twenty-five errors in the data out of those nine hundred compounds in the list.

HALLETT:

One of the major flaws in a lot of that data is that people don't measure the concentrations that they are testing. They calculate what should be and leave it at that. This can cause errors with two or three orders of magnitude. If you put that into a model, it is going to make it very difficult to find anything at all. So, when you are assembling data for a model you have to be very careful that you use data that are the most representative or otherwise you will never get a model to work right.

Comment:

I think it boils down to some degree to the fact that the modellers are years ahead of the experimentalists in creating data.

Comment:

I think the problem lies in standardizing the tests, the tests are still being standardized, yet you modellers are demanding data that are very comparable. This is years ahead of the labs that produce them, so unless your model has built in wide confidence limits, the data are useless.

LANDER:

Generally, the severely incorrect data will show up as outliers in a classification matrix.

Comment:

In some cases though, the correct data showed up as outliers.

LANDER:

Yes, it could be! We usually like to withhold as much data as we can for testing, but in this case we just weren't able to. Most of our models were not of homogeneous sets. That is another thing we are thinking about doing although a lot of times it comes down to the question of what group does this compound go into. In one way, what we've got is one big model, but really it is a series of small models.

DUNN:

That is contradictory. You treat them as a homogeneous set of chemical structures.

LANDER:

We treat the endpoint as homogeneous, as evenly distributed. We do not, however, treat the structures in most of our models as a homogeneous set because the substructural parameters are present only in certain classes of chemicals, hence are a set of "indicator" variables. In this (BOD) model, the independent parameters are continuous, and thus the data set is treated as a homogeneous one.

DUNN:

But you describe them, the molecules, from data that you assume to be homogeneous so you can't miss.

LANDER:

You hope that the variables that you have are going to be able to separate them.

KAISER:

I think this was was Rekker's complaint about them.

LANDER:

The only way we can respond to that is by withholding test compounds and so far we've been satisfied with those where we have had enough data, mainly the mutagenesis model.

HALLETT:

I think you had one series that you bascially got a much better correlation when you took out steroids. Did you look at that data set and examine that as to why?

LANDER:

We just assumed that the carcinogenic effect of steroids was not direct and there is probably no way to currently model that effect.

ENVIRONMENTAL HAZARD PROFILE – TEST RESULTS AS RELATED TO STRUCTURES AND TRANSLATION INTO THE ENVIRONMENT

Dieter Freitag, Jan P. Lay and Friedhelm Korte

Ges. für Strahlen- und Umweltforschung mbH München
Institut für Ökologische Chemie
Ingolstädter Landstr. 1, D – 8042 Neuherberg
FEDERAL REPUBLIC OF GERMANY

ABSTRACT

The "Environmental Hazard Profile" test scheme is described and relative bioaccumulation and biodegradation rates of approximately 100 chemicals belonging to various classes are given. Additional accumulation and degradation rates are given for representative benzene, benzoic acid, and phenol derivatives, and for selected aromatic amines, chlorobenzenes, and chlorobiphenyls. The "sphere fragmentation" of chemicals is shown on the examples of 3-methylbenzoic acid and acetaldehyde. The effects of 2,4,6-trichloro- and pentachlorophenol on the biota of outdoor experimental ponds are described.

INTRODUCTION

One problem frequently encountered in attempts to develop QSAR in the environmental field is a lack of data developed under identical experimental conditions. There is a vast number of such data from screening tests undertaken in industry. This includes hundreds of thousands of chemicals, but either the data are kept confidential or they are not reported in quantitative terms. The adoption of guidelines by the Organization for Economic Cooperation

K. L. E. Kaiser (ed.), QSAR in Environmental Toxicology, 111–136.

and Development (OECD) for testing of chemicals will improve the situation when so-called "existing chemicals" are tested according to these guidelines and the results will be published. In order to proceed, we developed a standardized test system and have started the evaluation of its results with respect to QSAR in cooperation with a university department, interested in tayloring chemicals. The first part of the results presented here, were obtained from a laboratory test system, which we call "Environmental-Hazard-Profile" (SCHEME 1).

Concept:	- Comparative assessment in a system of tests
Aim:	- Ranking of organic chemical substances - Screening for ecotoxicological potential - Priority list for further studies

SCHEME 1: The "Environmental Hazard Profile"

Our idea was to construct a simple, fast and inexpensive system of tests. The result should allow a ranking of environmental chemicals with respect to their accumulation and degradation potentials for various test species.

For this purpose, we used ^{14}C-labelled substances in order to facilitate the experimental procedure and to avoid analytical limitations. In general we used chemical concentrations of about 50 $\mu g \cdot L^{-1}$.

The "**Environmental Hazard Profile**" tests (FIGURE 1) comprise:

- bioaccumulation in algae (Clorella fusca) and fish (Goldorfe), representing aquatic organisms,
- retention, dispersion and excretion in rats (Wistar, male) as representative of warm-blooded animals,
- degradation, transformation and accumulation in activated sludge as microbial processes, and
- degradation and mineralization by photoirradiation, simulating abiotic processes in the atmosphere.

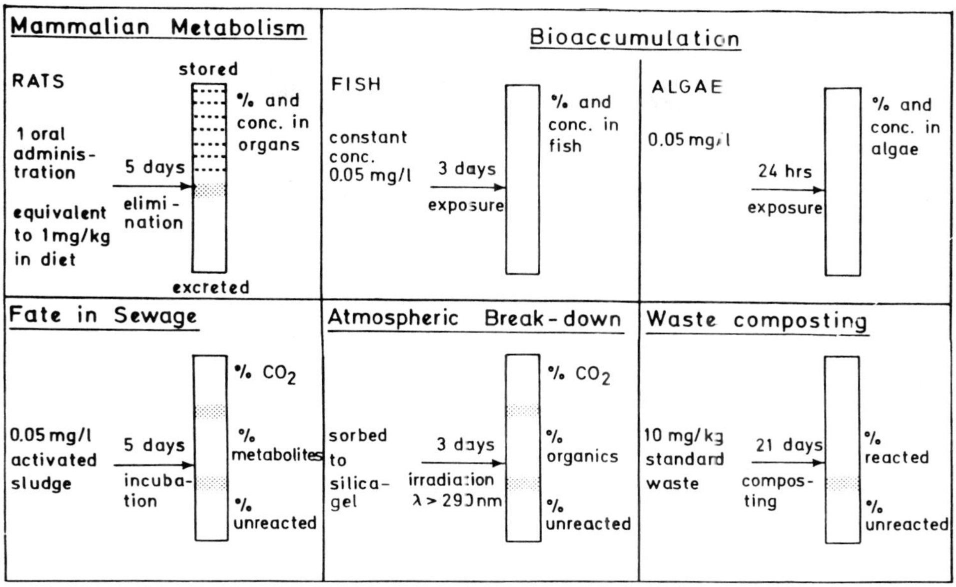

FIGURE 1: Test systems for the Environmental Hazard Profile analysis

RESULTS

General Observations

So far we have tested 100 organic chemicals with different structures covering a wide range of pollutants. TABLES 1 and 2 show the bioaccumulation and the mineralization data obtained from the activated sludge tests. The substances are listed from the highest to the lowest accumulation factors and mineralization rates, respectively. In general, highly chlorinated aromatic compounds show high accumulation rates (TABLE 1). Conversely, their mineralization rates, as shown in TABLE 2, decrease in the same order and are generally less than 1% CO_2. The bioaccumulation and biodegratation data of the first 35 substances tested, obtained from each of the five tests of the system, are also shown in FIGURE 2.

In FIGURE 2, the results of the single tests are contrasted. They are illustrated as bar graphs, representing the mentioned "Environmental Hazard Profile". The ranking is in order of increasing water solubility. It is evident that a negative correlation between accumulation factors and water solubilities exist.

TABLE 1: Relative bioaccumulation rates of organic chemicals in activated sludge; accumulation factor (BF), BF = [μg substrate·(g sludge)$^{-1}$]·[μg substrate·(g water)$^{-1}$]$^{-1}$

Compound	BF	Compound	BF
Pigment TK 10466	665300	Pentachlorophenol	1100
Pigment TK 10090	85730	Vinylchloride	1100
Dibenzanthracene	42800	Naphthaline	1000
Hexachlorbenzene	35000	Trichlorethylene	990
2,5,4'-Trichlorobiphenyl	32000	Cypermetrin	970
Acetic acid (Na-salt)	29100	Phenanthrene	930
2,4,6,2'4'-Pentachlorobiphenyl	27800	2,4,6-Trichloraniline	870
		Hydroquinone	870
Chlorhexidin	26700	γ-Hexachlorocyclohexane	820
Benzanthrancene	24400	Cortisonacetate	660
Perylene	22900	1,4-Dichlorobenzene	560
Aldrin	18000	Succinic anhydride	560
Dieldrin	17600	Aniline	500
Pentachlorobenzene	14300	Carbon tetrachloride	480
DDT	14000	Methanol	470
Docosane	10100	Phenylenediamine (hydrochloride)	460
Benzo-a-pyrene	10000		
Kepone	9900	2,4-Dichlorophenol	340
2,4'-Dichlorobiphenyl	9800	2,4-Dichloronitrobenzene	310
Ethylendiamine (hydrochloride)	8800	Cumaphos	290
		4-Chloroaniline	280
ADPA	6800	Maneb	250
Anthracene	6700	4-Isopropylnitrobenzene	240
2,4,6,2'-Tetrachlorobiphenyl	6500	4-tert-Butylphenol	240
2,2'-Dichlorobiphenyl	6300	Ethyleneglycol	200
Palmitic acid ethylester	5000	Urea	190
Quintozen	4500	Diethyleneglycol	180
Malonic acid diethylester	3700	Chlorferron	170
Tristearin	3600	4-Chlorobenzoic acid	170
Acetic acid ethylester	3300	Dodecylbenzene sulfonate (Na-salt)	140
Hexadecanol	3200		
3,3'-Dichlorobenzidine	3100	Zineb	130
Phthalic acid (2-ethyl) hexylester	3000	Thiourea	90
		2,6-Dichlorobenzonitrile	90
Maleic acid	2800	Carbaryl	90

TABLE 1: (Cont'd)

Compound	BF	Compound	BF
Palmitic acid	2800	Monolinuron	70
2,6-Di-tert-Butylphenol	2600	2-Nitropropane	70
Biphenyl	2600	2,4,6-Trichlorophenol	60
Hexachlorocyclopentadiene	2400	ETU	50
Phenol	2200	PTU	50
Toluene	1900	Atrazine	40
Chlorobenzene	1700	Nitrobenzene	40
Benzene	1700	N-Benzyl-N-Methylnitrosamine	30
ICM 2100	1500		
Bromobenzene	1500	2,6-Dichlorobenzamide	30
1,2,4-Trichlorobenzene	1400	Sencor	30
Coumarin	1400	4-Nitrophenol	30
Benzoic acid	1300	Captan	20
Dodecane	1300	4-Bromobenzoic acid	20
Belgard	1300	2,4-Dichlorophenoxy acetic acid	20
Benzidine	1200		
β-Hexachlorocyclohexane	1200	2,4-Dichlorobenzoic acid	10
3-Cresol	1100		

TABLE 2: Relative biodegradation rates of organic chemicals in activated sludge; data in percent of CO_2, relative to dose applied

Compound	CO_2	Compound	CO_2
Urea	85.9	Aldrin	1.5
Succinic anydride	78.3	PTU	1.3
Benzoic acid	65.4	Maneb	0.6
Acetic acid (Na-salt)	52.6	ADPA	0.6
Palmitic acid ethylester	44.0	4-Nitrophenol	0.5
Phenol	41.4	2,5,4'-Trichlorobiphenyl	0.5
Phenanthrene	39.6	Nitropropane	0.4
Cortisonacetate	38.5	Captan	0.4

TABLE 2: (Cont'd)

Compound	CO_2	Compound	CO_2
Dodecane	38.1	2,4,6-Trichloroaniline	0.4
Palmitic acid	37.8	Nitrobenzene	0.4
Methanol	37.0	Cypermetrin	0.4
3-Cresol	35.6	2,4,6-Trichlorophenol	0.3
Bromobenzene	34.8	Anthracene	0.3
Chlorobenzene	31.5	2,4,6,2'4'-Pentachlorobiphenyl	0.3
Acetic acid ethylester	30.3		
Carbaryl	29.7	Pentachlorophenol	0.2
Coumarine	29.5	4-Isopropylnitrobenzene	0.2
Benzene	29.2	Pigment TK 10466	0.2
Hexadecanol	28.0	γ-Hexachlorocyclohexane	0.1
Docosane	27.8	Sencor	0.1
Maleic acid	26.3	Chlorhexidine	0.1
Toluene	26.3	4-Bromobenzoic acid	<0.1
Ethylenediamine (hydrochloride)	25.0	2,4-Dichlorobenzoic acid	<0.1
		2,4-Dichlorophenoxy-acetic acid	<0.1
Tristearin	24.0		
ICM 2100	23.5	Perylene	<0.1
Vinylchloride	21.5	2,4,6,2'-Tetrachlorobiphenyl	<0.1
Aniline	20.5		
4-Chlorobenzoic acid	17.1	Quintozen	<0.1
4-Chloroaniline	14.5	1,4-Dichlorobenzene	<0.1
Thiourea	16.8	2,4-Dichloronitrobenzene	<0.1
Biphenyl	15.2	Benzidine	<0.1
Malonic acid diethylester	15.1	2,4'-Dichlorobiphenyl	<0.1
Naphthaline	9.0	2,2'-Dichlorobiphenyl	<0.1
Diethyleneglycol	8.4	Benzanthrancene	<0.1
Carbon tetrachloride	7.6	Atrazine	<0.1
Hydroquinone	7.5	β-Hexachlorocyclohexane	<0.1
Ethyleneglycol	5.7	Benzo-a-pyrene	<0.1
Chlorferron	5.2	Pentachlorobenzene	<0.1
Phenylenediamine (hydrochloride)	3.8	Pigment TK10090	<0.1
		2,6-Dichlorbenzonitrile	<0.1
2,6-Di-tert-Butylphenol	3.5	Dieldrin	<0.1
Trichloroethylene	3.4	DDT	<0.1
Belgard	3.1	Dibenzanthracene	<0.1
ETU	2.9	Kepone	<0.1

TABLE 2: (Cont'd)

Compound	CO_2	Compound	CO_2
2,4-Dichlorophenol	2.8	Hexachlorobenzene	<0.1
N-Benzyl-N-Methylnitrosamine	2.8	4-tert-Butylphenol	<0.1
3,3'-Dichlorobenzidine	2.7	1,2,4-Trichlorobenzene	<0.1
Dodecylbenzenesulfonate (Na-salt)	2.3		

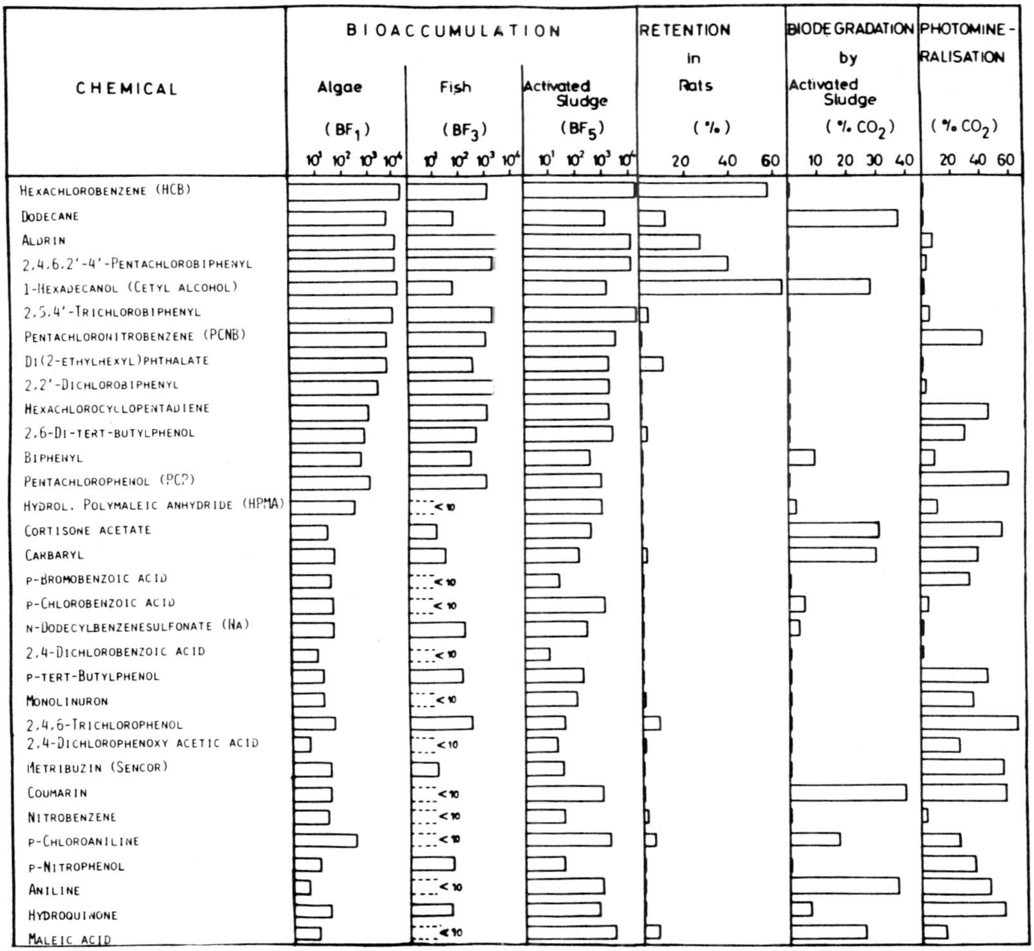

FIGURE 2: Bargraphs of the relative bioaccumulation rates of organic chemicals in algae, fish, and activated sludge, their retention in rats, their biodegradation by activated sludge and their photo-mineralization rates

Benzene Derivatives

From all the tested chemicals we selected some groups with different characteristics in their structure. The following TABLES show the results from the test system for different substituted benzenes, benzoic acids, anilines, nitrobenzenes, phenols, and chlorinated biphenyls and benzenes, respectively.

The set of substituted benzenes (TABLE 3) is listed in order of decreasing bioaccumulation factors in green algae. The ranking of the factors from the fish and sludge tests is comparable. In contrast, the mineralization rate by activated sludge (here expressed as % CO_2) shows a contrary course. Some of the data of the biodegradation and photomineralization are not comparable. This is also apparent from the data for the other groups of chemicals. From the data for the last three compounds of the benzene derivatives in TABLE 3, one can see that the introduction of other substituents to the nitrobenzene molecule leads to an increase of the accumulation factors.

TABLE 3: Relative bioaccumulation and degradation rates of benzene derivatives, benzoic acid derivatives, aniline derivatives, and phenol derivatives

Compound	Accumulation				Degradation		
						Photoirradiation	
	Algae BF_1	Fish BF_3	Rat % Retention	Activ. Sludge BF_5	Activ. Sludge % CO_2	% CO_2	% Org. Fragments
Benzene Derivatives							
Benzene	30	10	1.5	1,700	29.2	15.2	0.6
Biphenyl	540	281	0.4	2,600	14.5	9.5	0.1
Toluene	380	94	0.1	1,900	26.3	8.4	0.1
Phenol	200	20	0.3	2,200	41.4	32.5	0.2
Bromobenzene	190	48	0.7	1,500	34.8	19.7	0.3
Chlorobenzene	50	75	0.8	1,700	31.5	18.5	0.2
Nitrobenzene	20	2	-	40	0.4	6.7	0.1
Aniline	10	5	0.5	500	20.5	46.5	0.8
Benzoic acid	10	5	0.1	1,300	65.5	10.2	0.1

TABLE 3: (Cont'd)

Compound	Accumulation				Degradation		
						Photoirradiation	
	Algae BF_1	Fish BF_3	Rat % Retention	Activ. Sludge BF_5	Activ. Sludge % CO_2	% CO_2	% Org. Fragments
4-Isopropyl nitrobenzene	120	190	0.4	240	0.2	14.7	0.2
2,4-Dichloro-nitrobenzene	150	80	0.8	310	0.1	11.2	0.1
Pentachloro-nitrobènzene	3,100	1,140	0.4	4,500	0.1	43.0	0.1
Benzoic Acid Derivatives							
Benzoic acid	3	5.3	0.1	1,300	65.4	10.2	0.1
4-Chlorobenzoic acid	63	3.4	-	170	11.1	6.3	0.1
4-Bromobenzoic acid	25	6.5	1.3	11	1.3	36.4	0.1
2,6-Dichloro-benzoic acid	9	1.0	0.3	9	0.1	2.8	0.1
Aniline Derivatives							
Aniline	4	5	0.5	500	20.5	46.5	0.8
4-Chloroaniline	260	13	4.9	280	22.7	27.7	0.1
Phenylendiamine	450	6	0.8	460	3.8	53.7	0.1
Benzidine	850	83	0.7	1,210	0.1	40.8	0.1
2,4,6-Tri-chloroaniline	260	330	0.7	870	0.4	39.5	0.1
3,3'-Dichloro-benzidine	940	610	0.9	3,100	2.7	41.2	0.1
Phenol Derivatives							
Phenol	200	20	0.3	2,180	41.4	32.5	0.2
3-Cresol	40	17	0.2	1,120	35.6	33.5	0.2
Hydroquinone	40	40	0.5	870	7.5	57.4	0.1
4-Tert-butyl-phenol	30	118	-	240	0.1	46.8	0.1
4-Nitrophenol	20	51	0.3	30	0.5	39.5	0.1

TABLE 3: (Cont'd)

Compound	Accumulation				Degradation		
						Photoirradiation	
	Algae BF_1	Fish BF_3	Rat % Retention	Activ. Sludge BF_5	Activ. Sludge % CO_2	% CO_2	% Org. Fragments
Phenol Derivatives (cont'd)							
Pentachloro-phenol	1,250	260	1.0	1,100	0.2	50.0	0.1
2,6-Di-tert-butylphenol	880	660	2.9	2,600	3.5	29.5	0.1
2,4-Dichloro-phenol	260	100	0.2	340	2.8	50.4	0.1
2,4,6-Tri-chlorophenol	50	310	7.8	60	0.3	65.8	0.1

TABLE 3 further shows the corresponding bioaccumulation and degradation rates of substituted benzoic acids. The inductive and resonance effects of the para- and ortho-substituents could be responsible for the reduction of accumulation factors and degradation rates.

In the case of aromatic amines (TABLE 3), the data indicate a tendency to higher accumulation factors and lower mineralization rates. It is also apparent that the accumulation factors increase with an increase in chlorine substituents of the parent molecules.

Mono-substituted phenols (TABLE 3) show a pattern similar that of the benzoic acids as shown above in TABLE 3. The tert-butyl-substituents in position 4 causes a significant reduction of the bioaccumulation factors in the algae and sludge tests. The two-fold substitution in positions 2 and 6, however, causes a steric inhibition of the OH-group. Consequently, the accumulation factors of 2,6-di-tert-butylphenol are higher when compared to phenol itself and to 4-tert-butylphenol.

Chloroaromatics

Two groups of chemicals are presented, which consist of multi-chlorinated aromatic compounds. From these, a clear relationship between the degree of chlorination and the accumulation potentials, respectively, and the degradation rates can be seen.

TABLE 4 lists several chorobiphenyls in order of increasing substitution. It is evident that each of the accumulation rates in algae, fish and activated sludge increase in the same sequence. The mineralization values of biodegradation in sludge and by photoirradiation are decreasing in this order. However, the last two substances differ from this ranking. It can be assumed that the substitution in positions 2, 6 and 2' causes a certain steric inhibition of the internal rotation of the molecule, which results in the lower than expected accumulation rates in algae and fish.

TABLE 4: Relative bioaccumulation and degradation rates of chlorobiphenyl congeners and chlorobenzene congeners

Compound	Accumulation				Degradation		
						Photoirradiation	
	Algae BF_1	Fish BF_3	Rat % Retention	Activ. Sludge BF_5	Activ. Sludge % CO_2	% CO_2	% Org. Fragments
Chlorobiphenyl Congeners							
Biphenyl	540	280	0.4	2,600	15.2	9.5	0.1
2,2'-Dichloro-biphenyl	2,690	2,420	0.1	6,300	0.1	3.5	0.1
2,4'-Dichloro-biphenyl	6,720	3,550	0.6	9,800	0.1	2.9	0.1
2,5,4'-Tri-chlorobiphenyl	8,950	8,850	3.2	32,000	0.5	5.9	0.1
2,4,6,2'-Tetrachloro-biphenyl	18,300	3,150	2.7	6,500	0.1	3.9	0.1
2,4,6,2',4'-Pentachloro-biphenyl	11,500	2,320	39.8	27,800	0.3	5.2	0.1

TABLE 4: (Cont'd)

Compound	Accumulation				Degradation		
						Photoirradiation	
	Algae BF_1	Fish BF_3	Rat % Retention	Activ. Sludge BF_5	Activ. Sludge % CO_2	% CO_2	% Org. Fragments
Chlorobenzene Congeners							
Benzene	30	10	1.5	1,700	29.2	15.2	0.6
Chlorobenzene	50	70	0.8	1,700	31.5	18.5	0.2
1,4-Dichloro-benzene	100	50	0.4	560	0.1	5.1	0.1
1,2,4-Trichloro-benzene	250	490	1.0	1,400	0.1	9.8	0.1
Pentachloro-benzene	4,500	3,000	2.0	14,300	0.1	2.0	0.1
Hexachloro-benzene	24,800	2,600	55.7	35,000	0.1	1.5	0.1

The most significant relationship between structure and activity, can be seen from the the results for chlorinated benzenes (TABLE 4). The interpretation of all results is most comprehensible for simple aromatic compounds with subsequent substitution, as illustrated in this TABLE, showing a homologous series of chlorinated benzenes.

Structure Fragmentation

From the foregoing, it is apparent that quantitative structure-activity correlations (QSAR's) may be found in many homologous series of compounds. In contrast, series with substituents of different nature do not seem to be interpretable in a simple pragmatic way. In this situation we have initiated a coordinated research program with Dr. Schubert and Prof. Ugi at the Technical University of Munich. Some preliminary results to recognize structure-activity relationships by mathematical computer modelling are presented here.

The attributes of a chemical compound are based fundamentally on its structure. In terms of the chemical nature, the relationship between structure and attribute is comparatively obvious. The bioaccumulation, however, has to be considered as a complex function of structural parameters which cannot be recognized as easily. Therefore, it should be possible to elucidate structure-activity relationships with the aid of mathematical models.

A compound, which is to be analyzed for QSAR's is divided into substructures. Two different paths are possible for this procedure, as shown in FIGURES 3 and 4.

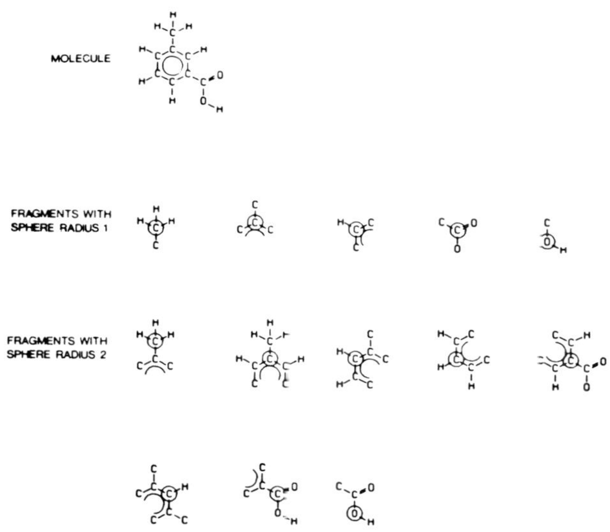

FIGURE 3: Sphere fragmentation of 3-Methylbenzoic acid. Central atoms are indicated by circles, duplicate structures are omitted

FIGURE 3 shows the so-called "sphere fragmentation" on the example of 3-methylbenzoic acid. Each atom, if at least two-fold coordinated, represents the centre of a sphere and is the principal of the spherefragmentation. By this method we get a series of substructures resembling the so-called functional groups, as mentioned above in the context of substituted compounds. The second method for dividing a compound, is the stepwise fragmentation (FIGURE 4). By this procedure the most distant atoms in the molecules will be removed stepwise. When calculating very large molecules, however, the number of fragments will cause an inordinate amount of mathematical work. In such cases, combination of both methods is indicated.

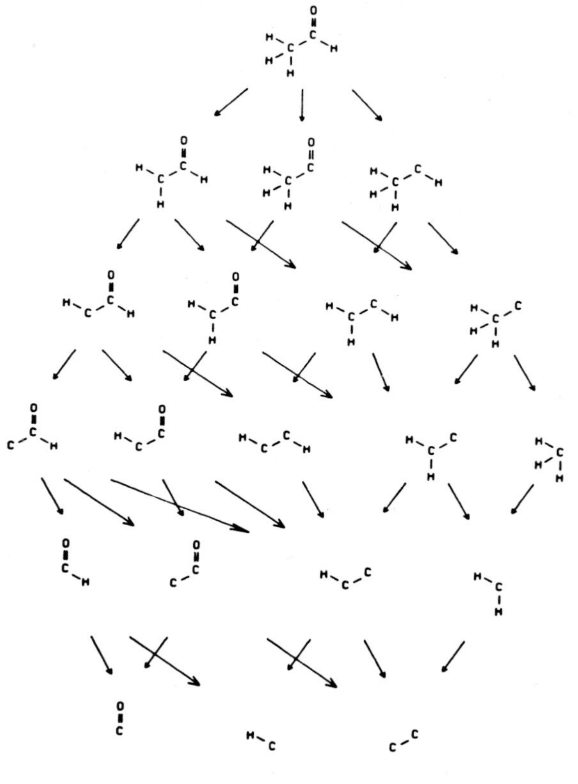

FIGURE 4: Stepwise fragmentation of acetaldehyde

In principle, each fragmentation leads to a network of structural elements of different properties. In this way we obtain a network of properties. The degree of overlap of the networks of structures and of properties represents the probability that a single chemical compound is endowed with a certain property.

The first model calculations, were carried out on the basis of fish accumulation data from Japan. Approximately 100 substances, consisting of polychlorinated aromatics, heterocyclics, alcohols and aliphatics, etc., were calculated.

FIGURE 5 shows the resultant structural elements responsible for the bioaccumulation. The numbers, given in the last column of FIGURE 5, represent the standard deviation. A high probability of accumulation is indicated by the structural elements F2 01, F2 02 and their combination. These results should be considered as a preliminary experiment only. The next step will be to use this mathematical model for the calculation of the data of the 100 substances, tested in our own laboratory test system. We hope that this computer modelling will be confirmed by our experimental derived data.

Effektor	ENS	ES	NES	NENS	ENS+NENS	$\frac{ENS}{ENS+NENS} \cdot \frac{E+NE}{E}$ 'F'
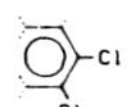 F3 02	9	12	2	82	91	0.495
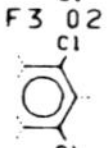 F3 01	6	15	1	83	89	0.337
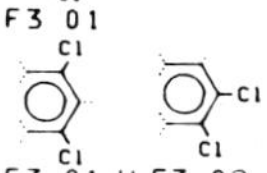 F3 01 ∪ F3 02	5	16	2	82	87	0.287
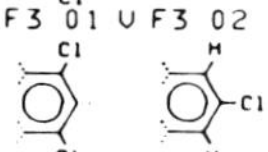 F3 01 ∪ F3 07	3	18	8	76	79	0.190
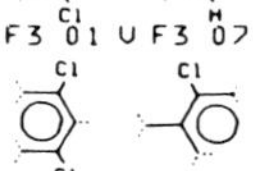 F3 01 ∪ F3 05	3	18	3	81	84	0.179
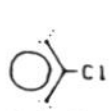 F2 01	2	19	13	71	73	0.137
 F2 02	16	5	2	82	96	0.083
 F2 01 ∪ F2 02	1	20	15	69	70	0.071

FIGURE 5: Stuctural elements found to be responsible for the bioaccumulation of 100 test substances

Chlorophenols

In the last chapter, some results are presented, concerning the effects of two chlorophenols on the biota of an aquatic outdoor system. Specifically, the effects of the chemicals 2,4,6-trichlorophenol (TCP) and pentachlorophenol (PCP) on the population dynamics of several species were studied in an experiment in separated compartments of an experimental outdoor pond.

For these investigations, a natural pond was divided into six compartments with tubes of 50 cm diameter resulting in a volume of about 200 L each. TCP and PCP were applied to two tubes each to result in water concentrations of 5 and 1 $mg \cdot L^{-1}$, respectively. These concentrations are equivalent to about three times the EC50 for the TCP and double the EC50 for PCP as determined in laboratory tests with *Daphnia magna*. The higher concentrations were chosen in consideration of the fact that natural systems have a higher buffer capacity than short-term laboratory test systems. Empty niches, for example, which were caused by chemical intoxication can be reoccupied by a more resistant species.

The disappearance of TCP and PCP from water is shown in FIGURE 6. The concentrations of both chemicals declined to about 10% of their initial values at 24 days after treatment. This decrease may be due to evaporation from water and to absorption by seston and the biota. It is evident that significant residues of the two chemicals remained in the aqueous layer over the whole experimental period of 24 days, and thus were available for bioaccumulation and may adversely affect planktonic organisms.

No significant differences in limnological conditions (temperature, pH, hardness, concentrations of electrolytes) could be found between treated and control compartments, with the exception of dissolved oxygen concentrations. In the second half of the investigative period, the treated water showed significantly lower oxygen-concentrations compared to the controls. The strong decrease of oxygen (FIGURE 7) in the water of the treated compartments is probably not directly caused by chemical reactions with either PCP or TCP, but rather may be interpreted as a secondary effect of the changes in biota.

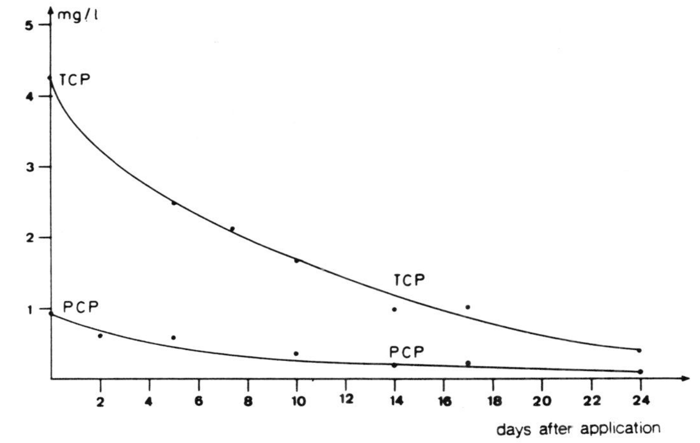

FIGURE 6: Disappearance of 2,4,6-trichlorophenol (TCP) and pentachlorophenol (PCP) from the water of an compartmentalized outdoor pond

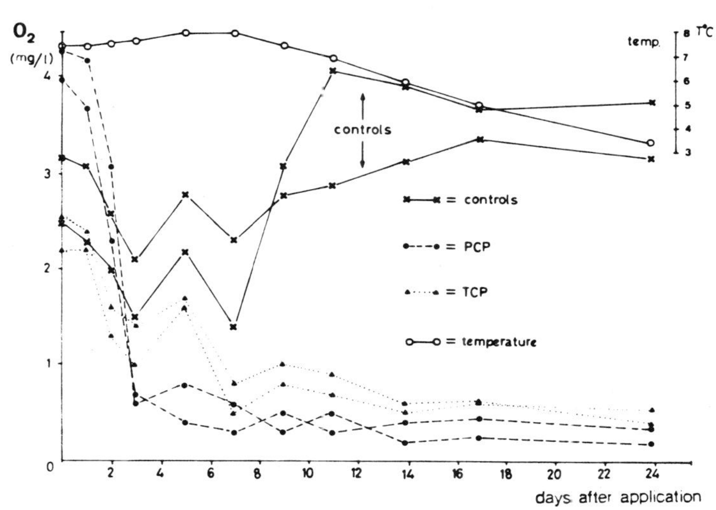

FIGURE 7: Dynamics of dissolved oxygen in the water of 2,4,6-trichlorophenol (TCP) and pentachlorophenol (PCP) treated, and of untreated compartments (controls)

As to be expected from the applied concentrations, both chemicals were highly lethal to *Daphnia pulex*. Three (PCP) and nine days (TCP) after application, the total daphnid populations declined to zero and did not recover. Considering the important intermediate role of daphnids in the food-chain as consumers of algae and detritus it can be assumed that their disappearence affected the stability of the phytoplankton and planktonic predator communities.

The population dynamics of three selected phytoplankton species are presented in FIGURES 8 and 9. *Chroococcus limneticus* (blue algae), *Trachelomonas hispida* (flagellate) and *Euglena acus*

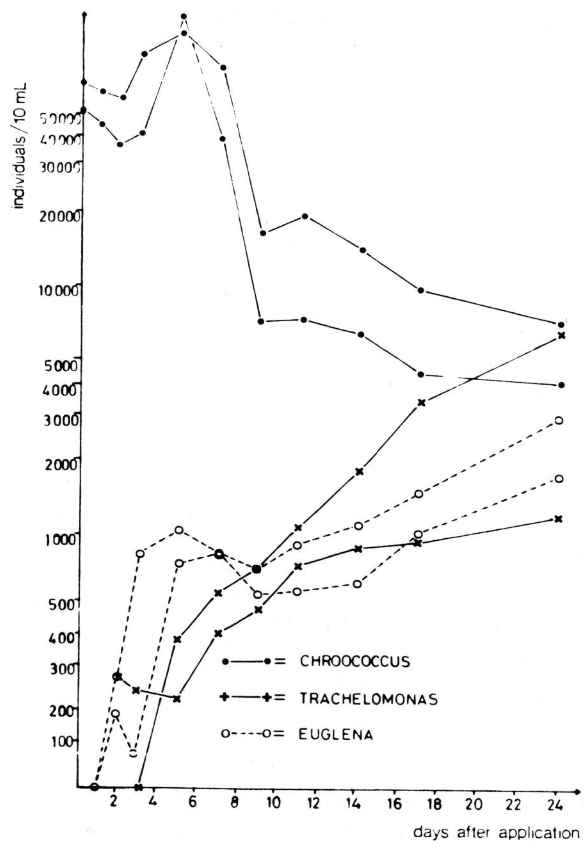

FIGURE 8: Dynamics of three plankton species in 2,4,6-trichlorophenol treated compartments

(flagellate) were selected as representative algae, because of their natural abundance and characteristic response to the chemical perturbation.

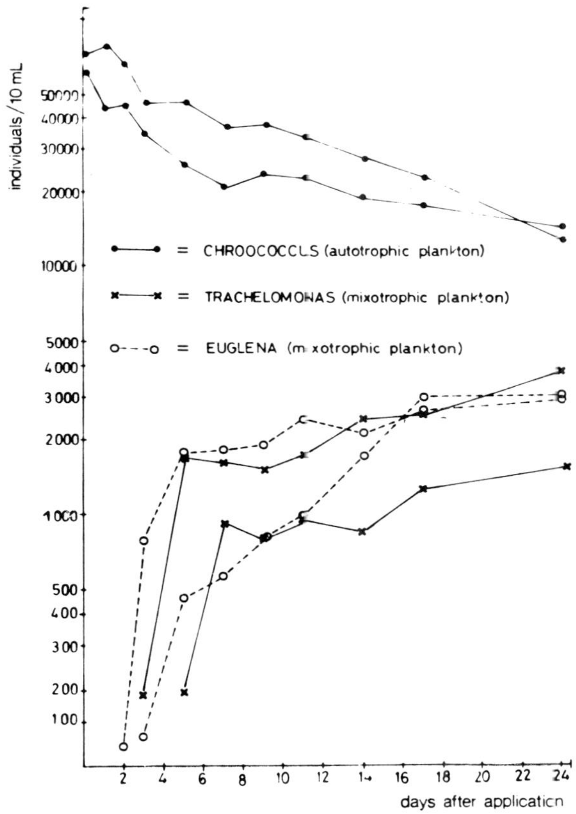

FIGURE 9: Dynamics of three plankton species in pentachlorophenol treated compartments

In the TCP (FIGURE 8) as well as in the PCP (FIGURE 9) treated compartments, the population densities of these autotrophic and mixotrophic plankton showed distinctive trends in response to chemical exposure. In comparison to their control population, Chroococcus limneticus decreased significantly demonstrating direct sensitivity to the chemical pollutant (primary effect), whereas the mixotrophic species showed a significant increase after chemical treatment. It is noteworthy, that both mixotrophic species (Trachelomonas and Euglena) could not be detected in the control compartments.

As shown in FIGURE 10, the presence of PCP and TCP in the water results in the complete elimination of the diatoms (Nitzschia acicularis) 11 and 17 days, respectively, after chemical application. The high sensitivity of diatoms to both chemicals is in agreement with the classification of these species as representatives of clean water associations according to water quality standards.

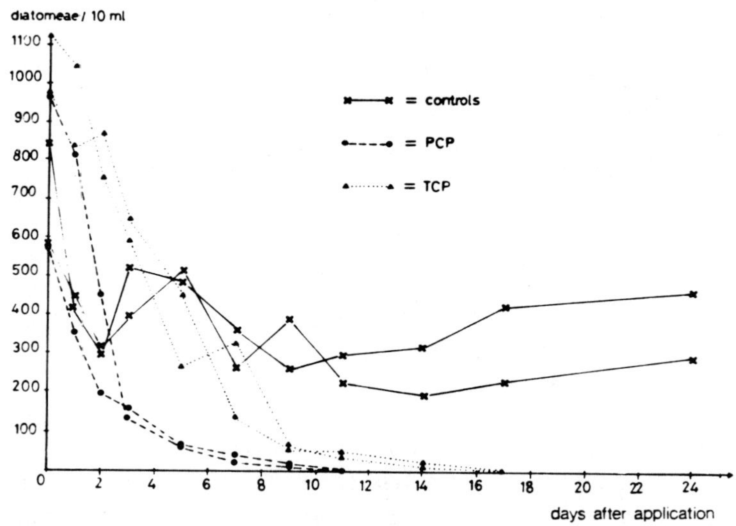

FIGURE 10: Dynamics of the diatom Nitzschia acicularis in 2,4,6-trichlorophenol (TCP) and pentachlorophenol (PCP) treated and in untreated compartments

FIGURE 11 illustrates the concentration dynamics of heterotrophic microorganisms in the water of the treated and untreated compartments. As can be seen from FIGURE 11, there is a sharp increase of numbers of bacteria in the contaminated water during the first week of investigation amounting to about the tenfold number of germs as counted in the samples of the control water. These data, however, should only be considered as additional information of the microbial activity for a week's time after application of the chlorophenols.

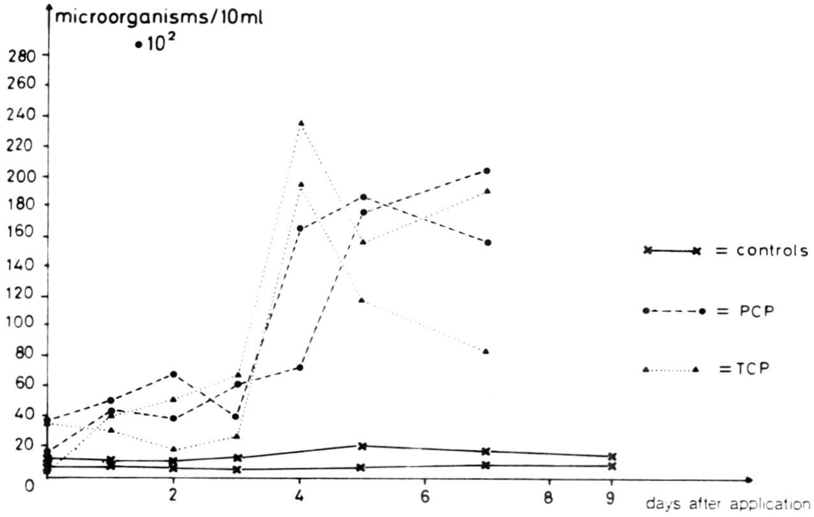

FIGURE 11: Dynamics of heterotrophic microorganisms in 2,4,6-trichlorophenol (TCP) and pentachlorophenol (PCP) treated and in untreated compartments

Summarizing the results of the effects of TCP and PCP upon the dynamics of planktonic organisms, it can be stated that single doses of 5 and 1 mg$\cdot$L^{-1}, respectively, adversely affected the plankton community. Autotrophic phytoplankters and daphnids were eliminated in part, while indicators of polluted water like flagellates and bacteria significantly increased. The five-fold higher concentration of TCP caused a faster reduction of blue algae and daphnids as compared to PCP, whereas the toxicity of TCP on the diatoms was weaker. The oxygen concentration (as a sum-parameter) in the TCP treated water was higher than in the PCP compartments, indicating a lesser toxic effect on the total mass of autotrophic phytoplankton. In general, it was established by this outdoor experiment that inspite of a five-fold lower dosage of pentachlorophenol, both TCP and PCP showed comparable declines in the water, and comparable influences on the physicochemistry and changes on the biota within the same order of magnitude.

WORKSHOP DISCUSSION

MACKAY:

Will this series of tests which you have described be used in Germany for assessing the hazards of chemicals or is it just your test?

FREITAG:

This is just our test. Our idea was to introduce a test system with ^{14}C labelled chemicals. We think that it is very easy to analyze the concentration in the organism and in the medium. Sometimes, we have problems too, but normally not in the algae test, because there is no metabolism within 24 hours, which is the duration in the algae test. In the fish test, which takes three days and the activated sludge test which takes five days, we sometimes have metabolism and we can see it when we calculate the regression for the relationship of the bioaccumulation with the water solubility or the K_{ow}. For example, in the case of 4-chloroaniline, the point is outside the line. Therefore, the radioactivity does not come from the chloroaniline, the original substance, but from the metabolized compound. However, for the graph, we used the water solubility of the original substance and the bioaccumulation of the metabolite.

MACKAY:

It was interesting to note on one of your tables that you had quite a low fish bioconcentration factor for hexachlorobenzene, lower than for pentachlorobenzene, which was exactly the same observation made by Barry Oliver. You must have encountered the same problem of kinetic control.

FREITAG:

Yes; the time for the fish test is three days and that is not enough to reach equilibrium. But, we have tested some chemicals for a longer time and we have seen that in most cases, a time of three days is enough for a screening test. This is very important; it is not exactly a quantitative result; it should be considered a screening test only and when we have the results, we can plan better for long-term experiments. The problem in the environment is the time parameter. We do not have enough time to investigate all the chemicals. So, I think, the short-term test system is very good for screening or pre-screening and after this you can make better long-term experiments based on these results.

McCARTY:

What sewage tests do you run for your degradation?

FREITAG:

Activated sludge.

Question:

Have you thought about modifying your system to take into account and to recognize compounds of extreme properties, such as very persistent or lipophilic substances.

KAISER:

Which particular part of the system?

McCARTY:

In light of what we have been talking about here, and in light of your system, have you set up some mechanism point out where some of these things might be in error, because obviously the three days are not sufficient time to screen things and to get an accurate estimate. Is there a cutoff point for certain compounds?

FREITAG:

The principle of this test system is not so much the single result from each test, but the overall result of the tests. For example, let's go to FIGURE 2 which shows that the single result of each test is not that important for the general test system, we must look at all the results together. For example, for hexachlorobenzene, you have a high accumulation in algae, in fish, in activated sludge, and in the rats, and no biodegradation and no photodegradation; together they give you a profile. For another substance such as coumarine you have varying degrees of degradation and the idea is to be able to say that the one is a very dangerous substance for the environment and the other is not. This is the important idea behind the system and not necessarily the single correct data.

KAISER:

I take it that even if all but one of the tests point to a non-hazardous compound, looking at the overall result one would still determine this to be a compound of concern. I think FIGURE 2 could be improved by taking the last two columns and using the inverted values. This way, you would get a large bar for zero degradation as you do for high accumulation and other, non-desirable properties.

McCARTY:

So, what you are saying, is that you have established a wide enough variety of tests to ensure that your interpretation is not based exclusively on one test?

FREITAG:

Yes.

SCHULTZ:

With the hundred chemicals that you have tested, have you tried to break those down into sub-groups based on classical organic chemistry to see if you can see some trends with increase in say, molecular size within a group or something like that concerning all these parameters that you have measured on these bar graphs?

FREITAG:

Yes, but only to the degree shown in the tables. We investigated chemicals which could be obtained in the labelled form which are usually very expensive. We took the "Ispra list" of dangerous chemicals and determined which chemical we could buy in the labelled form. This is why we have incomplete series of chlorobenzenes and so forth.

KAISER:

How are they labelled, always as ^{14}C?

FREITAG:

Yes, ^{14}C, generally in the ring, but the position for labelling is not so important for accumulation because in most cases, the complete molecule was accumulated. However, for the biodegradation test the label positions are very important. In the case of a benzoic acid derivative such as coumarine with ^{14}C in the carbonyl group, you will observe a high degradation, but it would present an incorrect impression of degradability. Therefore it is very important to use ring labelled compounds.

McCARTY:

In that particular case, did the other tests for that chemical compensate for that test, did you come to the same conclusion, eliminating the complicating factor of the incorrect result for this compound? In other words, is the test system so robust that even when it gets a wrong answer in one of the tests, the other tests will result in an overall correct evaluation?

KAISER:

Yes, I think so. You can probably judge for yourself from the data in FIGURE 2.

FREITAG:

For this type of compound, the degradation is always high and it stems from the carboxy group.

Comment:

But you can still come to the same conclusion without that piece of data. You know that it is wrong, but the rest of the tests will still give you the correct conclusion.

FREITAG:

Yes.

MACKAY:

This is an attempt to rank the hazard of chemicals. Did any chemicals which are regarded as environmentally troublesome, go through your system and appear to be harmless or is your system successful in identifying the environmentally troublesome chemicals?

FREITAG:

We have compared our monitoring data with those from outdoor experiments and we found a good correlation between the results from the outdoor experiments and the monitoring data. If we check a new chemical in our system, we can say, it has the profile similar to another chemical, with which we are familiar. Therefore we would know that it is a possible environmental hazard and we can start the right experiments for a detailed investigation.

MACKAY:

Unless a metabolite or a compound is formed on degradation of the chemical?

Question:

What is the photodegradation test you use?

FREITAG:

It is the so-called GSF test. The test is described in the annex to the European regulations. We put the substance on silica gel, in the vial and use a natural light lamp and the time for the test is seventeen hours. It could be fifteen or twenty but the point is, that we test the chemicals for the same time and seventeen hours is practical to work with. In the afternoon we start the test and in the morning we can work with the results. The air is pumped into the test system and we collect the organic fragments and the carbon dioxide. We have two chambers, so we can compare a known substance with an unknown substance and we can calculate relative degradation or we can have both chemicals in the two chambers to provide a replicate.

LANDE:

Is this being done in an oxygen atmosphere?

FREITAG:

No, it is done in normal room air.

PREDICTING THE ENVIRONMENTAL FATE OF TOXIC CONTAMINANTS IN LARGE LAKES: DATA REQUIREMENTS FOR MATHEMATICAL MODELS

Efraim Halfon

Aquatic Physics and System Division
National Water Research Institute
PO Box 5050, Burlington, Ontario L7R 4A6
CANADA

ABSTRACT

The relation between field research in ecotoxicology and modelling activities to predict the environmental fate of contaminants in large lakes is analyzed in terms of data needs. Modellers need data from research projects on pathways as well as concentration data. Are all these data needed and what are the alternatives? Examples are taken from research on the Great Lakes of North America, mainly Lake Ontario.

INTRODUCTION

Toxic substances in the aquatic environment are of great concern and the prediction of their fate and effects is an integral part of their risk analysis. Mathematical modelling and simulation have been offered as tools for that prediction (ASTM 1978). The question arising here is whether or not the sophistication in simulation techniques can be coupled with the regular monitoring of collecting and analyzing field samples. This paper analyses data needs for the prediction of the fate of known contaminants in large lakes.

K. L. E. Kaiser (ed.), QSAR in Environmental Toxicology, 137–151.

The Need for Communication

Models can be used to predict the environmental behaviour of known contaminants and especially of those compounds not yet in commerce or in the environment. Large lakes, such as the Great Lakes of North America, are spatially heterogeneous and many sampling stations are necessary to obtain a global picture. Fulfillment of this requirement is expensive, however it appears to be necesary. This paper seeks to encourage discussion with experimentalists about their views on model credibility, given the difficulties in model development. Communication is the key word rather than a dogmatic request for data since a credible model can only be developed within a collaborative and understanding framework.

Eutrophication Models

Mathematical models have been used extensively to model eutrophication processes (Patten 1971 to 1976). Agreement has been reached on the fact that eutrophication models are useful and have a role in the planning and analysis of field studies. Eutrophication models can be simple, or extremely complex (Vollenweider 1975; Park et al. 1974), but data exist, or can be collected, to determine their validity: Modelling the fate of toxic substances will take more time to reach the same credibility as eutrophication modelling as data become available.

Toxic Substance Models

TABLE 1 shows important topics for the research on modelling the fate and effects of toxic substances. Four areas of concern were identified by a group of scientists at the First International Conference on Modelling the Fate and Effects of Toxic Substances in Copenhagen, June 1983. These areas are model structure and testing, data requirements (this part is emphasized here), field and laboratory research programs associated with modelling needs, and fast and accurate communication of the results. The topics presented in TABLE 1 reflect the thoughts of the ecological modelling community present at the conference but they may not be all encompassing, even if the list appears quite extensive.

TABLE 1: Research topics and needs in the modelling area of fates and effects of toxic substances in large lakes

---

Model Structure and Testing

- Multiple compounds fate
- Spatial heterogeneity (linkages in terrestrial systems, soil profiles, succession, patchiness)
- Systems analysis techniques, model structure definition
- Statistical measures of magnitude and persistence
- Comparability of different models
- Evaluation of model structures and assumptions
- Quantitative measures of quality of fit
- Simple versus complex models
- Computational problems (stiffness of equations, numerical problems, initial conditions, "memory", history of exposure)
- Sediment transport models (estuaries, streams, lakes)
- Effects (acute effects, integrating the effects of compounds)
- Impacts on community structure (heterogeneity of populations, genetic potential)
- Equilibrium versus transient approach
- Importance of short term exposure, lag effects
- Combining process levels models (whole organisms, multi-age populations, intercommunity relations)
- Definition of goals and systems to be studied
- Artificial stress and response of a test system
- Baseline studies in natural ecosystems to build confidence in models
- Generality of fate models?
- Dangers of fossilizing ideas by models
- Dose/response in humans (Extrapolation from lake to human exposure and risk prediction)

Data

- Identification of past inputs
- Field calibration and verification of model
- Unsaturated transport
- Field, microcosm, laboratory relations

Table 1. (con't)

Specific Scientific Programs

Effects of landfills, waste dumps
Identification of chemicals
Definition of levels of pollution
Parent and daughter products
Effects of multiple compounds (synergism)
Fate of particulates (sedimentation, mixing within sediments, micellar states with colloids, filter feeding in sediments and organsim uptake, biological interactions)
Soil, sediment diffusion processes (varying water content)
Role of sediment degradation
Linkages between terrestrial and aquatic systems
Population dynamics, community structure
Effect level, chronic versus acute
Connection of body burden to mortality
Relation of LD50, body burden, mortality (long-term exposure)
Models of impacts on community structure (biomass, population dynamics)
Real indicators of system change
Bacterial dynamics - threshold levels
Terrestrial models

Dissemination and Communication

Transportability of computer code
Communication of approaches, cooperative studies, quick turnaround
Unification of studies, collection of case studies (systems, toxicants, priorities)
Wide surveys required to highlight broad relationships, e.g., eutrophication surveys
Acceptability to user
Understandability of model
Detailed documentation
Computer graphics presentation
Scientific consensus for unification of the applied model

Mathematical Models

Two kinds of modeling philosophies can be used in predicting the fate of contaminants in large water bodies; they are **Potential Environmental Distribution** (PED) and **Potential Environmental Concentration** (PEC). PED models describe only how the contaminant will become distributed between phases: air, water, sediments, soil, and so forth. Equilibrium kinetics are used and masses and concentrations provide only an indication of relative concentrations with no reference to the actual levels. The advantage of PED models is their ability to produce estimates where the contaminants will end up and whether accumulation in a component may be significant or not. One of the best known PED models is Mackay's fugacity model (Mackay 1979). EXAMS (Lassiter et al. 1979) can also be considered a PED model, since the original version predicts the ultimate steady state values in water; however, EXAMS has also been used as a PEC model (Halfon 1984) to predict Mirex concentrations in Lake Ontario. Normally, PEC models, or simulation models, are developed for specific environments and the parameters may be adjusted so to let the model simulations reproduce the actual environmental data. In some instances, uncertainty in the data is included in the model to originate probabilistic simulations (Halfon and Maguire 1983) but usually only deterministic simulations are produced.

Systems sensitivity analysis (Halfon 1984) can be employed to determine to which processes or parameters a model is most sensitive to. By developing generalized models of the fate of toxic substances in ecosystems, ecological modelers bring together many types of knowledge. The sensitivity analysis helps in determining which are the weak links, or where additional information should be acquired.

ANALYSIS OF DATA NEEDS

To predict the fate of a toxic substance in a large lake with a simulation model, the modeller needs information on several key factors: **Inputs,** i.e., how much of the mass of the substance enters the lake or river in a certain time interval? **Initial conditions,** i.e., if the lake is not completely clean, how much of the toxic substance is already there? **Physico-chemical parameters,** i.e., have we quantified all rates of transport, bioaccumulation, volatilization, photolysis, hydrolysis, biodegradation, sedimentation, burial, resuspension, etc., as functions of the compound itself and the environmental characteristics of the system of interest? **Data requirements,** i.e., can we measure the concentrations and quantities of the toxic substance in the bottom sediments, benthos, fish, plankton, suspended sediments, and water? Hence, can we estimate the total amount of the substance in the lake and follow its increase or decrease of the mass over time? Do we know the environmental pathways? The following will analyse these requirments in detail.

Inputs

Toxic contaminants may enter a lake from the atmosphere, or from tributaries or directly from effluents. For Lake Ontario, for example, the largest tributary is the Niagara River with a flow of $5 \cdot 10^8$ $m^3 \cdot day^{-1}$. Contaminants enter the river from industrial effluents, as leachates from dumps and from wastewater treatment plants. Even daily sampling may not be a representative mean, due to the frequent and large fluctuations in, for example, PCBs in suspended solids (Kuntz and Warry 1983).

Several chemicals, such as Mirex, PCBs and chlorobenzenes have been identified, but many other contaminants have not. Some of these chemicals are associated with suspended sediments of 1 μm size or less, possibly even with colloidal matter. Appropriate sampling and separation techniques to quantitatively measure the amount of contaminants associated with these small particles do not exist. Both filtration and centrifugation techniques are problematic because they modify the sample characteristics. Moreover, very large quantities of water need to be processed to obtain usable data. Sephadex columns used in phosphorus cycling research (Lean 1973) can not handle the large water volumes. Thus, the inputs can not be precisely monitored and the results may be erroneous by orders of magnitude, especially for compounds whose concentrations are near their analytical detection limits. Inputs from the atmosphere appear to be fluctuating less and can be measured with the appropriate dry and wet samplers at several locations (Strachan and Huneault 1979). Modellers would like to have input estimates of contaminants loadings over a period of several years, even if uncertain within a factor of one to four.

Initial Conditions

To run a model it is necessary to know how much toxic contaminant is in the water body at a determined time, called time zero. Time zero may be now, ten years ago or any other time in the past. To know the initial conditions one needs to have sufficient data collected at time zero. As it turns out, data for the initial conditions are usually wanting. In a model describing the environmental fate of a contaminant in a large lake, the usual state variables are contaminant concentrations in suspended sediments, fish, water, and bottom sediments. To these basic variables, we can add others such as concentrations in benthos, pore water, benthic fish, total plankton, or preferably subdivided into phytoplankton and zooplankton, detritus, dissolved organic matter, macrophytes, periphyton, etc. Some mathematical models require only a few state variables (e.g., EXAMS), others such as PEST (Park et al. 1980) require many. Whether time zero is now or years ago, one must have the data for the four main state variables mentioned above.

Water can be defined operationally as the fraction that filters through a given filter size and the remainder may be defined as suspended sediments. However, as noted above, there exist problems with filtering and the sampling of suspended sediments. The definition of the two compartments therefore may lead to problems. Bottom sediments can be sampled with a core sampler, segmented and dated. The bottom sediments therefore give an overview of the past behaviour of the contaminant. Sediment characteristics such as their mean grain size, water content, and organic carbon content, are also used in the models. For large lakes with large tributaries, such as Lake Ontario problems are encountered with the sedimentation rates. For example, the mean sedimentation rate in Lake Ontario is 1 to 2 mm$\cdot$year^{-1}. Near the mouth of the Niagara River, however, sedimentation rates are in the order of 1 to 2 cm$\cdot$year^{-1}. In other parts of the lake, they may be only 0.1 mm$\cdot$year^{-1}. The minimum core section which can consistently be cut is 0.5 cm. Therefore, for some zones of the lake, this layer corresponds to half a year, five years for the lake average, and fifty years for other parts of the lake. The reconstruction of the past loadings of a contaminant may therefore be difficult. The large lake surface, 18500 km^2, creates other problems. The lake must be sampled at many stations. Holdrinet et al. (1978) used an 8 km grid with a total of 212 stations but sampled the sediments with a grab sampler which does not provide for chronological information. A more practical number would be 30 to 40 sampling stations. Each sample must be dated and this operation may take up to two weeks (R.W. Durham, personal communication). The sediments must also be analyzed for contaminant contents and sediment properties. This analysis takes even more time. If this whole task is done correctly, we may have a mass estimate of different contaminants at time zero. We may have our initial conditions within a factor of two to four.

Physico-Chemical Parameters

Parameters describe environmental characteristics, chemical properties, and rates. Environmental characteristics include the lake bathymetry, sediment types, pH, light, bacterial concentrations, and so forth. Chemical properties are the contaminant molecular weight, vapour pressure, octanol/water partition coefficient (P or K_{ow}), Henry's constant, etc. Rates include both environmental rates, such as sedimentation rates and transport currents, and chemical rates, such as volatilization, photolysis, and hydrolysis. For a given environment and a given chemical, determination of these characteristics are the product of extensive laboratory work. If the complexity of the models is increased, for example by including food-chains, non-equilibrium chemical reactions, and sediment transport, the quantification of the parameters becomes much more difficult because _ad hoc_ experiments must be done for each particular situation. Experiments should lead to general formulations or empirical laws that can be used in models and applicable to a large number of chemical compounds. At present, only a limited number of "standard" chemical parameters have been used by chemists to describe compounds, e.g., the octanol/water partition coefficent, water solubility, melting point, organic carbon/water partition coefficient (K_{oc}), etc. These parameters might be sufficient for PED models, which are not site specific.

Models of large lakes require further knowledge of specific processes, such as food-chain transfers of contaminants. These needs can only be satisfied with specific experiments in controlled conditions, such as continuous cultures, microcosms and experimental ponds. Modellers may suggest experiments to be performed based on the ecological parameters or as a result of or following sensitivity analyses of preliminary fate models. Lassiter *et al.* (1979), who developed EXAMS, contracted out over millions of dollars in research funds over several years to obtain data for their process sub-models. When these kinds of funds are not readily available, modellers and ecotoxicologists must continue and enhance their dialogue to perform experiments which are to mutual benefit.

As mentioned above, the purpose should be one of developing mathematical equations of the processes studied and sub-models should be validated at the process level. As a result, model credibility will not be a function of how well model simulations fit the experimental field data, which in themselves may have a large variance, but from internal consistency. From eutrophication modelling we know that goodness of fit criteria are not sufficient to validate a model, however data from controlled experiments in microcosms can give the confidence that parts of the model are correct. For example, the question of fish uptake of contaminants from water or food-chain is continuously debated, but experimentation under dynamic conditions, for example using continuous cultures, should provide useful generalizations.

Data Requirements

The problems with data requirements are the same as with the initial conditions with the compounding effect that time series have to be collected. For some nonpersistent chemicals in ponds, the relevant dynamics may be short lived, i.e., a few days to a year. Persistent chemicals may have an environmental half life of decades. What then is a good data base for large lakes? As discussed above, contaminants data in large lakes, need not be very precise and extensive as long as ecotoxicologists understand that given the large variability, over several factors, goodness of fit should not be a requirement for them to believe in model results. As long as the model is internally consistent an agreement with data concentrations over a factor of one to four should be considered sufficient. With willingness and communication however, research may be focussed on processes, such as food-chain, contaminant recycling and flowthrough, while measures of concentrations are needed only to verify that model simulations are within the right order of magnitude.

CONCLUSIONS

Models need data that are collected for specific purposes and the system approach indicates which are the most important data that have to be collected. For example, to predict the fate of Mirex in Lake Ontario (Halfon 1984), a knowledge of the turbulence at the bottom sediment water interface is very important. In this case, modelling indicated the need for specific experiments to be undertaken. The model was also used to predict what percentage of Mirex was retained in the lake. It was shown that 72% of Mirex which entered from the Niagara and Oswego Rivers remained in the bottom sediments, in contrast, the model by Pickett and Dossett (1979), had computed a retention of only 50%. Clearly, an analysis of the ecosystem compartments' exposure to Mirex would be heavily influenced by which model was chosen. The question to be resolved is whether the data collected can or cannot be used for modelling purposes and for extrapolation to similar types of compounds. The knowledge of a mass balance is important for a good (PEC) model validation, but is it needed when we predict the fate of new contaminants? With a pathway analysis, we may estimate in which part of the aquatic environment (i.e,. water or bottom sediments), the contaminant will end up but concentrations can be predicted only if masses are known or can be estimated. It appears worthwhile to pool resources and to collect data for a selected number of compounds from a representative number of stations in the lake and to date and analyse them very carefully. Once a data set becomes available, which may take several years, good lake simulation models for the _chosen_ contaminants should be available. On the basis of quantitative structure-activity correlations and other exisitng models, we may then apply them to other chemicals with the knowledge that the real fate of a substance can be predicted within certain probability limits. Probabilities of model simulations can be computed using covariance analysis or Monte Carlo simulations (Halfon and Maguire 1983). Only with the mutual recognition of the various data requirements of both qualitative and quantitative nature, can the results of monitoring and modelling activities be integrated to allow reliable predictions on the impact of contaminants on the ecosystem compartments and on the assessment of contaminant hazards.

ACKNOWLEDGMENTS

I thank J.H. Carey, M.N. Charlton, R.W. Durham, F.C. Elder, A. El-Shaarawi, K.L.E. Kaiser, D.C.L. Lam, R.J. Maguire, F. Rosa, and R.L. Thomas for helpful comments.

REFERENCES

ASTM 1978. **Estimating the Hazard of Chemical Substances to Aquatic Life.** ASTM STP 657, American Society for Testing and Materials, Philadelphia.

Halfon, E. 1984. Error analysis and simulation of Mirex behaviour in Lake Ontario. In **Modelling the Distribution and Effects of Toxic Substances,** S.E. Jorgensen (Ed.), Elsevier, Amsterdam, in press.

Halfon, E. and Maguire, R.J. 1983. Distribution and transformation of fenitrothion sprayed on a pond: modelling under uncertainty. In **Uncertainty and Forecasting of Water Quality,** M.B. Beck and G. van Straten (Eds.), Springer Verlag, pp. 119-128.

Holdrinet, M., Frank, R., Thomas, R.L., and Hetling, L.J. 1978. Mirex in the sediments of Lake Ontario. **J. Great Lakes Res. 4:** 69-74.

Kuntz, K.W. and Warry, N.D. 1983. Chlorinated organic contaminants in water and suspended sediments of the lower Niagara River. **J. Great Lakes Res. 9:** 241-248.

Lassiter, R.R., Baughman, G.L., and Burns, L.A. 1979. Fate of toxic organic substances in the aquatic environment. In **State of the Art in Ecological Modelling,** S.E. Jorgensen (Ed.), International Society of Ecological Modelling, Copenhagen pp. 219-246.

Lean, D.R.S. 1973. Phosphorus dynamics in lake waters. **Science 179:** 678-680.

Mackay, D. 1979. Finding fugacity possible. **Environ. Sci. Technol. 13:** 1218-1223.

Park, R.A., O'Neill, R.V., Bloomfield, J.A., Shugart, H.H. Jr., Booth R.S., Goldstein, R.A., Mankin, J.B., Koonce, J.F., Scavia, D., Adams, M.S., Clesceri, L.S., Colon,, E.M., Dettmann, E.H., Hoopes, J.A., Huff, D.D., Katz, S., Kitchell, J.F., Kohberger, R.C., LaRow, E.J., McNaught, D.C., Peterson, J.L., Titus, J.E., Weiler, P.R., Wilkinson, J.W., Zahorcak, C.S. 1974. A generalized model for simulating lake ecosystems. **Simulation 23:** 33-50.

Park, R.A. et al. 1980. **Modelling Transport and Behaviour of Pesticides in Aquatic Environments,** CEM Report No. 7, Rensselaer Polytechnic Institute, Troy, New York, 163 p.

Patten, B.C. (Ed.) 1971 to 1976. **Systems Analysis and Simulation in Ecology,** Vol. I to IV, Academic Press, New York.

Pickett, R.L. and Dossett, D.A. 1979. Mirex and the circulation of Lake Ontario. **J. Phys. Ocean. 9:** 441-445.

Strachan, W.M.J. and Huneault, H. 1979. Polychlorinated biphenyls and organochlorine pesticides in Great Lakes precipitation. **J. Great Lakes Res. 5:** 61-68.

Vollenweider, R.A. 1975. Input output models with special reference to the phosphorus loading concept im limnology. **Schweiz. Z. Hydrol. 37:** 53-84.

WORKSHOP DISCUSSION

HALLETT:

You are asking the right questions but you are asking for the most expensive end, which we cannot afford and the whole direction of this symposium so far, is how to get more cheaper, and not just in dollars but also in terms of time. Because it takes a lot of time to develop these things as you well pointed out, you have a problem. To that end, in terms of modelling, I think you are going to have to concentrate more on the available data.

HALFON:

What you suggest is that we should go for simplicity in the models which may make them less predictive, and try to explain some of the existing data? In a sense, it would be in agreement with what is happening right now. But, how could you rely on the predictions if some data are complete and some incomplete?. That is the trouble with trying to assess reliability, finding some new data you can use and check the models with.

HALLETT:

I appreciate your comment about how to develop something that is predictive and I think one of the more useful things is that you have really laid out a whole series of questions and answers very competently. In terms of your focus and your modelling, just take your first questions about landfills and where all these compounds are coming from and I would think that you are looking at ten million dollars and five years down the road.

HALFON:

These are just questions that came out of the discussion in Copenhagen and we would like to have them commented on by ecologists.

WELLS:

Is there very much communication between the ecological modellers and others, for example with people like Don Mackay and his fugacity models?

HALFON:

Up until now, there hasn't been much communication. In fact, the general impression in Copenhagen was that we are a small group. I think that we are not only trying to do simulation, but we need a lot of assistance to understand the grouping of these chemicals into different compartments.

Question:

Do you have a sensitivity analysis to find out which parameter to follow or just one or two on the whole list to focus data collection on?

HALFON:

The necessity for sensitivity analysis is being described more and more, but what has replaced it is error analysis or uncertainty analysis that has a similar function and its assumptions are more general. I did that kind of error analysis and it showed that for a model like Mirex, the most important unknowns are related to processes in the sediments. It takes years to try and find out what chemical parameters and environmental parameters influence the uncertainty of the predictions.

McCARTY:

It seems to me that what you are asking for is that you want to have all the data so that you can have a 100% correct model all the time and instead of making several little forays and establish base camps up the mountain, you want to climb Mt. Everest all at once. The impression I get is that you are saying that modellers as a group are afraid to use models because if they have an imperfect model it is obvious people aren't going to believe them. Yet, the whole effort that we are trying to do here is to by and large make use of imperfect models, knowing that they are imperfect but taken as a first step and plotting out what kind of model you would like to see and then the next step is to improve the model making it a little bit better as we go along trying to take as much information as possible rather than make the perfect model first time out.

LANDE:

My question is how important is modelling for regulatory research?

HALFON:

When I introduced EXAMS at CCIW a couple of years ago, the initial framework was developed somewhere else, it was, in my opinion quite heavily criticized and thought to be incomplete. Management wanted something better, and did not accept the scientific aspects of it because it was imperfect and mainly because it was not representative of the large lakes.

Comment:

Toxicities can be predicted from QSAR models, but the environmental fate is more difficult to model. However, I think it is important to have good environmental models with which one can predict environmental concentrations and exposure.

MACKAY:

I sympathize with your problems. I think modellers should be a little more courageous and come out with models and give predictions with fuzziness in them like our fuzzy fugacity models and then, having done that, say to the managers, this is the result and these are the uncertainties and if you give me some more money I will shrink the uncertainties for you, if you want. The whole thing takes on a sense of that of an adult trying to explain to an eight year old what sex is all about, the whole things sounds very mysterious and vaguely disgusting. I think modellers are in the same situation in explaining the joys of modelling to other people.

THE NEED AND REQUIREMENTS FOR QUANTITATIVE STRUCTURE-ACTIVITY RELATIONS (QSAR) IN THE GREAT LAKES WATER QUALITY PROGRAM

Douglas J. Hallett

Ontario Region
Environment Canada
55 St. Clair Avenue West
Toronto, Ontario M4T 1M2
CANADA

INTRODUCTION

Under Annex 12 (Persistent Toxic Substances) of the Great Lakes Water Quality Agreement of 1978 between the United States and Canada, Section 5: Early Warning System, a) "Development and use of structure-activity correlations to predict environmental characteristics of chemicals", is a clearly defined requirement. The need for Quantitative Structure-Activity Relations (QSARs) and the complexity of the environmental chemical matrix present in the Great Lakes was well described by Dr. G. Veith of the U.S. Environmental Protection Agency to the Great Lakes community in a presentation at the annual meeting on Great Lakes Water Quality of the International Joint Commission in Windsor in July 1977.

Scientists in the Great Lakes area are developing tests to predict the toxicity of a compound based on the relationship of the toxicity of known compounds and their structures. Caution should be exercised in the reliability of these predictive tests since many of the compounds that are used today and occur as environmental contaminants had some lesser form of test applied to them that indicated they should not be harmful to the environment. There is therefore a great responsibility in developing these tests to ensure that they are

K. L. E. Kaiser (ed.), QSAR in Environmental Toxicology, 153–166.

accurate, predictive and that there is no danger that a chemical that appears environmentally sound is not oversold before it has been thoroughly tested. This cautious approach should not overshadow the need for developing faster and more accurate tests with which to establish QSARs. There are many compounds already present in the Great Lakes which require testing for toxic potential.

A framework for toxic chemicals management was first set out by the senior management committee of DOE. This scheme has also been adopted by the International Joint Commission (IJC) as shown in SCHEME 1 (First Report of the IJC Toxic Substances Committee 1980). The scheme involves the development of a hazard and risk assessment from an information base, control actions, and the evaluation of the effectiveness of that action. QSAR fits in at the information base, where there is the greatest expenditure of time, expertise and money. QSAR is one means of expediting the process of screening compounds which have potentially hazardous characteristics.

Herring gulls (*Larus argentatus*) and coho salmon (*Oncorhynchus kisutch*), both of which feed on alewives (*Alosa pseudoharengus*) and rainbow smelt (*Osmerus mordax*), provide an example of the ecosystem foodweb and bioaccumulation of toxic chemicals within Lake Ontario. TABLE 1 gives the concentrations of PCB's and other contaminants found in Lake Ontario herring gull eggs and coho salmon, and in alewives and smelt removed from their guts (Norstrom *et al.* 1978). PCB concentrations were a factor of approximately 20 lower in coho salmon and lower still in alewives and smelt when compared to concentrations found in herring gulls. It can be concluded that their common food source, alewives and smelt, are also the source of these contaminants for both coho salmon and herring gulls.

The results of studies on PCB concentrations in herring gulls (Struger *et al.* 1983), lake trout, rainbow smelt (GLWQB 1981; Whittle and Fitzsimons 1983) and spottail shiners (Ontario Ministry of the Environment, unpublished data, 1981) of Lake Ontario are summarized in FIGURE 1. Although an initial decline in PCB levels between 1974 and 1979 has been found, since that time, there has been a consistent increasing trend of PCB's in this ecosystem. Municipal and industrial effluents cannot account for the total loading to Lake Ontario. The major sources also include atmospheric deposition and contaminated groundwater in the Lake Ontario basin.

SCHEME 1: Toxic chemicals management framework. Connecting lines indicate interaction and feed-back in both directions between components. Numbers refer to the appropriate sections in Annex 12 of the Great Lakes Water Quality Agreement of 1978

Inventory

3(a) Identification of products and emissions
4(c) Disposal of wastes
5(b) Trends in production, import and use
5(h) Quantities in commerce

Characteristics

5(a) Structure-activity correlations
5(c) Testing of new chemicals
5(d) Toxicological research
5(h) Data bank

Environmental Measurements

3(b) Coordinated programs between air, water and solid wastes
4(a) Temporal and spatial trends
4(d) Emerging problems
5(f) Monitoring

Assessment and Prioritization

4(b) Effects on human health and other organisms
7(a) Significance of effects on human health and aquatic life (hazard and risk to ecosystem)
7(b) Interactive effects
7(c) Calculation of safe loading rates

Development of Plans

3(b) Establishment of coordination between air, water and solid waste management
3(c) Joint programs for disposal of hazardous waste
5 Determining priority substances for monitoring and regulation
7(c) Approaches to calculations of acceptable loadings

Implementation of Plans

2(a) Regulatory strategies for controlling input
2(b) Rehabilitation programs
3(a) Development of comprehensive inventories
3(b) Implementation of coordinated programs between air, water and solid wastes
3(c) Implementation of waste disposal programs

Evaluation of Effectiveness

TABLE 1: Levels relative to DDE of selected organochlorine compounds in coho salmon muscle, liver, and their pooled stomach contents (alewives and smelt) and in herring gull eggs in $mg \cdot kg^{-1}/mg \cdot kg^{-1}$ (DDE)

Species and Tissue	PCB's	Mirex	Photomirex	HCB[a]	p,p'-DDD'
Alewives and smelt	4.70	0.19	0.06	0.05	0.10
Coho salmon muscle	5.95	0.24	0.11	0.10	0.11
Coho salmon liver	5.63	0.24	0.10	0.16	0.18
Herring gull eggs	7.64	0.25	0.09	0.03	0.01

a Hexachlorobenzene.

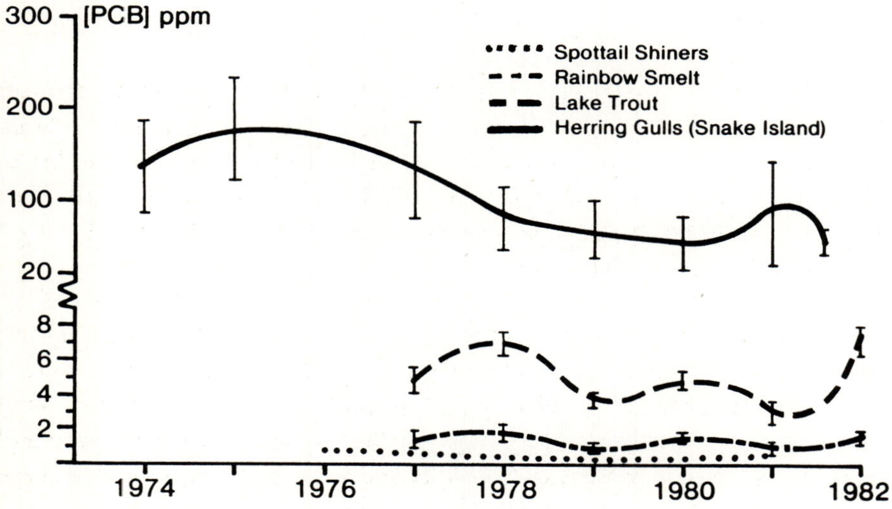

FIGURE 1: PCB concentrations in Lake Ontario biota

Similar trends are indicated for mirex, photomirex, and the DDT family of compounds in the Lake Ontario basin. This not only exemplifies the need for increased effort in toxic chemical control, but also the utility of mutually supporting studies on various trophic levels within the same ecosystem.

Another way of looking at PCB's and other toxic chemicals in the Great Lakes and perhaps the best method of looking at a spatial trend is to examine the sediment. A series of sediment profiles can be found in the final report of the International Reference Group on Great Lakes Pollution from Land Use Activities (PLUARG 1978). The highest concentrations of PCB's and other contaminants such as lead are found at the mouth of the Detroit River in Lake Erie and at the mouth of the Niagara River in Lake Ontario. Sediment transfer and movement reflects currents and deposition in the deeper basins of all lakes.

Mirex and photomirex are the third and fourth major toxic chemicals in the Lake Ontario ecosystem. Sources of mirex occur in the Niagara and Oswego Rivers (Kaiser 1978). Photomirex is not found in the sediment nor in the suspended sediment in the Niagara River but it does show up in the rest of the ecosystem in higher levels (Hallett _et al_. 1976; Norstrom _et al_. 1980).

In reality, it is not just a mirex or photomirex problem in Lake Ontario; we are looking at six mirex related compounds. Compared to the toxicity of mirex, little is known regarding the toxicity of photomirex and the other related compounds. We begin here to see the need for deriving toxicity correlations for very similar, but in this case very complex, chemical structures.

Another major toxic chemical source in the Great Lakes is the atmosphere. The incomplete combustion of organic compounds will either volatilize those compounds or produce other compounds through side reactions, particularly higher molecular weight polynuclear compounds (PNA's) and condensation products. For example, incomplete combustion of polyvinylchloride produces chlorobenzenes, some PCB isomers and octachlorostyrene (Ahling _et al_. 1978).

Large concentrations of chlorobenzenes are found in gulls and there are specific isomers which are common to fish, water, and gulls (Hallett _et al_. 1982). The same types or isomers of chlorobenzenes have been found in certain PCB Askerels that have been contaminated with chlorobenzenes (Rappe _et al_. 1983). There are several theories of how octachlorostyrene enters the Great Lakes. The most plausible theory involves the use of graphite electrodes in the production of chlorine. One of the predominant chlorostyrenes that has been found basin-wide is heptachlorostyrene. Three isomers of chlorostyrenes have been reported to occur in Great Lakes biota (Brecher and Hallett 1984). Limited data on the toxicity of octachlorostyrene has been generated only recently (Chu _et al_. 1982; Villeneuve _et al_. 1982; Strik and Koeman 1976). The bioconcentration factor of octachlorostyrene in fish has been estimated at 33,000 (Veith _et al_. 1978). Perhaps some QSAR considerations are possible for chlorinated styrenes and chlorinated benzenes.

The types of PNA's found in the Great Lakes are those occurring from combustion or from coking processes common to the iron and steel industry which are different from those typically found in crude, diesel, and bunker oils (Hallett and Brecher 1984). There are compounds with structures similar to PNA's which may be significant, particularly those compounds with methyl substitutions. Very little research has been done on the toxicology of methyl-substituted or nitro-substituted PNA's.

It is interesting to note that although PNA's occur in vast quantitities in the Great Lakes and there are tremendous source inputs, they have not bio-accumulated to the top of the food chain on a broad scale (Hallett and Brecher 1984). Although PNA's have the lipophilic properties which allow bioaccumulation, once they reach an organism with a well-developed arylhydrocarbon hydroxylase (AHH) system, they are metabolized very quickly and excreted. Studies on the benthic organisms in the Great Lakes have found bioaccumulation occurring in the benthos (Eadie _et al_. 1982) while higher fish with well-developed AHH systems have lower levels of PNA's.

An important consideration for QSAR is the study of enzymes such as AHH, the levels of AHH locus activators such as PNA's in various species and determining the activity of those enzymes. Information on detoxification and excretion of metabolites of more labile compounds will be valuable in predicting what might happen in ecosystems.

FIGURE 2 shows the pattern of 2,3,7,8-tetrachloro-p-dibenzodioxin (TCDD) which has been found in the Great Lakes (Norstrom _et al_. 1982). There are particularly high levels at Saginaw Bay which is associated with the Dow Chemical Company, and in the Detroit River where the source is undetermined. The high levels in gull eggs from

Lake Ontario is probably associated with waste dumpsites and landfills in the Niagara River area. The Love Canal and the Hyde Park landfills are likely sources on the Niagara River. There was a downward trend in dioxin between 1974 and 1980. The 1974 inflection point in the curve corresponds with the capping of the Hyde Park dump which had a direct input through Bloody Run Creek to Lake Ontario. There are also vast quantities of mirex in the dump. It is interesting that the trend curve for both TCDD and mirex in gulls parallel each other exactly.

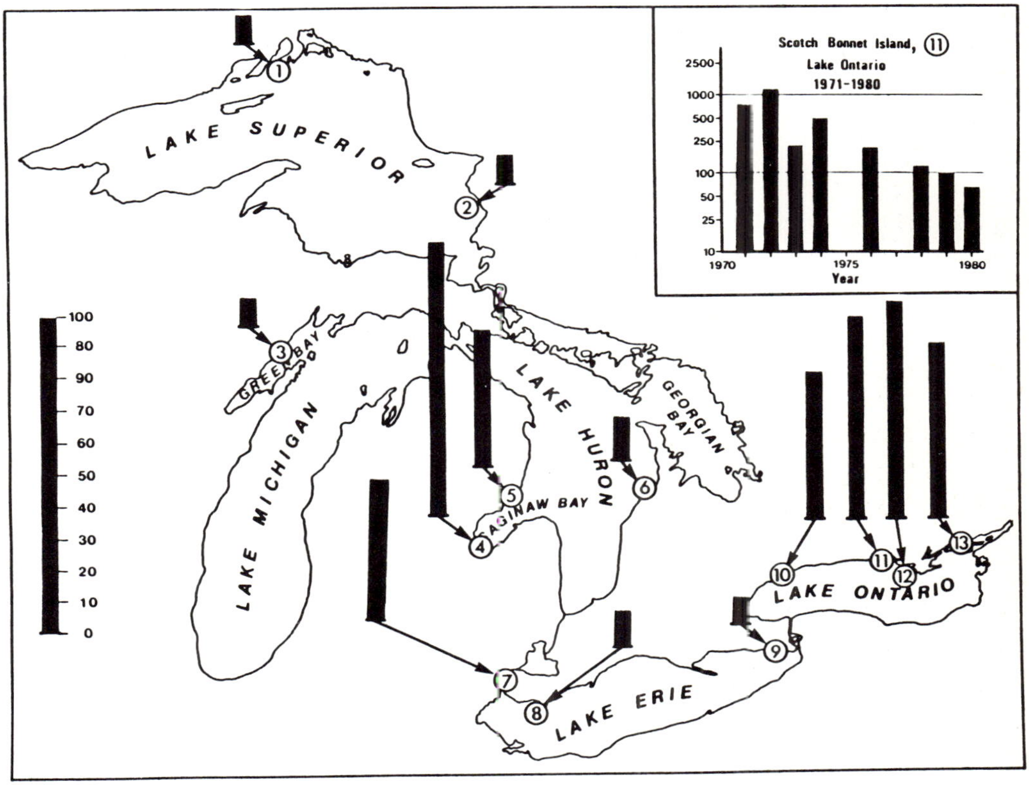

FIGURE 2: Concentrations of 2,3,7,8-tetrachlorodibenzo-p-dioxin in herring gull eggs from various colonies in the Great Lakes basin in 1980, concentrations in ng•kg^{-1}; from GLWQB 1981

TABLE 2 shows the other compounds in the dioxin class which are presently under study including pentachloro-, hexachloro-, heptachloro-, and octachlorodioxin, which have been found in the Detroit River at the Fighting Island herring gull colony. Isomers of polychlorinated dibenzofurans are also present. Dioxin and furan isomers with chlorine substituents in the 2,3,7,8 positions are highly toxic. Clues as to the similarity of toxic effects of structural stereoisomers of chlorinated dioxins, furans, and biphenyls as well as polychlorinated naphthalenes and chlorinated azo and azoxy benzenes have been reviewed by Safe (1983).

TABLE 2: Polychlorinated dibenzofurans (PCDF's) and polychlorinated dibenzo-p-dioxins (PCDD's) in herring gull eggs from Fighting Island, 1979, in $\mu g \cdot kg^{-1}$

Isomer[a]	PCDF's	PCDD's
Tetrachloro-		
2,3,7,8-	15	70
Pentachloro-	2	nd[b]
2,3,4,7,8-	48	
Hexachloro-	40	88
Heptachloro-	18	nd[b]
Octachloro-	5	28
Total	128	186

a Concentrations corrected for recovery of ^{13}C tetrachloro-PCDD, but not for other PCDD's or PCDF's.

b not detected.

It is becoming increasingly obvious that we are dealing with a multitude of chemicals in the environment for which we know little or nothing of their toxicity. However, within those classes and to a lesser extent also between classes, toxicological similarities are beginning to be realized. Further research to address these gaps and lack of information, through the synthesis and testing of key compounds, development of a battery of relevant biological tests, and the analysis of such data by QSAR methods will allow us to assess the hazards of these compounds.

REFERENCES

Ahling, B., Bjorseth, A. and Lunde, G. 1978. Formation of chlorinated hydrocarbons during combustion of polyvinylchloride. **Chemosphere 7:** 799-806

Brecher, R.W. and Hallett, D.J. 1984. Incidence of chlorinated styrenes in Great Lakes herring gulls. **Chemosphere,** submitted.

Chu, I., Secours, V.E., Villeneuve, D.C. and Valli, V.E. 1982. Acute and subacute toxicity of octachlorostyrene in the rat. **J. Tox. Environ. Health 10:** 285-296.

Eadie, B.J., Faust, W., Gardner, W.S. and Nalepa, T. 1982. Polycyclic aromatic hydrocarbons in sediments and associated benthos in Lake Erie. **Chemosphere 11:** 185-191.

GLWQB. 1981. Great Lakes Water Quality Board. **Report on Great Lakes Water Quality, Appendix B, Great Lakes Surveillance.** International Joint Commission, Windsor, Ontario, 174 p.

Hallett, D.J., Norstrom, R.J., Onuska, F.I., Comba, M.E. and Sampson, R. 1976. Mass spectral confirmation and analysis by the Hall detector of mirex and photomirex in herring gulls from Lake Ontario. **J. Agric. Food Chem. 24:** 1189-1193.

Hallett, D.J., Norstrom, R.J., Onuska, F.I. and Comba, M.E. 1982. Incidence of chlorinated benzenes and chlorinated ethylenes in Lake Ontario herring gulls. **Chemosphere 11:** 277-285.

Hallett, D.J. and Brecher, R.W. 1984. Cycling of polynuclear aromatic hydrocarbons in the Great Lakes. In **Toxic Contaminants in the Great Lakes,** J.O. Nriagu (Ed.), John Wiley & Sons, New York, pp. 127-137.

Kaiser, K.L.E. 1978. The rise and fall of mirex. **Environ. Sci. Technol. 12:** 520-528.

Norstrom, R.J., Hallett and D.J. Sonstegard, R.A. 1978. Coho salmon and herring gulls as indicators of organochlorine contamination in Lake Ontario. **J. Fish. Res. Board Can. 35:** 1401-1409.

Norstrom, R.J., Hallett, D.J., Onuska, F.I. and Comba, M.E. 1980. Mirex and its degradation products in Great lakes herring gulls. **Environ. Sci. Technol. 14:** 860-866.

Norstrom, R.J., Hallett, D.J., Simon, M. and Mulvihill, M.J. 1982. Analysis of Great Lakes herring gull eggs for tetrachlorodibenzo-p-dioxins. In **Chlorinated Dioxins and Related Compounds, Impact on the Environment,** O. Hutzinger, R.W. Free, E. Merian and F. Pocchiari (Eds.), Pergamon Press, Toronto, Ontario, pp. 173-181.

PLUARG. 1978. International Reference Group on Great Lakes Pollution from Land Use Activities. **Environmental Management Strategies for the Great Lakes System.** Final Report to the International Joint Commission, July 1978, Windsor, Ontario, 115 p.

Rappe, C., Marklund, S., Bergqvist, P.-A., Kjeller, L.-O. and Hansson, M. 1983. Composition of PCDF's formed in PCB fires. Preprint Extended Abstract No. 93. Presented before the **Division of Environmental Chemistry, American Chemical Society,** 186th National Meeting, Washington, D.C., August 1983, Vol. 23, No. 2, pp. 252-255.

Safe, S. 1983. 2,3,7,8-TCDD - Biochemical effects. **Chemosphere 12:** 447-451.

Strik, J.J.T.W.A. and Koeman, J.H. 1976. Porphyrinorganic action of hexachlorobenzene and octachlorostyrene. In **Porphyrins in Human Diseases,** M. Doss (Ed.), Karger Press, New York, pp. 418-423.

Struger, J., Weseloh, D.V., Mineau, P. and Hallett, D.J. 1983. Levels and trends of organochorines in herring culls in the Great Lakes, 1974 to 1981. Presented at the **International Association for Great Lakes Research,** 26th Annual Conference, Oswego, New York, May 1983.

Veith, G.D., Kuehl, D.W., Leonard, E.N., Puglisi, F.A. and Lemke, A.E. 1978. Polychlorinated biphenyls and other organic chemical residues in fish from major watersheds of the United States, 1976. **Pestic. Monitor. J. 13:** 1-11.

Villeneuve, D.C., Chu, I., Secours, V., Valli, V.E.O. and Becking, G.C. 1982. Subchronic toxicity of octanochlorostyrene in the rat. Proc. Ann. Conf. Soc. Toxicol., **The Toxicologist 2:** Abstract No. 651.

Whittle, D.M. and Fitzsimons, J.D. 1983. The influence of the Niagara River on contaminant burdens of Lake Ontario biota. **J. Great Lakes Res. 9:** 195-302.

WORKSHOP DISCUSSION

KAISER:

What is the most urgent need in terms of research on environmental contaminants, is it determining toxicity, identification of contaminants, or what?

HALLETT:

There are two major thrusts going on right now; one is towards hazard and risk assessment, and QSAR will be very, very useful in predicting the hazards of compounds so that we can take action on them. We are dealing with literally thousands of compounds in the Great Lakes and I emphasize, that we are not so concerned about new compounds on the market as we are about existing compounds that people are already exposed to, that our ecosystem is already exposed to and we keep finding out that there are more and more of them already there and we are not going to keep them off the market. Once something is there, it is hard to get rid of it and I feel that we've got to go and take some strides towards controlling, not on a compound by compound basis, but maybe on a class by class basis or source by source basis. Usually, on an individual source you have very similar compounds coming out or maybe one, two or three classes - the coke industry would be a good example of polynuclear aromatics and the like and I think QSAR will be useful in predicting what are the compounds we want to eliminate from a waste water stream for instance and what remedial action may be taken to get rid of that class of chemicals.

I would emphasize to you that it is applied research and you are trying to apply it to a current problem. You have a tremendous responsibility in that your predictive tests really have to predict what the reality is; you can't be wrong this time or else we are going to be with the problem for quite some time yet.

DUNN:

One thing I tend to get from your presentation, I guess somewhat consistent with my view of how these problems should be treated, from the QSAR point of view is that it is certainly a most variable problem. We are dealing with very complex mixtures and we are going to have to use the appropriate methods to treat the data. If you want to optimize your predictability with the treatment of such problems you have to use multi-varied techniques.

HALLETT:

The way we have tried to manage chemicals in the past is on a compound by compound basis and you can see that with the vast number of compounds that, even if we had all the money to spend, we would not have the time. It would take probably 20 to 50 years to develop the required data base. We have to shorten the time frame or else live with the problem and accept it. I think we have to decide what are the key variables and we have to draw this out, particularly with respect to control.

SCHULTZ:

Your point that we are dealing with a very complex mixture of contaminants is well taken and we have been interested in that same problem for several years and are using chemical fractionations. The problem is that chemical fractionation divides chemicals along classic chemical lines, whether they be aromatics, hydrocarbons, or nitrogen substituted heterocyclics, or whatever. But when you look at the biology, the fish do not divide the compounds into chemical classes based on classic organic chemistry, they break them down on toxic response based on how they are able to handle that with the enzyme system they happen to have.

HALLETT:

Such as the arylhydrocarbon hydroxylase activity?

SCHULTZ:

Exactly. As you well pointed out, those beasts tend to be very active and have a well developed liver and a lot of enzyme activity. They seem to be able to handle those things while the domestic animals tend to be a little "slower" and tend to accumulate.

HALLETT:

We have recently written a tentative objective we are trying out for polynuclear hydrocarbons, dealing with those polyaromatics which have high arylhydrocarbon hydroxylase activity and in developing that objective by reviewing a great deal of literature and there was an excellent table in one review paper which had all the compounds, it was almost like a list of the key compounds we are concerned about in the Great Lakes. They all had a very high cytochrome P-450 activity. I think the genetic aspect is something I have not seen people develop at all or be concerned with, it is a very well developed science but in terms of QSAR no one is looking at genetic aspects.

Again, there is a long series of compounds that should be correlated with that type of activity and some which are inactive. Also, there are anomalies and that is another thing I think you should be looking for compounds that don't fit the norm, what do they fit?

Question:

You said that we must not be wrong this time with QSAR. What level of confidence would a regulatory agency be willing to accept?

HALLETT:

I don't think the regulatory agencies know that and really, the onus is on the scientists in this case. I see developing QSAR and developing a very good correlation as important but to do that, we are going to have to use a vast battery of tests to begin with, and really test out your correlations.

Comment:

You are never going to get 99% accuracy as you will in the bioassay but that doesn't mean to ignore our predictions.

HALLETT:

I am just concerned. Some models are coming out which are very simple and people are developing systems such as the Michigan Department of Natural Resources System, to prioritize chemicals with a very simplistic approach, useful as a first cut.

ZITKO:

Do you have an example of where did we go wrong using QSAR?

HALLETT:

Did I say that QSAR failed? I didn't mean that, but basically predictive tests have gone wrong - essentially many of the compounds on the market have been tested to some degree and were found to be acceptable and we now have a great deal of environmental contaminants. On top of it, we are fighting a huge lobby from industry which basically says, "there is the test, and everything looks fine", and then ten years down the road you may be finding a high level of it in the environment.

ZITKO:

I think that most of the chemicals that are causing problems now, have never really been tested with respect to the environment. When PCB's came on the market, nobody knew about them, when DDT came on the market and did a great job of killing mosquitos, no one was concerned about biodegradation; we just didn't ask the right questions.

HALLETT:

That's because there was no one in the position to ask the question.

ASSESSMENT OF MUTAGENIC EFFECTS IN AMPHIBIAN EMBRYOS

Donald R. Hart

IEC Beak Consultants Ltd.
6870 Goreway Drive
Mississauga, Ontario L4V 1P1
CANADA

ABSTRACT

Adult male South African clawed frogs, *Xenopus laevis*, were mutagenized by three-day immersion in aqueous solutions of ethyl methanesulfonate (EMS), ethyl nitrosourea (ENU) or diethyl nitrosamine (DEN), or by acute gamma irradiation. Dominant lethal effects were obtained with EMS or gamma radiation. Recessive lethal effects were uncovered in haploid embryos after paternal treatment with EMS, ENU or gamma radiation. DEN was not an effecitve *in vivo* mutagen.

INTRODUCTION

The traditional mutagen test systems are less than ideal for evaluation of hazards from environmental contaminants. The Ames test is rapid and sensitive, but involves the rather tenuous extrapolation of results from a bacterial to a vertebrate system. The mouse or rat dominant lethal test minimizes the extrapolation problem but does not detect recessive mutations. The specific locus test measures recessive mutations but low mutation rates at specific loci and small litter sizes necessitate treatment of a prohibitive number of animals.

The sex-linked recessive lethal approach can be used to measure recessive effects over a larger portion of the genome. Male

K. L. E. Kaiser (ed.), QSAR in Environmental Toxicology, 167–177.

fruit flies, which lack a homologue for the X chromosome, will express recessive mutations induced anywhere on that chromosome. The probability of uncovering recessive effects is thus increased by sampling a whole chromosome rather than a few specific loci.

This principle can be extended to the entire genome by production of androgenetic haploid progeny from mutagenized males. Such haploids carry no maternal genes to mask induced recessive mutations. Armstrong and Fletcher (1983) recently used this technique to uncover recessive lethals in salamander embryos. I am using it with the aquatic from *Xenopus laevis*, as part of a test system for dominant and recessive lethal effects.

Haploids are easily produced in this frog. Large numbers of eggs can be obtained on a year-round basis and their development is entirely external so that induced embryonic abnormalities are easily observed and characterized.

METHODS

1. Adult male frogs were mutagenized by three-day immersion in aqueous solutions of various alkylating agents or by gamma irradiation at 73 $R \cdot min^{-1}$.
2. Males were spawned with untreated females one week later and thereafter at two-week intervals.
3. Spawning was induced by injection of human chorionic gonadotropin.
4. Eggs were collected in plastic petri dishes and covered with dechlorinated tap water.
5. Androgenetic haploids were produced by UV-inactivation of egg pronuclei following Gurdon's (1960) methodology.
6. Normal haploid and diploid blastulae were separated and transferred to finger bowls where they developed through hatching at 20°C.
7. Haploid and diploid survival to hatching was recorded and embryonic abnormalities noted.
8. Reduction of diploid survival from pre-mutagenesis levels was used to assess dominant lethal effects.
9. Reduction of haploid survival was used to assess total induced lethality, including both dominant and recessive effects.

RESULTS AND DISCUSSION

FIGURE 1 shows the reduction in embryonic survival after paternal treatment with ethyl methanesulfonate (EMS) at 400 mg•L^{-1} (73% of the adult LC50). Two adult males received this mutagen treatment.

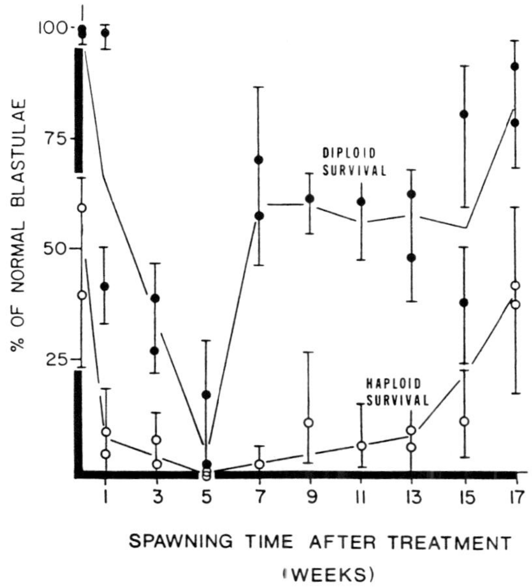

FIGURE 1: Survival of diploid and haploid male South African clawed frog embryos after paternal treatment with the mutagen ethyl methane-sulfonate (EMS) at 400 mg•L^{-1}

In spawnings prior to mutagenesis, both males produced highly viable offspring. Survival of diploid embryos approached 100%. Haploid survival was about 50%. Haploids never survive as well as diploids and this has been attributed to the uncovering of natural genetic load.

Spawning after mutagenesis produced less viable offspring. Diploid survival was reduced to about 4% in spawnings done five weeks after paternal EMS treatment. Thereafter diploid survival improved, indicating that the spermatogenic stage of maximum EMS sensitivity takes about five weeks to mature and fertilize eggs.

The spermatid stage is most sensitive to dominant lethal damage by EMS in mice (Cattanach *et al.* 1968; Ehling *et al.* 1968) and this is the stage most strongly labelled by tritiated EMS in both mammals (Sega and Owens 1978) and amphibians (Miltner 1982).

Haploid survival was reduced over a broader time period, from one to thirteen weeks after treatment of the male. This unique haploid pattern can be attributed to induced recessive lethal effects which do not show up in diploids. The pattern suggests that recessive lethal mutations are less specific than dominant ones to a particular spermatogenic stage, and may arise by different mechanisms.

The ratio of pre- and post-mutagenesis survival can be used to calculate induced embryonic lethality. FIGURE 2 shows induced lethality in haploid and diploid embryos at different paternal dose levels. Dominant lethal induction in diploids was minimal up to 250 $mg \cdot L^{-1}$ EMS. At this dose most diploid embryos hatched, though abnormalities at hatching (dashed line) were induced. The dose response for induced lethality in haploids was more linear. I suspect that this is due to induction and expression of recessive lethal mutations at low dose levels.

Haploid larvae develop a characteristic haploid syndrome shortly after hatching. The syndrome includes edema, microcephaly and an upturned tail, and is unrelated to paternal mutagenesis. Since it leads to heavy mortality approximately one week after hatching, hatching must be used as the end point in all survival estimates.

Diploid larvae are rarely abnormal in the absence of mutagenesis. Most abnormalities at low EMS dose levels result in embryonic death at around the time of hatching. Abnormalities include spina bifida, scoliosis, missing eyes and other gross deformities.

At high paternal dose levels abnormalities are more severe, and generally result in embryonic death at an earlier developmental stage. Exogastrulation is common. A large yolk plug is excluded from the gasterocoel, and the yolk cells may eventually undergo cytolysis. Abnormalities do not appear prior to gastrulation.

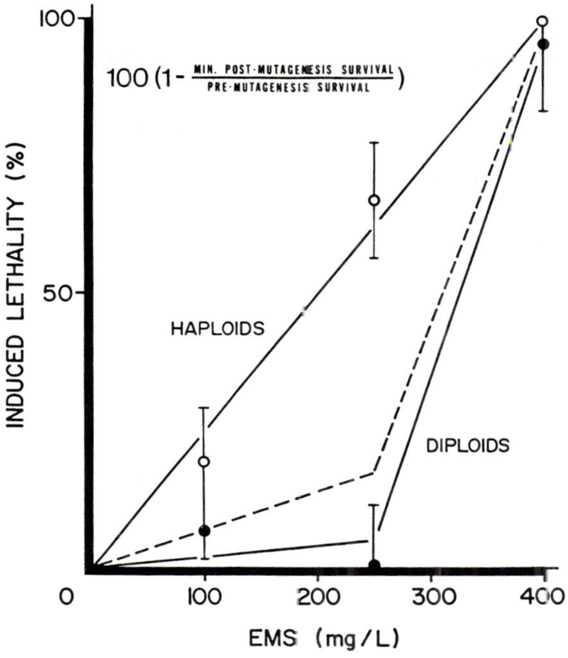

FIGURE 2: Ethyl methanesulfonate (EMS) induced lethality in haploid and diploid South African clawed frog embryos versus treatment concentration

FIGURE 3 shows a reduction in haploid survival after paternal treatment with ethyl nitrosourea (ENU) at 75 $mg \cdot L^{-1}$ (83% of the adult LC50). Diploids survived well to hatching, though some hatchlings were abnormal in the first few spawnings after mutagenesis. The results are similar to those obtained with EMS at 45% of the LC50. Thus, it appears that treatments nearer the lethal range are required to mutagenize with ENU as opposed to EMS.

The comparison between ENU and EMS is interesting because these are alkylating agents near opposite ends of the Swain and Scott (1953) scale. ENU acts mainly at the O(6) site of the guanine base in DNA, while EMS acts mainly at more nucleophilic sites such as N(7). Vogel and Natarajan (1979) found in fruit flies that mutagens near the ENU end of the scale are more toxic than those at the other end and also less likely to produce translocations.

The action of ENU can also be compared to that of diethyl nitrosamine (DEN). This mutagen is considered to form identical alkylating species, but requires metabolic conversion. I have been unable to obtain a mutagenic response with DEN. Negative *in vivo* results in mice (Propping *et al.* 1972) have been attributed to a short-lived intermediate metabolite which fails to survive transport from the liver, where it is formed, to the testis. I suspect that similar problems prevent DEN mutagenesis in frogs.

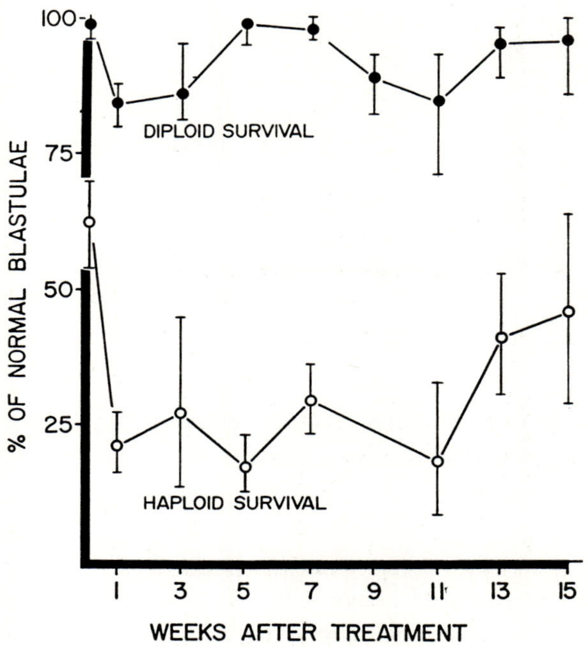

FIGURE 3: Survival of diploid and haploid male South African clawed frog embryos after paternal treatment with the mutagen ethyl nitrosourea (ENU) at 75 mg$\cdot$L^{-1}

FIGURE 4 shows a reduction in both haploid and diploid survival after paternal treatment with gamma radiation at 1500 R (approximately 1455 rad; 75% of the adult LD50). The effect on haploids was similar to that observed at 400 mg$\cdot$L^{-1} EMS, with reduced embryonic survival in spawnings from one to thirteen weeks after mutagenesis. However, permanent sterility at 15 weeks prevented recovery. This is likely due to spermatogonial killing.

Peak dominant lethality, as indicated by reduced diploid survival, occurred from one to seven weeks after paternal treatment. The width of this survival trough suggests that dominant lethal effects of radiation are less specific than those of EMS to a particular spermatogenic stage. Similar results have been obtained with mice (Ehling 1971).

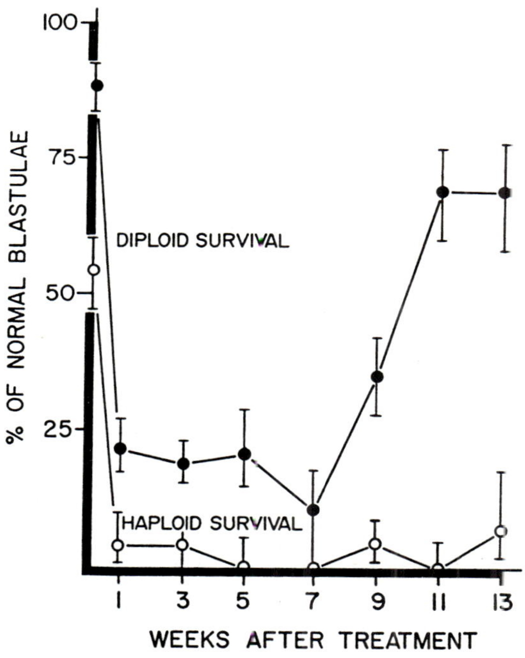

FIGURE 4: Survival of diploid and haploid male South African clawed frog embryos after paternal treatment with gamma radiation at 1500 R (approximately 1455 rad)

Mutagenic responses in _Xenopus_ have so far been qualitatively similar to those obtained with mice, though EMS treatments nearer the lethal range are required to produce dominant lethality. Dominant lethality reaches 50% in _Xenopus_ at 325 mg$\cdot L^{-1}$ EMS (60% of the LC50). The EC50 for dominant lethality in mice is about 40% of the LD50 by intraperitoneal injection.

Comparison of McKinnell's (McKinnell *et al.* 1980) *Xenopus* data with those of Mathews *et al.* (1978) for *Poecilia* indicate that fish may also be somewhat more sensitive to dominant lethal induction by triethylenemelamine (TEM). However, the genetic response of the frog can be greatly enhanced by production of haploid embryos for assessment of recessive lethal damage. For this reason the frog system may have potential as a vertebrate test for mutagenicity in aquatic environments.

ACKNOWLEDGEMENTS

I wish to thank J.B. Armstrong for valuable advice on all aspects of this research, and the Department of National Defense for providing gamma irradiation facilities. The work was supported by the National Sciences and Engineering Research Council of Canada.

REFERENCES

Armstrong, J.B. and Fletcher, W.S. 1983. Assessment of mutagenic damage following ethyl methanesulfonate mutagenesis in the axolotl (*Ambystoma mexicanum*). **J. Exp. Zool. 226:** 333-340.

Cattanach, B.M., Pollard, C.E. and Issacson, J.H. 1968. Ethyl methanesulfonate-induced chromosome breakage in the mouse. **Mut. Res. 6:** 297-307.

Ehling, U.H. 1971. Comparison of radiation- and chemically-induced dominant lethal mutations in male mice. **Mut. Res. 11:** 35-44.

Ehling, U.H., Cumming, R.B. and Malling, H.V. 1968. Induction of dominant lethal mutations by alkylating agents in male mice. **Mut. Res. 5:** 417-428.

Gurdon, J.B. 1960. The effects of ultraviolet irradiation on uncleaved eggs of *Xenopus laevis*. **Quart. J. Microscop. Sci. 101:** 299-311.

Mathews, J.G., Favor, J.B. and Crenshaw, J.W. 1978. Dominant lethal effects of triethylenemelamine in the guppy Poecilia reticulata. **Mut. Res. 54:** 149-157.

McKinnell, R.G., Kren, B.T., Bergad, R., Schultheis M., Byrne, T. and Schaad, J.W. 1980. Dominant lethality in Xenopus laevis induced with triethylenemelamine (TEM). **Teratogen. Carcinogen. and Mutagen. 1:** 283-294.

Miltner, M.J. 1982. Spermatogenesis in the axolotl, Ambystoma mexicanum. M.Sc. Thesis, University of Ottawa, 96 p.

Propping, P., Rohrborn, G. and Buselmaier, W. 1972. Comparative investigations on the chemical induction of point mutations and dominant lethal mutations in mice. **Molec. gen. Genet. 117:** 197-209.

Sega, G.A. and Owens, J.G. 1978. Ethylation of DNA and protamine by ethylmethanesulfonate in the germ cells of male mice and the relevancy of these molecular targets to the induction of dominant lethals. **Mut. Res. 52:** 87-106.

Swain, C.G. and Scott, C.B. 1953. Quantitative correlations of relative rates: Comparison of hydroxide ion with other nucleophilic reagents toward alkyl halides, esters, epoxides and acyl halides. **J. Am. Chem. Soc. 75:** 141-147.

Vogel, E. and Natarajan, A.T. 1979. The relation between reaction kinetics and mutagenic action of mono-functional alkylating agents in higher eukaryotic systems I. Recessive lethal mutations and translocations in Drosophila. **Mut. Res. 62:** 51-100.

WORKSHOP DISCUSSION

KAISER:

What is the significance of the African frog; could you use another frog?

HART:

One of the problems you run into using most of our local species is that there is seasonal reproduction. Unless you want to go to a great deal of trouble you can only do your test in spring when the frog is in a reproductive condition. Now, you can maintain them under artificial laboratory conditions, that is keeping them in the cold and bringing them out and warming them up when you want them to spawn. This is a lot of trouble, at least initially, to verify that a frog system will work. I think that Xenopus is the organism of choice because of the year round reproduction.

SCHULTZ:

Also, there is the definitive work by Neikoop and Saber. They have done all the biology for you and you can get right into the testing.

HART:

Yes, there is a nice compendium of embryonic stages so you can easily determine at any point whether you have a normal or an abnormal embryo.

SCHULTZ:

What did you use for your water?

HART:

It was ordinary dechlorinated tap water. We did put for the haploids 100 $mg \cdot L^{-1}$ of Penicillin in the water and we did that for the diploids as well. It is not necessary for the diploids but for the haploids it improves the survival during the onset of this haploid syndrome. In other words, it gets you safely beyond your hatching point before the haploid syndrome sets in.

SCHULTZ:

Have you done any water quality testing, specifically for magnesium, is Xenopus sensitive to the magnesium content?

HART:

No, we don't have data on magnesium content of the water we are using.

HODSON:

What sort of approach do you take for water insoluble compounds?

HART:

We haven't used those. I think that what you would want to do for those, would be to use an appropriate solvent and inject the material intraperitoneally. I think that would be the way to maximize concentrations in adults. We haven't done that yet. I did a little work trying to mutagenize with benzo-a-pyrene by immersing, using DMSO as solvent and that was unsuccessful; we couldn't get high enough concentrations in the water.

LETHAL DOSE VERSUS LETHAL CONCENTRATION AS INDICATOR OF CONTAMINANT TOXICITY TO FISH

Peter V. Hodson[1], D. George Dixon[2] and Klaus L.E. Kaiser[3]

[1] Great Lakes Fisheries Research Branch
Department of Fisheries and Oceans
PO Box 5050, Burlington, Ontario L7R 4A6
CANADA

[2] Department of Biology
University of Waterloo
Waterloo, Ontario N2L 3G1
CANADA

[3] Environmental Contaminants Division
National Water Research Institute
PO Box 5050, Burlington, Ontario L7R 4A6
CANADA

ABSTRACT

Median lethal dose determinations by intraperitoneal injection (IPLD50) provide a rapid and reliable indication of the lethal effects of chemicals on fish. The IPLD50s are correlated with oral lethal dose (OILD50) as well as aqueous lethal concentration (LC50), the traditional measure of toxicity. The speed and efficiency of toxicity testing by IPLD50 makes it a valuable tool for structure-activity studies of toxic substances.

K. L. E. Kaiser (ed.), QSAR in Environmental Toxicology, 179–187.

INTRODUCTION

Quantitative Structure-Activity Relationships (QSARs) have been proposed as a method to predict the toxicity and environmental impact of chemicals in aquatic ecosystems. While this approach would significantly reduce the time and effort required to screen large numbers of chemicals, QSARs will only be useful if they are based on rapid and inexpensive toxicity tests of representative chemicals.

Traditionally, the toxicity of chemicals to fish has been estimated by measuring their LC50s, the median lethal concentrations in water. This bioassay is time consuming and expensive. Complex diluting systems are needed as well as frequent measurements of the chemicals in water to characterize the exposure. LC50s are limited to those chemicals that are soluble in water; when solubility is low, organic solvents such as acetone must be used for stock solutions and their presence can confound the results. A final and more fundamental problem is that 'dose received' by the fish is totally unknown. Hence, QSARs of LC50s may describe factors affecting 'dose received' (e.g., solubility) rather than variability in inherent toxicity.

Therefore, we adapted the measurement of the LD50, or median lethal dose, from mammalian toxicology. LD50s were determined by intraperitoneal injection of fish with test chemicals.

EXPERIMENTAL

The toxicity of chemicals to fish was measured by three methods: intraperitoneal injection (IPLD50); oral intubation (OILD50s); and aqueous exposure (LC50). The chemicals used included para-substituted phenols, chlorinated phenols, chlorinated benzenes and anilines. The three measures of toxicity were compared by multiple stepwise regression techniques (Steel and Torrie 1960) to verify whether IPLD50s were a relevant measure of the toxicity of chemicals in the diet (OILD50s) or water (LC50s). All bioassays were repeated in triplicate so that all regression analyses were based on means.

IPLD50s were measured by dissolving the test chemicals in either (a) a solution of 5% ethanol in 0.9% saline (water soluble compounds), or (b) cod liver oil (lipid soluble compounds). The concentrations were adjusted so that, at any given dose each fish was injected with 1.0 mL of solution per 100 g of body weight.

Rainbow trout (Salmo gairdneri) weighing from 10 to 40 g each were purchased from a commercial supplier of 'specific-pathogen free' fish and acclimated to the laboratory for at least one week at 15°C. They were anaesthetized in MS-222, weighed, and injected with an appropriate volume of test solution. Glass gas-tight or Tuberculin syringes were used with 22 ga needles. The needle was inserted in the mid-ventral line of the gut, behind the pectoral fins. Care was taken to ensure that the solutions were not injected into the stomach or other organs, but rather into the peritoneal cavity. Ten fish were injected at each of five test doses plus injected and uninjected controls. The fish recovered in clean, flowing water and mortality was recorded for 96-hour. The results of the assays were calculated by probit analysis (Finney 1971) or by graphical interpolation for assays giving no partial responses (Litchfield and Wilcoxon 1949).

OILD50s were determined by the same method as IPLD50s with the exception that the solutions were placed in the fish's stomach by gavage using either 3 mm OD gelatin capsules or by injection through a 1 mm i.d. flexible tube.

LC50s were determined using recommended aqueous bioassay techniques (Sprague 1973) with 1 to 5 g rainbow trout. Toxicants were added to water in aqueous solution by a syringe pump and a Mount-Brungs diluter (Mount and Brungs 1967). In cases of low water solubility, chemicals were first dissolved in dimethylformamide. The stock concentration was high enough that only 10 to 15 μL of stock were added to 1.08 L of water. Therefore, the highest concentration of DMF was about 15 $mg \cdot L^{-1}$, well below the measured LC50 of >6400 $mg \cdot L^{-1}$ (Hodson, unpublished data). Chemicals in water were measured daily by either ultraviolet absorption or gas chromatography. Ten fish were tested at each of five concentrations plus control, and results were analyzed as described above.

RESULTS AND DISCUSSION

The measurements of toxicity in this experiment were fairly precise. The standard deviations of the three replicate bioassays, expressed as a proportion of the mean, averaged 10.7% for IPLD50s, 9.9% for OI LD50s, and 6.2% for LC50s.

IPLD50s were linearly correlated to OILD50s. The relationship could be described by:

$$\text{OILD50} = -0.186 + 1.255\ \text{IPLD50} \quad (1)$$
$$n = 12;\ r^2 = 0.99$$

Regression (1) is highly significant and the two measures of toxicity appeared equivalent. However, they were not equal since the slope was significantly different from one ($P < 0.05$). OILD50s were consistently higher than IPLD50s, perhaps due to the slower uptake of chemicals from the gut compared to chemicals injected IP (Dixon _et al_. 1984). In any case, these data demonstrate that IPLD50s can be used to accurately estimate the toxicity to fish of chemical ingested orally.

The relationship of IPLD50s to LC50s was curvilinear: IPLD50s increased with the square of LC50s. The equation relating them was:

$$\text{IPLD50} = 1.971 + 1.14 \times 10^{-4}\ (\text{LC50})^2 \quad (2)$$
$$n = 13;\ r^2 = 0.45$$

While this regression was significant and allows a rough prediction of LC50s, the variability about the regression line was very high and the regression was strongly influenced by extreme results for aniline and benzene.

The high observed variability about the regression line may be related to the interactions between the solubility of compounds in water and their uptake across lipid membranes. The partitioning of compounds from water to lipids can be described by their octanol/water partition coefficients (P) (Hansch and Leo 1979). A good correlation has been found between log P and the negative log of LC50s, suggesting that water/lipid partitioning may control the aqueous toxicity of compounds (Veith _et al_. 1983). Therefore, 'correcting' either the IPLD50s or LC50s by log P may improve the relationship between the two if variations in partitioning have affected the toxicity measurements.

An improved relationship between IPLD50s and LC50s was found when IPLD50s were divided by log P (or LC50s multiplied by log P). The relationship was still curvilinear but much stronger than before:

$$\text{IPLD50}/\log P = 0.348 + 1.03 \times 10^{-4} (\text{LC50})^2 \quad (3)$$
$$n = 12;\ r^2 = 0.82$$

The improvement in the correlation coefficient was statistically significant ($P < 0.05$) and there was a large increase in precision of the equation: the standard error of the estimate decreased from 5.92 to 2.31. Therefore, by correcting for the lipophilicity of compounds, LC50s and IPLD50s can be interconverted with reasonable accuracy. OILD50s could also be related to LC50s by a curvilinear equation:

$$\log \text{OILD50} = -0.193 + 4.18 \times 10^{-3}\ \text{LC50} \quad (4)$$
$$n = 7;\ r^2 = 0.69$$

Equation (4) differs from that of the IPLD50s (equation 3). Curvilinearity is described by the log of the OILD50 rather than the square of LC50s; as well, the correlation coefficient is higher. These differences are probably due to the smaller number of data rather than to any real difference in toxicity; with fewer data, there were fewer extreme data points to generate error. While this equation indicates that OILD50s and LC50s are interconvertable, more data are required to accurately define the relationship.

The above results have demonstrated that IPLD50s provide a suitable estimate of the toxicity of chemicals to fish. Furthermore, our experiments showed considerable increases in efficiency of LD50 bioassays over LC50 bioassays. We were able to complete about three times as many LD50s as LC50s for equivalent time and manpower. These results also provided a suitable basis for calculation of QSARs (Kaiser et al. 1984) so they meet our requirements for QSAR research: efficiency, relevance, and applicability.

REFERENCES

Dixon, D.G., Hodson, P.V. and Kaiser, K.L.E. 1984. Mobilization and tissue distribution of radiolabelled benzene, hexachlorobenzene, phenol and para-nitrophenol following dietary and intraperitoneal dosing of rainbow trout. In preparation.

Finney, D.J. 1971. **Probit Analysis.** 3rd ed. Cambridge University Press, Cambridge, 333 p.

Hansch, C., and Leo, A.J. 1979. **Substituent Constants for Correlation Analysis in Chemistry and Biology.** John Wiley & Sons, New York.

Kaiser, K.L.E., Dixon, D.G. and Hodson, P.V. 1984. QSAR studies on chlorophenols, chlorobenzenes and para-substituted phenols. In **QSAR in Environmental Toxicology,** K.L.E. Kaiser (Ed.), D. Reidel, Publ. Co., Dordrecht, Holland, pp. 189-206.

Litchfield, J.T., and Wilcoxon, F. 1949. A simplified method of evaluating dose-effect experiments. **J. Pharm. Exper. Therapeut. 95:** 99-113.

Mount, D.I. and Brungs, W.A. 1967. A simplified dosing apparatus for fish toxicology studies. **Water Res. 1:** 21-29.

Sprague, J.B. 1973. The ABC's of pollutant bioassay using fish. In **Biological Methods for the Assessment of Water Quality. ASTM STP 528.** American Society for Testing and Materials, Philadelphia, Pennsylvania, pp. 6-30.

Steel, R.G.D., and Torrie, J.H. 1960. **Principles and Procedures of Statistics.** McGraw-Hill, New York.

Veith, G.D., Call, D.J. and Brooke, L.T. 1983. Structure-activity relationships for the fathead minnow Pimephales promelas: Narcotic industrial chemicals. **Can. J. Fish. Aquat. Sci. 40:** 743-748.

WORKSHOP DISCUSSION

MACKAY:

What Alice Bobra said towards the end of her presentation was that it was the concentration of the substance in the organism which was important and this is a reflection of the concentration in the water, divided by the water solubility or multiplied by K_{ow}. Mortality occurs at a relatively constant concentration within the fish, regardless of the hydrocarbon. Your data indicated, I think, that the IPLD50, which is also the amount of material in the fish, was relatively constant, it was a much more constant quantity than were the LC50s, which agrees entirely with Alice Bobra's work. It is the amount of material which is in the organism that matters. Whether it gets there by injection or whether it gets there by partitioning, does not matter a great deal. The amount is what controls the toxicity.

HODSON:

Presumably, the observed variations in toxicity would then be due either to partitioning within the organism, which takes us back to log P, or some change in the inherent toxicity.

MACKAY:

What you described earlier is really elegant and it is entirely complimentary to the water concentration uptake times LC50s product. It is another piece in the jigsaw puzzle.

SCHULTZ:

Would you care to comment about your outlier?

HODSON:

Para-hydroxybenzoic acid? There are several possibilities, one of which is that it breaks down very rapidly. Even though we measure it in the water, the animal may metabolize it so quickly that its concentration never becomes constant unless you add a very large amount, very suddenly. It is a strange situation.

KAISER:

In addition, it is probably fully ionized and just doesn't get into the tissues, doesn't get taken up very much.

SCHULTZ:

We had to look at some carboxyl substituted compounds and we had the same thing. You need to really load up the system to get it into the stomach and we are not looking at the carboxy group, but would that have to be ionized?

KAISER:

Our lake pH is about 8 to 8.5.

ZITKO:

If you leave the aniline out, you have no outlier but also no relation.

HODSON:

That is true; that is why we left it in. We have some more data. If I remember correctly, we are still trying to verify the benzene point because we don't have all the water chemistry to back it up yet. If you accepted the nominal concentrations in water that we added, the benzene data fell on the line up near the aniline, so I think benzene is going to reinforce the aniline data.

HALLETT:

What knocks out the fish with those chemicals?

HODSON:

We will be giving some data on Thursday which suggests that it is liver damage, but there is also a suggestion in the literature that it may be neurotoxicity. We have definitely seen sub-lethal liver damage caused by short-term exposures.

KWASNIEWSKA:

Did you look at the accumulation of toxicant either in whole fish or in specific organs?

HODSON:

We have not looked at accumulation from water in these tests. We have only looked at the tissue distribution of a single dose intraperitoneal of two compounds.

KWASNIEWSKA:

Did you look for the age of the fish as, on some days they are supposed to be growing and at certain times they are more sensitive or less sensitive or if they are old they may be less sensitive to the toxicants?

HODSON:

That is a good point. The LC50s were measured on small fish, generally less than 2 g and, as you point out, the other fish in the LD50 tests were a good ten times larger and represent much older fish. The relationship between age of fish and toxicity is a function of body weight and metabolic rate, and hence the variation of toxicity is a variation of uptake. It is just a further wrinkle on the animal's ability to take up materials. That is one advantage of the IPD50s, you get away from that complication.

Question:

Does the tissue distribution vary or have you measured that?

HODSON:

The data that I presented on blood is part of a study in which we looked at two separate radiolabelled compounds. The data are not completely analyzed yet but we will be looking at any variation of the distribution of the compound with the method of dosing.

QSAR STUDIES ON CHLOROPHENOLS, CHLOROBENZENES AND PARA-SUBSTITUTED PHENOLS

Klaus L.E. Kaiser[1], D. George Dixon[2] and Peter V. Hodson[3]

[1] Environmental Contaminants Division
National Water Research Institute
PO Box 5050, Burlington, Ontario L7R 4A6
CANADA

[2] Department of Biology
University of Waterloo
Waterloo, Ontario N2L 3G1
CANADA

[3] Great Lakes Fisheries Research Branch
Department of Fisheries and Oceans
PO Box 5050, Burlington, Ontario L7R 4A6
CANADA

ABSTRACT

The acute toxicities of chlorophenols (CP's) chlorobenzenes (CB's) and para-substituted phenols (PP's) to rainbow trout (*Salmo gairdneri*) as intraperitoneal injection lethality, serum sorbitol dehydrogenase activity, and 96-hr lethal concentration are investigated in terms of quantitative structure-activity correlations (QSARs). The effects of CP's and CB's are primarily related to the chemicals' octanol/water partition coefficients (log P). The slopes of the QSARs of CP's for rainbow trout are nearly parallel to those of acute and semichronic CP toxicities to bluegill (*Lepomis macrochirus*), guppy (*Poecilia reticulata*), waterflea (*Daphnia magna*), and the marine organism *Photobacterium phosphoreum*, indicating similar mechanisms of toxic effects. The toxicites of PP's are dependent on both log P and

K. L. E. Kaiser (ed.), QSAR in Environmental Toxicology, 189–206.

other molecular descriptors, such as molar refractivity and field and resonance constants of the para-substituents.

INTRODUCTION

Several hundred organo-chlorine compounds are found to be present in many aquatic and terrestrial samples from around the world. Apart from pesticides such as DDT and its derivatives, which were deliberately introduced to the environment, many other environmental contaminants are by-products and wastes from manufacturing processes or come from unknown sources. For example, phenol derivatives such as pentachlorophenol (Jones 1981) and several chlorobenzenes are commonly found in water, sediment and fish samples from the Great Lakes. Recent findings of a host of other contaminants (Kuehl *et al.* 1983; Kaminsky *et al.* 1983), some of which are not even manufactured anywhere on the continent, are liable to heighten the public concern about the safety of our drinking water and the edibility of fish from the lakes (Allan *et al.* 1983).

As a general rule, toxicological data on specialty chemicals are non-existent and even for some compounds produced in large quantities, little information is available on their sublethal effects on aquatic and terrestrial species, including man. Obviously, not every chemical will be of concern. However, in the absence of sufficient information as to their safety, how can hazardous compounds be screened out from those which are safe? Given the large number of chemicals and formulations potentially involved, and the enormous resources and time necessary to actually test each, it is evident that routes must be found to predict at least the type and magnitude of hazard associated with their use and release to the environment. For this purpose, the field of quantitative structure-activity correlations (QSARs) has shown great promise.

In theory, its potential to predict toxicities, environmental pathways, bioaccumulation factors, persistence and other ecological parameters, is unlimited. In practice, only small, yet promising beginnings, have been made. The work reported here is a minute addition to the rapidly increasing base of knowledge on the significance of molecular structure and a variety of physico-chemical parameters as descriptions of the biological effects of some common classes of contaminants.

DATA AND METHODS

Fish toxicity data for this investigation are those from ongoing collaborative studies by the Great Lakes Fisheries Research Branch, Department of Fisheries and Oceans, Burlington, Ontario, the Departments of Biology and Pharmacology at the University of Waterloo, Waterloo, Ontario, and the Environmental Contaminants Division, National Water Research Institute, Burlington, Ontario. The microbiological toxicity data are those from other studies at the Environmental Contaminants Division. In addition, selected literature data on toxic effects of these sets of compounds were used.

Where necessary, the data were converted from weight ($mg \cdot L^{-1}$ or $mg \cdot kg^{-1}$) to the inverse molar concentrations ($L \cdot mmol^{-1}$ or $kg \cdot mmol^{-1}$) and are given in logarithmic form. This provides for a consistent presentation of the toxicity data on the basis of functional groups, such as the phenolic hydroxy group, and shows higher toxicity with more positive or higher values. All regression analyses were performed with Hewlett-Packard calculators and pre-recorded statistical programs for linear and multiple linear regression analysis.

BACKGROUND

When this study was at its beginning stage, about three years ago, it became obvious that a close collaboration between chemists and biologists had to be the foundation of this work. Chemists rarely understand the complexities and vagaries of biological work and biologists generally have little chemical inclination. One of the first tasks to be solved was the selection of compounds to be studied. This might seem easy, but actually it isn't. After all there are roughly 2×10^6 chemicals to choose from and we could test only a few per year. So, it became necessary to develop criteria for this selection process. We used the criteria shown in SCHEME 1. Based on these criteria, we arrived at three sets of contaminants for the initial phase of investigation as shown in SCHEME 2. In fact, however, these numbers have been somewhat reduced due to several reasons. For example, several chlorophenols and chlorobenzenes were not available in 100+ g quantities and reasonably pure form at a reasonable cost. Other compounds could not be purified at the necessary quantities due to laboratory limitations. Yet others were found difficult to analyze by routine methods (e.g., p-chlorophenol), or were insufficiently stable due to photodegradation or hydrolysis (e.g., p-methylaminophenol). Also, some were found to be insufficiently soluble or difficult to prepare as stable solutions due to high lipophilicity and adsorption onto diluter and bioassay system walls.

1) Relevance to aquatic ecosystem (e.g., Great Lakes), i.e., present or suspected environmental contaminant?

2) Relevance in toxicological respect, i.e., known or suspected to have toxic effects on biota?

3) Relevance to industrial processes, i.e., manufactured or imported and used in large quantities?

4) Relevance to other research, i.e., are other biological data available?

5) Relevance in form of chemical considerations, i.e., are structural parameters known (for entire set)?

6) Practicability to work with, i.e., are compounds
   (i) sufficiently soluble,
   (ii) of too high acute toxicity,
   (iii) carcinogens, mutagens, tumorigens,
   (iv) stable under test conditions,
   (v) amenable to analytical control?

7) Availability, i.e., can compounds be purchased, made or refined at the required
   (i) quantity,
   (ii) quality,
   (iii) cost?

SCHEME 1: Criteria for the selection of compounds for QSAR studies

1) Para-substituted phenols, ≈15.
2) Chlorophenols, 20.
3) Chlorobenzenes, 13.

SCHEME 2: Sets of compounds selected and number of compounds per set for initial QSAR research phase

Even then, further problems may restrict either the experimental or theoretical QSAR work with each set. For example, a very important physico-chemical property, such as the octanol/water partition coefficient, has only recently been measured for all chlorophenols and chlorobenzenes by scientists interested in their environmental behaviour and toxicology. For the set of choroanilines described elsewhere in this volume (Kwasniewska and Kaiser 1984; Ribo and Kaiser 1984), octanol/water partition coefficients (log P) of only six of the possible 20 isomers are given in the large compilation by Hansch and Leo (1979). Furthermore, of the 210 possible polychlorobiphenyl isomers of which approximately 100 compounds are routinely found in environmental samples, measured log P values for only 10 isomers are reported in the literature. Naturally, there are methods to estimate partition coefficients, such as from π constants (Hansch and Leo 1979) and from fragment values (Rekker 1977), but for highly lipophilic compounds, such as hexachlorobenzene, the estimated partition coefficients are one to two orders of magnitude higher than most measured values. Only very recently has a new method for the estimation of log P of polychloro-substituted aromatic hydrocarbons become available (Kaiser 1983).

Of course, partition coefficients are only one of several physico-chemical parameters which are useful for structure-activity correlations (Purcell _et al_. 1973). A variety of other factors and properties, such as the connectivity index (Kier and Hall 1981), field and resonance constants of substituent groups (Hansch and Leo 1979), sigma constants and acid dissociation constants are just a few of many. However, as with log P, measured values are to be found in the literature for only a few compounds in almost any coherent set; for example, pKa data exist for only 13 of the possible 20 chlorophenol isomers. Moreover, the modern scientific view is to regard the laborious measurement of physico-chemical constants no longer as science, hence little effort is directed towards the determination of basic constants of common chemicals.

Notwithstanding this lack of available data for many of the chlorophenols, chlorobenzenes, and para-substituted phenols, these sets were chosen for the purpose of QSAR investigations of their acute and sublethal effects to aquatic biota. TABLES 1, 2, and 3 summarize the most important physico-chemical data for these compounds for this purpose and also identify existing gaps in that regard.

TABLE 1: Summary of physico-chemical constants for the thirteen possible chlorobenzene isomers

Substituent Position	log P[a]	mp[b]	dpm[c]	IP[d]
-	2.14	5.5	0.15	9.24
Cl	2.86	-45.6	1.59	9.09
1,2-Cl_2	3.55	-17.0	2.27	-
1,3-Cl_2	3.60	-24.7	1.4	-
1,4-Cl_2	3.62	53.5	0.3	9.15
1,2,3-Cl_3	4.11	53	2.33	-
1,2,4-Cl_3	3.93	17.0	1.26	-
1,3,5-Cl_3	4.15	64	0.28	9.62
1,2,3,4-Cl_4	4.46	47.5	0.65	9.56
1,2,3,5-Cl_4	4.50	54.5	0.73	9.56
1,2,4,5-Cl_4	4.52	140	0.88	-
Cl_5	4.88	86	-	-
Cl_6	5.00	231	0	9.90

a From Könemann *et al*. (1979).
b Melting point (°C), from MI (1976), CRC (1978), and other sources.
c Dipole moment (D) in benzene, from McClellan (1963).
d Vertical ionization potential (eV) from various sources.

RESULTS AND DISCUSSION

Chlorobenzenes

A summary of acute toxicity and semi-chronic threshold data of the chlorobenzene isomers to several species of aquatic organisms is given in TABLE 4. The data include acute lethality of chlorobenzenes to rainbow trout (*Salmo gairdneri*) by way of intraperitoneal injection in oil base (IPLD50), serum serbitol dehydrogenase threshold (SSDH) in rainbow trout, acute toxicity to bluegill (*Lepomis macrochirus*) (LC50), semi-chronic toxicity to guppy (*Poecilia reticulata*) (LC50), and toxicities to *Photobacterium phosphoreum* (30EC50) as determined by the Microtox[R] test.

TABLE 2: Summary of physico-chemical constants for twenty possible chlorophenol isomers

Substituent Position	log P[a]	pK_a [b]	mp[c]	dpm[d]	$E_{\frac{1}{2}}$ [e]
Phenol	1.48	9.82	41.5	1.55	0.633
2-Cl	2.17	8.29	9.0	1.33	0.625
3-Cl	2.50	8.78	36	2.10	0.734
4-Cl	2.35	9.14	43.7	2.11	0.653
2,3-Cl_2	3.19	7.70	59	-	-
2,4-Cl_2	3.08	7.89	45	2.0	-
2,5-Cl_2	3.20	7.51	59	-	0.695
2,6-Cl_2	2.86	6.79	69	-	-
3,4-Cl_2	3.37	8.59	68	-	-
3,5-Cl_2	3.52	8.19	68	-	-
2,3,4-Cl_3	4.07	~6.5	83.5	-	-
2,3,5-Cl_3	4.21	~6.5	62	-	-
2,3,6-Cl_3	3.88	~6.5	58	-	-
2,4,5-Cl_3	3.72	~7.0	70	-	-
2,4,6-Cl_3	3.69	~6.1	69.5	1.6	-
3,4,5-Cl_3	4.39	~6.5	101.0	-	-
2,3,4,5-Cl_4	4.95	~5.3	117.0	-	-
2,3,4,6-Cl_4	4.10	~5.3	70	-	-
2,3,5,6-Cl_4	4.90	~5.3	115	-	-
Cl_5	5.01	~4.8	191	2.1	-

a From Hansch and Leo (1979).

b Acid dissociation constant, from Sillén and Martell (1971).

c Melting point (°C) from CRC (1978) and other sources.

d Dipole moment (D), from McClellan (1963).

e Redox potential (V), from Suatoni *et al.* (1961).

TABLE 3: Summary of physico-chemical constants for a selection of para-substituted phenols

Substituent	log P	F^a	R^b	MR^c	mp	$E_{\frac{1}{2}}^d$	pK_a^e
NH_2	0.04	0.02	-0.68	5.42	190.2	-	-
CH_2OH	0.25	0.00	0.00	7.19	112	-	-
OH	0.59	0.29	-0.64	2.85	171	-	-
NO	~1.3	0.50	0.45	5.20	~144	-	-
OCH_3	1.34	0.26	-0.51	7.87	57	0.406	10.21
CHO	1.35	0.31	0.13	6.88	119	-	7.41
$COCH_3$	1.35	0.32	0.20	11.18	108	0.791	-
H	1.48	0.00	0.00	1.03	41.5	-	9.82
COOH	1.58	0.33	0.15	6.93	214	0.716	4.58
CN	1.60	0.51	0.19	6.33	113	-	7.70
NO_2	1.91	0.67	0.16	7.36	114		6.90
CH_3	1.94	-0.04	-0.13	5.65	35.5	0.543	-
CF_3	~2.3	0.38	0.19	5.02	47	-	8.68
Cl	2.35	0.41	-0.15	6.03	43.7		9.14
COC_6H_5	3.07	0.30	0.16	30.33	135	-	-
C_6H_5	3.20	0.08	-0.08	25.36	167	0.534	-
OC_6H_5	3.5	0.34	-0.35	27.68	85	-	-

a Swain and Lupton field constant of substituent, from Hansch and Leo (1979).

b Swain and Lupton resonance constant of substituent, from Hansch and Leo (1979).

c Molar refractivity of substituent, from Hansch and Leo (1979).

d Redox potential (in V), from Suatoni *et al*. (1961).

e From Sillén and Martell (1971).

The toxicity data were analysed by linear regression analysis with some of the physico-chemical parameters (TABLE 1) and the resulting QSAR equations are summarized in TABLE 5. It is quite obvious that there are several significant correlations with log P, such as for the rainbow trout IPLD50, bluegill 24LC50, guppy 14DLC50, and Microtox 30EC50 values. In contrast the trout SSDH and *Daphnia* 48LC50 data did not give meaningful correlations with log P. It is also of interest to note that the slopes of the toxicity dependence on log P is close to one; this indicates that these toxicities to bluegill and guppy are significantly influenced by the chlorobenzene's ability to pass through one water/lipid barrier.

TABLE 4: Concentrations of chlorobenzenes causing fifty percent lethality to rainbow trout (Salmo gairdneri) by intraperitoneal injection in oil base (IPLD50), of the serum sorbitol dehydrogenase threshold (SSDH), of acute toxicity to bluegill (Lepomis macrochirus), (24LC50), of semichronic toxicity to guppy (Poecilia reticulata), (14DLC50), acute toxicity to Daphnia magna (48LC50), and Microtox[R], 30 min effective concentrations (30EC50)

Compound	Trout[a] IPLD50	Trout[a] SSDH	Bluegill[b] 24LC50	Guppy[c] 14DLC50	Daphnia[d] 48LC50	Microtox[e] 30EC50
Benzene	-1.41	-0.72	-	0.09	-	0.02
Cl	-0.98	-0.15	0.85	0.77	0.12	1.00
1,2-Cl_2	-0.87	-	1.42	1.40	1.79	1.56
1,3-Cl_2	-1.00	-	1.47	1.30	0.72	1.46
1,4-Cl_2	-1.01	-0.42	1.53	1.57	1.13	1.44
1,2,3-Cl_3	-0.95	-	1.73	1.89	-	1.76
1,2,4-Cl_3	-0.99	-0.59	-	1.88	0.56	1.66
1,3,5-Cl_3	-1.48	-	-	1.74	-	1.11
1,2,3,4-Cl_4	-0.69	-0.58<	-	2.43	-	1.73
1,2,3,5-Cl_4	-0.89	-	1.53	2.43	1.34	1.94
1,2,4,5-Cl_4	-1.37<	-	2.13	2.85	-0.39	1.68
Cl_5	-0.76	-1.25	3.00	3.15	1.67	-
Cl_6	-1.24<	-	-	-	-	-

a From Dixon, Hodson and Kaiser, unpublished data.
b From Buccafusco *et al*. (1981).
c From Könemann (1981).
d From LeBlanc (1980).
e From Ribo and Kaiser (1983).

Chlorophenols

Available toxicity data are summarized in TABLE 6. The data include acute lethal doses of chlorophenols to rainbow trout (*Salmo gairdneri*) by way of intraperitoneal injection (IPLD50), serum sorbitol dehydrogenase thresholds (SSDH) to rainbow trout, acute toxicity to bluegill (Lepomis macrochirus) (24LC50), semi-chronic toxicity to guppy (Poecilia reticulata) (14DLC50), lethal threshold to shrimp (Crangon septemspinosa) (96LT), and acute toxicity in the Microtox[R] test (Photobacterium phosphoreum) (30EC50).

TABLE 5: Summary of chlorobenzene structure-activity correlations of several fish and bacteria toxicities with the octanol/water partition coefficients

(1)	Trout IPLD50	=	1.71 + 0.20 log P
	n	=	11; r^2 = 0.78
(2a)	Trout SSDH	=	0.06 - 0.19 log P
	n	=	6; r^2 = 0.27
(2b)	SSDH	=	0.17 - 0.23 log P
	n	=	5 (ex 1.2.3.4-Cl_4); r^2 = 0.35
(3a)	Bluegill 24LC50	=	-1.51 + 0.81 log P
	n	=	8; r^2 = 0.75
(3b)	Bluegill 24LC50	=	-1.96 - 0.95 log P
	n	=	7 (ex 1.2.3.5-Cl_4); r^2 = 0.92
(4)	Guppy 14DLC50	=	-2.38 + 1.08 log P
	n	=	12; r^2 = 0.95
(5)	Daphnia magna 48LC50	=	-0.04 + 0.23 log P
	n	=	8; r^2 = 0.04
(6)	Microtox 30EC50	=	-0.98 + 0.63 log P
	n	=	11 (Ribo and Kaiser 1983); r^2 = 0.76

Analyses of the data by linear regression showed several significant correlations which are summarized in TABLE 7. As can be seen from the correlation coefficients for these equations, there is in general a very strong dependence of the observed toxicity values on the octanol/water partition coefficient. As the chlorophenols are weak acids which dissociate in water as a function of its pH, it can be expected that multiple linear correlations with log P and pKa show even better correlations. Such investigations have already been reported, for example, on the shrimp data by McLeese *et al.* (1979). TABLE 8 shows a summary of the correlation equations so obtained.

TABLE 6: Toxic concentrations of chlorophenols to rainbow trout (*Salmo gairdneri*), bluegill (*Lepomis macrochirus*), guppy (*Poecilia reticulata*), in the Microtox Test (*Photobacterium phosphoreum*), and to shrimp (*Crangon septemspinosa*). All values given as logarithms of the inverse molar concentrations (log L·mmol^{-1})

Compound	Trout[a] IPLD50	Bluegill[b] 24LC50	Guppy[c] 14DLC50	Shrimp[d] 96LT	Microtox[e] 30EC50
Phenol	-0.65	-	0.50	1.10	0.42
2-Cl	-	1.25	1.06	1.39	0.58
3-Cl	-	-	1.30	-	0.96
4-Cl	-0.53	1.51	-	1.45	1.19
2,3-Cl_2	-	-	-	-	1.52
2,4-Cl_2	-0.60	1.54	1.59	-	1.47
2,5-Cl_2	-	-	-	-	1.24
2,6-Cl_2	-0.95	-	-	0.93	1.09
3,4-Cl_2	-	-	-	-	2.00
3,5-Cl_2	-	-	1.78	2.04	1.77
2,3,4-Cl_3	-	-	-	2.00	2.20
2,3,5-Cl_3	-	-	2.10	-	2.25
2,3,6-Cl_3	-	-	1.59	1.86	1.19
2,4,5-Cl_3	-	2.51	-	-	2.19
2,4,6-Cl_3	-	2.44	-	-	1.41
3,4,5-Cl_3	-	-	2.24	-	2.74
2,3,4,5-Cl_4	-	-	-	-	3.12
2,3,4,6-Cl_4	-	3.09	-	1.29	2.26
2,3,5,6-Cl_4	-	2.76	2.23	-	2.02
Cl_5	0.85	-	2.85	1.91	2.71

a From Hodson, Dixon and Kaiser, unpublished data.
b From Buccafusco *et al.* (1981).
c From Könemann (1981).
d From McLeese *et al.* (1979).
e From Ribo and Kaiser (1983).

TABLE 7: Summary of chlorophenol structure-activity correlations of several aquatic species with the octanol/water partition coefficient

(1)	Trout IPLD50	=	−1.70 + 0.45 log P
	n	=	5; r^2 = 0.69
(2)	Bluegill 24LC50	=	−0.12 + 0.66 log P
	n	=	7; r^2 = 0.81
(3)	Guppy 14DLC50	=	−0.18 + 0.54 log P
	n	=	10; r^2 = 0.93
(4)	*Daphnia magna* 48LC50	=	0.39 + 0.45 log P
	n	=	9; r^2 = 0.73;
(5)	Shrimp 96LT	=	0.78 + 0.24 log P
	n	=	9 (McLeese *et al.* 1979); r^2 = 0.42
(6)	Microtox 30EC50	=	−0.88 + 0.76 log P
	n	=	18 (Ribo and Kaiser 1983); r^2 = 0.91

TABLE 8: Summary of QSAR equations of chlorophenol toxicities with octanol/water partition coefficients and the acid dissociation constants

(1)	Trout IPLD50	=	−6.98 + 1.17 log P + 0.40 pKa
	n	=	5; r^2 = 0.98
(2)	Bluegill 24LC50	=	3.44 + 0.23 log P + −0.30 pKa
	n	=	7; r^2 = 0.87
(3)	Guppy 14DLC50	=	−0.38 + 0.56 log P + 0.02 pKa
	n	=	10; r^2 = 0.93
(4)	*Daphnia magna* 48LC50	=	3.39 + 0.14 log P + 0.22 pKa
	n	=	9; r^2 = 0.76

Para-Substituted Phenols

The para-substituted phenols available for our investigations are fundamentally different from the previously mentioned two sets of chlorobenzenes and phenols. As shown in TABLE 3, (para)

TABLE 9: Matrix of correlation coefficients (r^2) of molecular descriptors for para-substituted phenols (n = 11)

Descriptor	log P	F	R	MR
log P	1.00	0.21	0.12	0.39
F		1.00	0.12	0.04
R			1.00	0.02

TABLE 10: Toxic concentrations of para-substituted phenols to rainbow trout (Salmo gairdneri) by intraperitoneal injection (IPLD50), oral lethal dose (OLD50), serum sorbitol dehydrogenase activity (SSDH), acute lethal concentrations (LC50), and cell attachment bioassay [a] (EC50)

Substituent	IPLD50	OLD50	SSDH	LC50	EC50
H	-0.65[b]	-0.75[c]	0.77	0.99	-1.62
CH_3O	-0.46	-0.52		0.64	-1.76
C_6H_5O	0.06		0.42		
Cl	0.04		1.36	1.83	-0.96
CH_3	0.15	0.06	0.82	1.16	-1.47
CH_2OH	0.16				
$NH_2CH_3{}^+$	0.25	0.14		2.62	-0.45
NO_2	0.27	0.34	1.87(?)	1.24	-1.11
CH_3	0.56				
CN	0.74			0.72	-1.04
NH_2				1.96	
OH	0.66[b]	-0.32(?)		2.21	

a N.C. Bols, Dept. of Biology, University of Waterloo, unpublished data.

b Outlier, based on equation 1, TABLE 11.

c Outlier, based on equation 2, TABLE 11.

substituent parameters, such as field (F) and resonance (R) constants vary independently of each other, the octanol/water partition coefficient (log P), the molar refractivity (MR), the acid dissociation constant (pK_a), and other physico-chemical parameters, such as the redox potential ($E_{\frac{1}{2}}$). This is also evident from the matrix of linear correlation coefficients of those compounds tested (TABLE 9). Due to experimental limitations, in particular, solubility and stability problems encountered, not all phenol derivatives could be tested.

A summary of the observed toxicities to rainbow trout by intraperitoneal injection (IPLD50), oral lethal dose (OLD50), serum sorbitol dehydrogenase activity (SSDH), acute lethal concentration (LC50), and a trout cell attachment bioassay (N.C. Bols, unpublished data) is given in TABLE 10. Except for the OLD50 and SSDH data, where the number of data points are inconclusive, good structure-activity correlations were obtained with log P, MR, F, and R, as shown in TABLE 11. Except for equation 1 in TABLE 11, all correlations show a

TABLE 11: Summary of structure-activity correlations of para-substituted phenols with log P and other parameters

(1)	Trout IPLD50	=	0.21 + 0.012 MR + 1.54 R
	n	=	8; r^2 = 0.94; s.d. = 0.09
(2)	Trout OLD50	=	-2.85 + 1.35 log P + 0.007 MR
	n	=	4; r^2 = 0.97
(3)	Trout SSDH	=	1.71 - 0.51 log P + 1.73 F
	n	=	5; r^2 = 0.93
(4)	Trout 96LC50	=	2.04 - 2.15 log P + 0.88 $(\log P)^2$
	n	=	7; r^2 = 0.95;
(5)	Trout cell attachment bioassay[a] EC50	=	-2.23 + 0.53 log P
	n	=	8; r^2 = 0.94; s.d. = 0.15

a Regression includes toxicity data for 2,4-dichlorophenol (-0.76), pentachlorophenol (0.40), and aniline (-1.51).

significant dependence on log P. This indicates the general importance of the water/lipid partitioning step for the exertion of toxic effects of these phenols, irrespective of the mode of application. It may be expected, however, that some of the observed relationships may change if more data become available. In particular, expansion of the chemical sets to include compounds with other functional groups or heterocyclic compounds such as aniline, pyridine, furan, and dioxin derivatives would be useful.

ACKNOWLEDGEMENTS

We thank the Great Lakes Water Quality Program for substantial support of this study.

REFERENCES

Allan, R.J., Mudroch, A. and Sudar, A. 1983. An introduction to the Niagara River/Lake Ontario pollution problem. **J. Great Lakes Res. 9:** 111-117.

Buccafusco, R.J., Ells, S.J. and LeBlanc, G.A. 1981. Acute toxicity of priority pollutants to bluegill (Lepomis macrochirus). **Bull. Environ. Contam. Toxicol. 26:** 446-452.

CRC. 1978. **CRC Handbook of Chemistry and Physics**, 58th ed., R.C. Weast (Ed.), CRC Press, Inc., West Palm Beach, Florida.

Hansch, C. and Leo, A.J. 1979. **Substituent Constants for Correlation Analysis in Chemistry and Biology.** John Wiley & Sons, New York.

Jones, P.A. 1981. Chlorophenols and their impurities in the Canadian environment. **Environment Canada, Economic and Technical Review,** Report EPS 3-EC-81-2, 434 p.

Kaiser, K.L.E. 1983. A non-linear function for the approximation of octanol/water partition coefficients of aromatic compounds with multiple chlorine substitution. **Chemosphere 12:** 1159-1167.

Kaminsky, R., Kaiser, K.L.E. and Hites, R.A. 1983. Fates of organic compounds from Niagara Falls dumpsites in Lake Ontario. **J. Great Lakes Res. 9:** 183-189.

Kier, L.B. and Hall, L.H. 1981. Derivation and significance of valence molecular connectivity. **J. Pharm. Sci. 70:** 583-589.

Könemann, H. 1980. Structure-activity relationships and additivity in fish toxicities of environmental pollutants. **Ecotoxicol. Environ. Safety 4:** 415-421.

Könemann, H., Zelle, R., Busser, F. and Hammers, W.E. 1979. Determination of log P_{OCT} values of chloro-substituted benzenes, toluenes, and anilines by high-performance liquid chromatography on ODS-silica. **J. Chrom. 178:** 559-565.

Kuehl, D.W., Leonard, E.N., Butterworth, B.C. and Johnson, K.L. 1983. Polychlorinated chemical residues in fish from major watersheds near the Great Lakes, 1979. **Environment International 9:** 293-299.

Kwasniewska, K. and Kaiser, K.L.E. 1984. Toxicities of selected chloroanilines to four strains of yeast. In **QSAR in Environmental Toxicology,** K.L.E. Kaiser (Ed.), D. Reidel Publ. Co., Dordrecht, Holland, pp. 223-233.

LeBlanc, G.A. 1980. Acute toxicity of priority pollutants to water flea (Daphnia magna). **Bull. Environ. Contam. Toxicol. 24:** 684-691.

McClellan, A.L. 1963. **Tables of Experimental Dipole Moments.** W.H. Freeman and Co., San Francisco, California.

McLeese, D.W., Zitko, V. and Peterson, M.R. 1979. Structure-lethality relationships for phenols, anilines and other aromatic compounds in shrimp and clams. **Chemosphere 2:** 53-57.

MI. 1976. **The Merck Index,** 9th ed., Merck & Co., Inc., Rahway, New Jersey.

Purcell, W.P., Bass, G.E. and Clayton, J.M. 1973. **Strategy of Drug Design: A Guide to Biological Activity.** John Wiley & Sons, New York.

Rekker, R.F. 1977. **The Hydrophobic Fragmental Constant.** Elsevier Scientific Publ. Co., Amsterdam.

Ribo, J.M. and Kaiser, K.L.E. 1983. Effects of selected chemicals to photoluminescent bacteria and their correlations with acute and sublethal effects on other organisms. **Chemosphere 12:** 1421-1442.

Ribo, J.M. and Kaiser, K.L.E. 1984. Toxicities of aniline derivatives to Photobacterium phosphoreum and their correlations with effects on other organisms and structural parameters. In **QSAR in Environmental Toxicology,** K.L.E. Kaiser (Ed.), D. Reidel Publ. Co., Dordrecht, Holland, pp. 319-332.

Sillén, L.G. and Martell, A.E. 1971. **Stability Constants of Metal-Ion Complexes.** Supplement No. 1, Spec. Publ. No. 25, The Chemical Society, London.

Suatoni, J.C., Snyder, R.E. and Clark, R.O. 1961. Voltammetric studies of phenol and aniline ring substitution. **Anal. Chem. 33:** 1894-1897.

WORKSHOP DISCUSSION

MACKAY:

I agree entirely about the importance of measuring solubilities. In fact, they are easier to measure than toxicity. There is a need for fundamental physical chemistry. The National Bureau of Standards and similar agencies need to support and foster this type of research; it is demanding work but it is not difficult.

KAISER:

I think the scientific community at large doesn't seem to care too much about it and I think that we will find out sooner or later that we have to go back to some of the old-fashioned chemistry.

BOLS:

If the correlations with log P are going to be of value at all, accurate log P's are necessary.

MACKAY:

What often happens with these correlations is what a few points away to the right of the diagram are calculated log P values of 7, 8 or as high as 13. They drag the slope of the line down and it has repercussions at the other end where other strange things are happening so it really fouls up the correlation. It is very important to get that tied down accurately.

Comment:

I think that a lot of people are calculating log P inadequately.

KAISER:

I would say that it depends on the type of compound.

LANDE:

I would like to comment on the question of molecular configuration and on the use of HPLC versus the direct measurement. HPLC is not a direct measurement of the partition coefficient. It is just a correlation that says the partitioning between the solvent phase and the solid or stationary phase in the LC is related to the octanol/water partitioning. So, I think one is taking a chance when using log P values from LC rather than the actual physical measurement of log P. Secondly, there is the question of the effect of configuration on physical properties. For example, hydrocarbon chains of twelve carbons are probably folding over and I think that any computation would not, at this point, be able to account for it.

Comment:

That is why I say you have to use calculated log P values within their limitations.

QUANTITATIVE STRUCTURE-ACTIVITY RELATIONSHIPS IN ECOTOXICOLOGY: POSSIBILITIES AND LIMITS

Rainer Koch

Institute of Hygiene
Department of Toxicology
Burgstrasse 2, DDR-6500-Gera
GERMAN DEMOCRATIC REPUBLIC

ABSTRACT

The application of quantitative structure-activity relationships to environmental contaminants is reviewed in general terms. In particular, the successes and limitations of structure-activitiy correlations for important environmental parameters, such as bioconcentration factors, sediment sorption and toxic effects of chemicals, are shown. Quantitative relationships of these parameters with molecular connectivity indices, parachors, octanol/water partition coefficients, and negative entropies are given.

INTRODUCTION

Man's activities frequently result in qualitative and quantitative, locally and temporally differentiated chemical, physical and biological changes in the hydrosphere, the atmosphere and most significantly, the biosphere. The recognition and registration of pollutants through environmental monitoring and an overall and generally applicable ecotoxicological evaluation, including the provision of measures for amelioration of contaminations, as well as setting of priorities for testing based on a classification guide, are all part of the strategy towards the determination and elimination of

K. L. E. Kaiser (ed.), QSAR in Environmental Toxicology 207–222.

alterations in the viability of the ecosystems and their effect on human beings.

Compared to several years ago, much more data are available on environmental chemistry and toxicology and on the fate and toxicity of contaminants. However, there is often no real qualitative growth of the knowledge based on new hypotheses and scientific theories for interpreting measured data as well as to recognize fundamental connections and relationships for predicting the environmental fate and toxicity of contaminants on the basis of findings from other sciences.

The idea that there is a connection between molecular structure and effects is rather old. The basis of this recognition is the fundamental empirical law of sciences that any changes of the composition and structure of systems or their parts are connected with alterations of their properties and functions. From this point of view, the relationships between molecular structure, physicochemical properties, chemical reactivity and biological activity are a fundamental principle of chemical compounds.

On the basis of these fundamental regularities, physicochemical and biochemical related processes and reactions of environmental chemicals can be characterized as interactions between various molecular structures. The mode of interaction will be determined by the chemical structure and physicochemical properties of pollutant molecules as well as the molecular structure of biotic and abiotic environmental compartments, their relative abundance and interactions and the physicochemical properties of the total system. For that reason, it is of major importance to know both the molecular structures and properties of pollutants and those of the target structures in the environment.

STRUCTURE-ACTIVITY RELATIONSHIPS

As a consequence of the complexity and great variability of the composition of biotic and abiotic environmental systems, the receptor site of the interaction between a contaminant and its environment is still a "black box". One of the tools for getting some light into this "black box" are the theoretical methods used in the analysis of relationships between molecular structure or physicochemical properties of contaminants and their biological activities in the broadest sense.

These methods, called Quantitative-Structure-Activity-Relationships (QSAR) are structure-analogy-concepts, and are established principles in pharmacology and drug development. For the latter, the aim is to predict the biological activites of non-tested compounds in drug design. Conversely, if we transform the basic thoughts of these models in ecotoxicology it is probably possible to explain the observed environmental effects of certain classes of chemicals as a function of both the molecular structure and their changes caused by different substituents. The basic item of this theoretical-methodological concept in ecotoxicology is the description of interactions between chemicals and biotic as well as abiotic environmental structures under application of different molecular structure parameters, physicochemical properties related to fundamental physicochemical, quantum mechanical and biochemical or biophysical regularities without consideration of the complexities and variabilities of environmental structures.

Therefore, one of the main goals of this concept is to characterize quantitatively the molecular structure and therefore those controlling molecular properties by a representative numerical descriptor. On the basis of a logical physicochemical and biochemical hypothesis, which is of great importance in this concept, it is the aim to recognize quantitative relationships between these numerical descriptors of molecular structures and the fate as well as the toxicity or biological activity of environmental pollutants as a basis for predicting the fate and toxicity of non-tested chemicals and for interpreting these environmental parameters in a qualitative sense.

Applications

One of the major problems in using structure-analogy-models in ecotoxicology is the ignorance of the molecular organization and accessibility of environmental target molecules or structures as well as the lack of reproducible and comparable data as there are few standardized tests. Furthermore, it must be emphasized that the quality of a structure-analogy model depends largely on the accuracy of the experimental environmental data. Therefore, it is often difficult to explain and to interpret quantitative structure-activity relationships in a logical physicochemical and biochemical sense. However, these difficulties become less with simpler environmental systems with simpler processes and reactions.

The two methods of quantitative structure-activity relationships which have received the most application in the past 20 years in pharmacology and drug design are the Hansch multiple parameter method or linear free energy relationship (LFER-method) and the Free-Wilson approach or additive method. In the last few years, pattern recognition and related classification methods have been introduced as a third type of methodological approach.

Hansch's Method

According to the linear free energy related model (Hansch-analysis), effects of a homologous series of chemicals are related to a set of theoretical or empirical selected molecular parameters, which determine various changes in physicochemical properties of compounds (Hansch and Leo 1979).

The basic thought of the Hansch approach is the assumption that different magnitudes of biological effects within a homologous series correspond to changes in the free energy of the compounds which occur on interaction with biological macromolecules or on participation in biochemical reactions. Besides quantum mechanical parameters, mainly empirical parameters related to electronic, steric and hydrophobic features of the total molecule or of the functional moieties only are used in these models. Relationships between the activity in its broadest sense and electronic, steric and hydrophobic parameters of molecules can be expressed by the general equation 1:

$$\log A = f(X_h) + f(X_e) + f(X_s) + \text{const.} \tag{1}$$

where A is a quantitative measure of biological activity, usually a chemical concentration causing a certain effect, and X_h, X_e, X_s are molecular parameters. This linear combination of different molecular structure parameters shows the additivity as a fundamental principle in QSAR. This prinicple as well as the basic thought of the

Hansch approach show that the usefulness of this method in general is limited to series of structurally related compounds.

The most important variable for characterizing electronic effects are Hammetts' σ constants, pK_a-values, electronic polarizability as well as parameters which can be derived from spectrometric data such as NMR, IR, UV and MS data. Parameters for representing steric effects are Taft's E_s, and molar refractivity (MR). The hydrophobicity (lipophilicity) can be expressed by Hansch's π, Kubinyi's hydrophobic fragmental constant, octanol/water partition coefficient or related parameters such as water solubility, parachor, molvolume and the total surface area of molecules.

Generally, in QSAR models the influence of lipophilicity on biological activity is described either by the log P values of the total molecule or by the lipophilic influence of different substituents on the parent molecule (π). Other empirical parameters used to represent changes in partition behaviour and with that in lipophilicity are the (R_m)-values obtained from the reverse-phase thin layer chromatography and the (R_t)-values from the reverse-phase high pressure liquid chromatography. It is evident from the literature that lipophilicity, mainly expressed as log P, correlates reasonably well with the bioavailability, transport and distribution pattern as well as bioactivity of diverse series of congeners and, therefore, with processes of ecotoxicological relevance. A number of studies show that the magnitude of log P is dependent on molecular size and shape, such as the van der Waals volume or total surface area of molecules. However, it should be pointed out that the molecular volume is not well defined because of the probability nature of electron density based on quantum mechanical considerations. Moreover, it is known that not only the geometric properties of molecules are related to their physicochemical properties but also the non-geometric properties, such as bonding type, electron density and branching patterns. Therefore, also such parameters influence the physicochemical properties, hence chemical reactivity and biological activity.

Topological Indices

Recently, a number of non-geometric molecular parameters called topological indices were developed which characterize the bonding type, molecular size and shape and the branching pattern. Several of these indices, especially molecular connectivity indices of different order, are well correlated with molecular parameters or physicochemical properties such as lipophilicity (log P), molar refractivity (MR), parachor, molar volume, water solubility, boiling point, diamagnetic susceptibility and the energy of the highest occupied molecular orbital (E_{HOMO}) or of the lowest unoccupied molecular orbital (E_{LUMO}).

The fundamental thought in developing topological indices is to express the chemical's structural formula as a mathematical graph. Graphs representing molecules are usually called chemical graphs. A chemical graph may be characterized by a matrix or by a numerical index (topological index). With that the molecular structure can be translated into a unique numerical descriptor. For example, Bonchev and Trinajstic (1977) applied information theory to the problem of characterizing molecular structure in a numerical index. The application of information theory is based on the recognition that it is possible to construct a finite probability scheme for every system or structure. Basak _et al_. (1982) developed a numerical descriptor to measure the total topological information contained in a molecule with the application of information theory. Gombar and Wadhwa (1982) calculated the information content of molecules using Shannon and Weaver's (1949) equation based upon the probability of choice among atoms classified into groups of topologically and chemically equivalent atoms. In information theory the structural and functional information content of a molecule can be expressed as a numerical descriptor in terms of negative entropy or negentropy. Altogether, these considerations show that besides log P and Hansch's π values, a number of different ways exist to translate molecular structures and with that physicochemical properties into specific, molecule characteristic numerical descriptors which may be used in structure-activity correlations.

Free-Wilson Method

The second fundamental method used in QSAR analysis is the Free-Wilson (Kubinyi and Kehrhan 1976) approach. The basic thought of this model is the assumption that each group in a molecule contributes additively to the overall biological activity of the total molecule.

In addition, it is presumed that the activity of each of the molecule fragments is constant and independent from any structural changes of the total molecule. Because of this tenuous condition, the usefulness of this model will be very limited if one is dealing with structurally unrelated chemicals. In addition, problems which cannot be solved by the Free-Wilson approach are nonlinear relationships and the prediction of activities of chemicals with substituents which are not part of the training set. As evident from the literature and as a logical consequence of biochemical and biophysical facts relating to partitioning processes and the transport through biological membranes, linearity between structural parameters and molecular activities exist only in a limited range. The classification of chemicals into groups of different biological activity on the basis of molecular structure parameters using non-elementary discriminant analysis, is the third type of quantitative approach in QSAR.

ENVIRONMENTAL PARAMETERS

In principle, the LFER-model, the Free-Wilson approach and the classification methods are useful in ecotoxicology with the assumption that the fundamental processes of interaction between a contaminant and the environment are related to molecular structure.

Chemicals discharged into the environment may go to biotic or abiotic compartments of water, sediment, soil and air. Generally, the environmental fate of chemicals may be characterized by physicochemical, especially thermodynamical and biochemical/biophysical regularities. Various processes and reactions affect their movement into and between the liquid, solid, gaseous and biological phases, such as transport and distribution processes and related transformation reactions. When a chemical is distributed between two immiscible phases, physicochemical equilibrium conditions require that the chemical potential or fugacity of the chemical be equal in the two phases at equilibrium. These processes are based upon thermodynamic regularities such as Nernst's Partition Law and Henry's or Rault's Law. The molecular diffusion of chemicals in soil structures can be described by Fick's Law and the adsorption processes can mathematically be characterized by Freundlich or Langmuir relationships. With that it seems to be probable that various processes of environmental relevance can be described by few fundamental thermodynamic regularities. The parameters of these regularities are often related to molecular structure or physicochemical properties of chemicals. For example, the transfer of environmental chemicals between water and sediment, water and soil, water and air, and water and biota or biological fluids can be characterized as partitioning processes. The primary physical properties of the chemical which determine how a chemical will become distributed is the hydrophobic nature and its vapor pressure. Therefore, the tendency of distribution, their rate and degree are mainly based on molecular structure parameters related to lipophilicity.

In context with the following considerations it seems to be useful to characterize chemical distribution processes between immiscible phases in terms of system parameters. Sorption behaviour and bioconcentration tendency represent such environmentally related system parameters which are of major importance for evaluating environmental contaminants and for assessing environmental and human risk. Soil or sediment sorption coefficients and bioconcentration factors are the numerical descriptors of these system parameters.

Bioconcentration Factors

Bioconcentration factor is a term to describe the tendency of a chemical to concentrate in biological structures or fluids. This process is a special case of the distribution of hydrophobic chemicals between an aqueous phase and a lipophilic phase. Bioconcentration factor is an equilibrium constant, calculated as the ratio of the chemical's concentrations in the biological system and its aqueous environment. As it depends on the specific biological system, it is difficult to exactly compare bioconcentration factors of different species. Given the fact that bioconcentration processes are basically related to water-lipophilic phase partitionings it is not surprising that relationships exist between bioconcentration factors and molecular structure parameters such as molar refractivity, parachor, molar volume and topological indices. A selection of regression equations which were calculated on the basis of measured bioconcentration factors is given in TABLE 1.

TABLE 1: Selected regression equations for the prediction of bioconcentration factors (BCF) from molecular connectivities ($'X^V$) for Chlorella sp. and Goldorfen (Leuciscus idus melanotus [L])

---

log BCF various sp.[a] $= -0.048\ (^1X^V)^3 + 0.359\ (^1X^V)^2 + 0.388\ (^1X^V) - 0.815$

n = 13 (PCB's, chlor. pesticides); r = 0.999; s = 0.1778

log BCF various sp.[a] $= 0.789\ (^1X^V) - 0.147$

n = 13 (PCB's, chlor. pesticides); r = 0.957; s = 0.132

log BCF Goldorfen[b] $= -0.106\ (^1X^V)^2 + 1.499\ (^1X^V) - 1.746$

n = 15 (halog. organics); r = 0.936; s = 0.297

log BCF Chlorella[c] $= -0.016\ (^1X^V)^2 + 0.985\ (^1X^V) - 1.133$

n = 9 (organic compounds); r = 0.867; s = 0.642

---

a Data from Kenaga and Goring (1980).

b Data from Korte et al. (1978).

c Data from Geyer et al. (1981).

Sediment Sorption

Apart from bioconcentration processes, the distribution characteristics of organic hydrophobic pollutants between water and sediment or soil are also of major importance for evaluating their fate in environmental compartments. Although there are different types of molecular interactions between a chemical and sediment or soil structures making sorption complex process, some fundamental characteristics in the sorption of organic nonpolar chemicals have recently been recognized. For example, the soil or sediment-water partition coefficients were found to be directly related to organic matter or organic carbon content. Chiou (1981) pointed out that, as a result of thermodynamic considerations, the sorption of hydrophobic organic chemicals from water onto sediment is mainly determined by the available organic matter, readily expressed as the organic carbon content. This recognition is of great importance in context with the processes. On that basis then it is not surprising that sorption coefficients of non-polar organic pollutants (as an equilibrium constant) correlate well with physicochemical properties such as log P, water solubility, molar volume, total surface area of molecules and molecular structure parameters such as molar refractivity, parachor, and molecular connectivity indices. Selected QSAR equations describing sediment sorption as function of parachor and molar refractivity are given in TABLE 2.

TABLE 2: Selected regression equations for the prediction of sediment sorption coefficients (BS) from parachors (Pc), molar refractivities (MR), and molecular connectivities ($^{1}X^{v}$)[a]

$$\log BS = -0.916 + 0.009\ Pc$$
n = 7 (PCB's); r = 0.998; s = 0.028

$$\log BS = -0.268 + 0.077\ MR$$
n = 7 (PCB's); r = 0.996; s = 0.025

$$\log BS = 0.284 + 0.010\ Pc$$
n = 6 (halogenated alkanes); r = 0.996; s = 0.144

$$\log BS = 0.445 + 0.673\ (^{1}X^{v})$$
n = 19 (PCB's chlorobenzenes, chloroalkanes); r = 0.974; s = 0.211

---

a Data from Koch (1981).

It should be pointed out that correlations between partitioning like processes, such as bioconcentration and sediment sorption and molecular structure parameters are not of universal nature. Generally, the more closely related the chemicals are the better are the correlations. Vice versa, the more varied the groups of compounds are from which these correlations are determined, the weaker are the correlations. Furthermore, the complexity of partition processes between water phase and natural structures does probably not allow predictions based on these parameters such as water solubility and log P. The best empirical or semi-empirical relationships for predicting bioconcentration and sorption are often with those parameters which are directly based on the molecular structure of the chemicals. For example, it is well known from Langmuir's equation that adsorption isotherms are linear only at low surface concentration but not throughout the whole concentration range. Similarly, the partitioning of a solute between two phases is in general only constant at sufficiently low bulk concentration.

Biological Effects

One of the major properties of an environmental contaminant is its bioavailability. Not only the exposure concentration but also the rate of absorption and the total amount absorbed are of great importance for the resultant biological effects.

It has been recognized that bioavailability depends on such molecular properties as hydrophobicity/lipophilicity, polarity, molar volume, degree of ionization, molecular shape and size. Using the results of pharmacology, it may be assumed that in moving from the site of absorption to the receptor site an environmental contaminant passes several alternating hydrophilic and hydrophobic barriers. These processes can be expressed as a series of partitioning steps each of which is related to molecular parameters. Furthermore, the last of these processes, namely the binding compound to the receptor may also be viewed as a distribution step which is dependent on the lipophilic properties or molecular structure of the compound and the receptor site. On the basis of such simplifications, it appears likely that relationships exist between molecular activity and structural parameters not only for drugs and related compounds but also for environmental contaminants.

With biological activity data expressed by different toxicity parameters such as LD50 or LC50 values, and molecular structure parameters characterized by electronic, hydrophobic and topological properties of the compounds, many excellent correlations have been obtained. TABLE 3 gives a selection of QSAR regression equations for acute toxicities of several classes of contaminant to various species.

On the basis of structure-analogy/models it seems possible to classify different environmental chemicals into groups based on discriminant analysis as classification rule. For example, we have

TABLE 3: Selected regression equations for the prediction of acute toxicities of chlorophenols, chlorinated aliphatics and benzenes, styrene oxide derivatives and related compounds to guppy (*Poecilia reticulata*); mice; *Daphnia* sp., *Salmonella typh.*, and to sheepshead minnows (*Cyprinodon variegatus*) with connectivity indices ($^1X^v$) ($^2X^v$), octanol/water partition coefficients (log P) and negentropy (log N) values

---

Chlorophenols[a]

log LC50 guppy $= -0.6719\ (^2X^v) + 3.257$
$n = 8;\ r = -0.982;\ s = 0.284$

Chlorinated Aliphatics and Benzenes[a]

log LC50 guppy $= -0.835 \log P - 0.167 \log N + 0.103\ (^2X^v) + 4.16$
$n = 27;\ r = -0.952;\ s = 0.252$

Chlorinated Aliphatics, Benzenes and Phenols[a]

log LC50 guppy $= -0.748 \log P - 0.590 \log N + 0.165\ (^2X^v) + 4.725$
$n = 36;\ r = -0.917;\ s = 0.3003$

Styrene Oxide Derivatives

log LD50 mice[b] $= -3.435\ (^2X^v) + 5.643 \log N + 0.847 - 4.115\ \alpha$
$n = 7;\ r = -0.938;\ s = 0.257$

log LC50 *Salmonella t.*[c] $= -0.899\ (^1X^v) + 4.865$
$n = 7;\ r = 0.838;\ s = 0.256$

Various Halogenated Aliphatics and Aromatics

log LC50 *Daphnia*[d] $= 0.0728\ (^1X^v)^2 - 1.3\ (^1X^v) + 5.484$
$n = 14;\ r = -0.9014;\ s = 0.303$

log LC50 *Cyprinodon*[e] $= 0.106\ (^1X^v)^2 - 1.419\ (^1X^v) + 4.581$
$n = 14;\ r = -0.832;\ s = 0.563$

---

a Data from Könemann (1979).
b Data from N. Tamura (personal communication 1983).
c Data from Tamura *et al.* (1982).
d Data from Bringmann and Kühn (1977).
e Data from Heitmuller *et al.* (1981).

selected 48 chemical compounds and classified them into groups of highly toxic and less toxic substances, on the basis of their acute toxicities. Using molecular structure parameters for describing the acute toxicities we have then computed the following equation for their classification:

$$D = 0.402({}^1X^V) - 2.525({}^2X^V) + 0.041\ Pc + 0.999\ MR - 2.728\ \alpha \quad (2)$$

D	-	Value of the discriminance function
$({}^1X^V)$	-	First order valence molecular connectivity index
$({}^2X^V)$	-	Second order valence molecular connectivity index
Pc	-	Parachor
MR	-	Molar refractivity
α	-	Polarizability of electrons

With this algorithm (2) for the classification, it would to be possible to set priorities for experimental examinations of the toxicities of non-tested industrial chemicals or environmental contaminants.

CONCLUSION

In summary, it is evident that the most important hypothesis of these structure-analogy-models is that the ecotoxicological fate and biological activity of chemicals may be characterized by various simple, yet fundamental, physicochemical and biochemical processes. These processes are related to molecular structures both of global (related to the total molecule such as parachor, molar refractivity, polarizability of electrons, topological indices) and local nature (parameters characterized by functional groups such as Hansch's π value, Hammett constant) and can be described by fundamental thermodynamic regularities.

As shown above, good correlations can be calculated between molecular structure parameters and processes of major ecotoxicological relevance such as soil/sediment sorption behaviour, bioconcentration tendency and biological activity. In this context, it is also of interest to note that sorption and bioconcentration can be characterized quite well as simple partition processes between the water phase and an organic phase or structure. These processes may be described on the basis of various simple regression equations under application of descriptors of the total molecular structure. Because biological activity is the result of very complex and differentiated processes of absorption, distribution, transformation (metabolism) and excretion of chemicals, the situation is more complicated for describing acute toxicity parameters using molecular structure. It is evident from calculating LD50 values that successful correlations can only be computed under application of both the structure parameters which characterize the total molecule and the active parts of a molecule which may be responsible for a particular biological effect.

Altogether, the regression equations show that it is impossible to compute the "best" structure-analogy-model for predicting ecotoxicological parameters of non-tested chemicals. Therefore, the main topic in this concept is to select from among different regression equations one for the best model using a physicochemical, quantum mechanical and biological probable hypothesis of the mode of action of the congeneric series of chemical compounds. Furthermore, it should be noted that because physicochemical properties depend on molecular structures, the use of log P or Hansch's π only places an indirect relationship between molecular structure and activity. On the other hand, non-geometrical properties expressed by topological indices place a direct connection between the total structure of molecule and activity. The disadvantage of using topological indices is that it is difficult to interpret exactly the physicochemical sense of computed regression equations.

Finally, it should be pointed out that much more work is needed for using this theoretical-methodological concept of structure-activity in environmental toxicology. However, it may be a successful concept for predicting ecotoxicological parameters of non-tested chemicals and for recognizing various connections between molecular structure and activity. Both QSAR analysis and experimental research are needed for evaluating ecotoxicological and human risk of chemicals.

REFERENCES

Basak, S.C., Gieschen, D.P., Magnuson, V.R. and Harriss, D.K. 1982. Structure-activity relationships and pharmacokinetics: a comparative study of hydrophobicity, van der Waals' volume and topological parameters. **IRCS Med. Sci. Libr. Compend. 10:** 619-620.

Bonchev, D. and Trinajstic, N. 1977. Information theory, distance matrix, and molecular branching. **J. Chem. Phys. 67:** 4517-4533.

Bringmann, G. and Kühn, R. 1977. Results of the damaging effect of water pollutants on Daphnia magna, Z. **Wasser-, Abwasser Forsch. 10:** 161-166.

Chiou, C.T. 1981. Partition coefficient and water solubiity in environmental chemistry. In **Hazard Assessment of Chemicals: Current Developments. Vol. 1.** J. Saxena and F. Fisher (Eds.), Academic Press, New York.

Geyer, H., Viswanathan, R., Freitag, D. and Korte, F. 1981. Relationship between water solubiity of organic chemicals and their bioaccumulation by the algae Chlorella. **Chemosphere 10:** 1307-1313.

Gombar, V.K. and Wadhwa, L. 1982. Quantitative structure-activity relationships. IV. Use of molecular negentropy as structural parameter. **Arzneim.-Forsch. 32:** 715-718.

Hansch, C. and Leo, A.J. 1979. **Substituent Constants for Correlation Analysis in Chemistry and Biology.** J. Wiley & Sons, New York.

Heitmuller, P.T., Hollister, T.A. and Parrish, P.R. 1981. Acute toxicity of 54 industrial chemicals to sheepshead minnows (Cyprinodon variegatus). **Bull. Environ. Contam. Toxicol. 27:** 596-604.

Kenaga, E.E. and Goring, C.A.I. 1980. Relationship between water solubility, soil sorption, octanol/water partitioning, and concentration of chemicals in biota. In **Aquatic Toxicology,** J.G. Eaton, P.R. Parrish and A.C. Hendricks (Eds.), American Society for Testing and Materials, ASTM STP 707, pp. 78-115.

Koch, R. 1981. **Zu Problemen der Analytik, Bewertung und Normierung organischer Wasserschadstoffe.** Dissertation B, Technische Universität Dresden, Sektion Wasserwesen, Bereich Hydrochemie, Dresden.

Könemann, H. 1979. **Quantitative structure-activity relationships for kinetics and toxicity of aquatic pollutants and their mixtures in fish.** Thesis. Rijksuniversiteit Utrecht.

Korte, F., Freitag, D., Geyer, H., Klein, W., Kraus, A.G. and Lahaniatis, E. 1978. Ecotoxicologic profile analysis. **Chemosphere 7:** 79-102.

Kubinyi, H. and Kehrhan, O.H. 1976. Quantitative structure-activity relationships. I. The modified Free-Wilson approach. **J. Med. Chem. 19:** 578-586.

Shannon, C.E. and Weaver, W. 1949. **The Mathematical Theory of Communication.** University of Illinois Press, Urbana, Illinois.

Tamura, N., Takahashi, K., Shirai, N. and Kawazoe, Y. 1982. Studies on chemical carcinogens. XXI. Quantitative structure-mutagenicity relationship among substituted styrene oxides. **Chem. Pharm. Bull. 30:** 1393-1400.

TOXICITIES OF SELECTED CHLOROANILINES TO FOUR STRAINS OF YEAST

Kazimiera Kwasniewska and Klaus L.E. Kaiser

Environmental Contaminants Division
National Water Research Institute
PO Box 5050, Burlington, Ontario L7R 4A6
CANADA

ABSTRACT

The growth inhibitions of the yeast strains *Pichia* sp., *Rhodotorula* sp., *Rhodotorula rubra* and *Saccharomyces cerevisiae* (baker) by aniline (A), and the chloroanilines 4-Cl-A; 2,4-Cl_2-A; 2,6-Cl_2-A; 3,5-Cl_2-A; 2,3,4-Cl_3-A; 2,3,5,6-Cl_4-A; and 2,3,4,5,6-Cl_5-A were determined. In all cases, the toxicities were found to increase in the order A < 4-Cl-1 < 2,6-Cl_2-A < 2,4-Cl_2-A < 3,5 Cl_2-A < 2,3,4-Cl_3-A. Linear structure-activity correlations were obtained for the toxicites as dependent on the octanol/water partition coefficients. The chloroaniline toxicities are also linearly correlated with those to *Photobacterium phosphoreum* and with the growth inhibitions by the corresponding chlorophenols to the same yeast species.

INTRODUCTION

Determinations of toxicity and biodegradability of existing and new chemicals are necessary for the assessment of their impact on the ecosystem. In particular, the aquatic sphere is subjected, and frequently especially sensitive to environmental contaminants because of bioaccumulation of certain types of compounds in the food chain. For that reason, current research efforts are directed towards the examination of the effects of contaminants on important aquatic species.

K. L. E. Kaiser (ed.), QSAR in Environmental Toxicology, 223–233.

Among the groups of chemicals of high interest in this regard are polychloro derivatives of phenol, many of which are present in Great Lakes' waters and fish (Jones 1981). Recently, we reported on the toxicities of halogenated phenols to a *Bacillus* sp. (Liu *et al.* 1982), to yeast (Kwasniewska and Kaiser 1983), to *Photobacterium Phosph.* (Ribo and Kaiser 1983), and to rainbow trout (*Salmo gairdneri*) (Hodson *et al.* 1983) were reported. Other groups of contaminants include aromatic amines (Diachenko 1979; Nelson and Hites 1980), and chloro derivatives of benzene (Oliver and Nicol 1982), biphenyl, styrene and other polynuclear aromatic hydrocarbons (Kaminsky *et al.* 1983). For such reasons and as part of our ongoing studies on structure-activity correlations of hazardous chemicals (Kaiser 1983; Ribo and Kaiser 1983), we report here on the toxicities of selected chloroanilines to four strains of yeast.

MATERIALS AND METHODS

Materials: Aniline and seven chloroanilines were purchased at the highest degree of purity available from FLUKA Chem. Corp., and were used without further purification. Of each compound, a 1% stock solution in water (aniline) or in dimethyl sulfoxide (chloroanilines) was prepared which served for the preparation of the test solutions.

Yeast strains: The investigated yeast strains were obtained from several sources, namely *Pichia* sp. and *Rhodotorula* sp. from activated sewage sludge from a municipal sewage treatment plant in Burlington, *Rhodotorula rubra* from the shore of Lake Ontario at Burlington, and *Saccharomyces cerevisiae* (baker) from a grocery store.

Cultures: Each strain was cultured in nutrient broth medium (Difco Co.) with the addition of 2% glucose and was agitated in a 125 mL Erlenmeyer flask on a gyratory shaker at 120 strokes$\cdot$min^{-1} at 22°C for 24 hours. These culture stocks were used to seed the test solutions. Each inoculum was added aseptically to a test flask to give an initial reading of 10 Klett Summerson units using a red filter. Test flasks were of 250 mL size with sidearms and also contained 20 mL of the nutrient broth, and 0, 5, 10, 20, 50, 100, and 300 mg$\cdot$L^{-1} of the test substance. The flasks were incubated on a gyratory shaker as for the yeast cultures. At selected time intervals, the turbidity in each flask was measured in the sidearm which provided a direct measurement of the yeast growth. Each species was tested for each chemical at various concentrations and the yeast growth determined in relation to the controls without substrate. For tetra- and pentachloroaniline, opalescent turbidity was observed.

RESULTS AND DISCUSSION

By definition, yeasts are fungi in which the unicellular form predominates. They are therefore essentially similar to higher eukaryotes (such as mammals) in both cellular structure and metabolic behavior. As a result, they offer certain biochemical, metabolic and genetic advantages over bacteria. Furthermore, yeasts or yeast-like fungi are ubiquitous organisms and have been shown to possess mechanisms to cope with various physical and chemical stresses. Such characteristics make yeasts useful and important organisms for the study of contaminant toxicity and biodegradation.

The scope of this study was to determine the growth inhibition of both fermentative and oxidative yeast by exposure to aniline and several chloroanilines.

The fermentative strain, *Pichia* sp., was the one previously used for chlorophenols (Kwasniewska and Kaiser 1983). FIGURE 1 shows the growth inhibition by aniline (A), 4-chloroaniline (4-Cl), 2,4-dichloroaniline (2,4-Cl_2), 2,6-dichloroaniline (2,6-Cl_2), 3,5-dichloroaniline (3,5-Cl_2), 2,3,4-trichloroaniline (2,3,4,-Cl_3), 2,3,5,6-tetrachloroaniline (Cl_4), and 2,3,4,5,6-pentachloroaniline (Cl_5). Unsubstituted aniline (A) was the least toxic compound. From the shape of the dose response curve, it is evident that the fermentative *Pichia* sp. was essentially unaffected by aniline concentrations up to 1.0 $mmol \cdot L^{-1}$ (lower scale, LS, in FIGURE 1). With increasing chlorine substitution of the aniline, particularly in the meta and para positions, the inhibitory effects of the chloroanilines increase strongly. The toxicities, defined as the concentration at which a 50% growth inhibition is observed, increase in the order:

$$A < 4\text{-Cl} < 2,6\text{-Cl}_2 < 2,4\text{-Cl}_2 < 3,5\text{-Cl}_2 < 2,3,4\text{-Cl}_3$$

In contrast to the positive slope of the dose of response curves for the aforementioned compounds, negative slopes are evident for both Cl^4

and Cl_5, with Cl_4 effecting a higher apparent toxicity than either Cl_3 or Cl_5. A similar decrease of toxicity was noted for the effects of a trichlorophenol and the pentachlorophenol (Kwasniewska and Kaiser 1983) on the growth of Pichia sp. and has also been found for pentachlorophenol toxicity in the Microtox test (Ribo and Kaiser 1983). However, the negative slope of the Cl_4 and Cl_5 dose/response curves suggest the possibility of inhibitory effect limitations of these compounds due to their low solubility. Such an effect has also been discussed for the comparatively low toxicity of Cl_5 to Photobacterium phosph. (Ribo and Kaiser 1984) and is also found for the other yeasts reported in here.

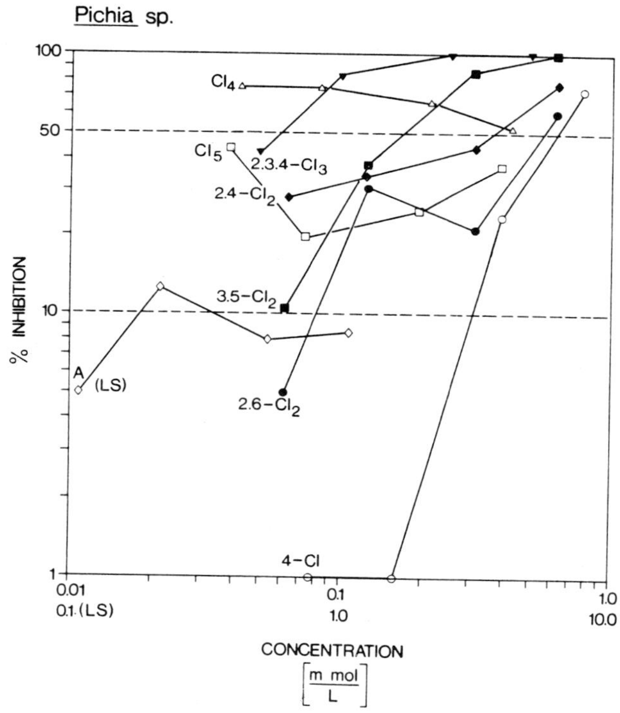

FIGURE 1: Growth inhibition of Pichia sp. versus concentrations of aniline and chloroanilines

For the second fermentative yeast strain reported in here, Saccharomyces cerevisiae (baker), dose/response curves very similar to those for Pichia sp. were obtained (FIGURE 2). Also, an identical order of increasing toxicity is evident with unsubstituted aniline the least and 2,3,4-Cl_3 the most toxic compound of the series. Again, as with Pichia sp. (FIGURE 1), both Cl_4 and Cl_5 show negative slope curves for Saccharomyces cerevisiae (FIGURE 2), with Cl_4 producing a somewhat stronger inhibition than Cl_5. As mentioned above, we interpret the Cl_4 and Cl_5 results as being solubility limited which is further supported by the comparitive little change of growth inhibition with the substrate concentration. For example, the tenfold increase of 2,6-Cl_2 results in a change of growth inhibition from 2% to 100%, while the same order of concentration change of Cl_5 changes the inhibition from 19% to 17%.

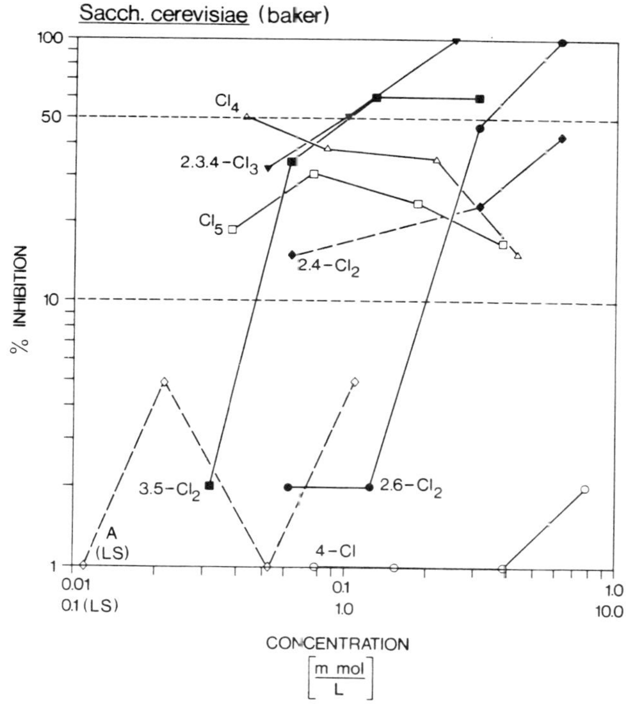

FIGURE 2: Growth inhibition of Saccharomyces cerevisiae (baker) versus concentrations of aniline and chloroanilines

For the two oxidative yeast species investigated, Rhodotorula sp. (FIGURE 3) and Rhodotorula rubra (FIGURE 4), the dose/response curves obtained are basically similar to those of the

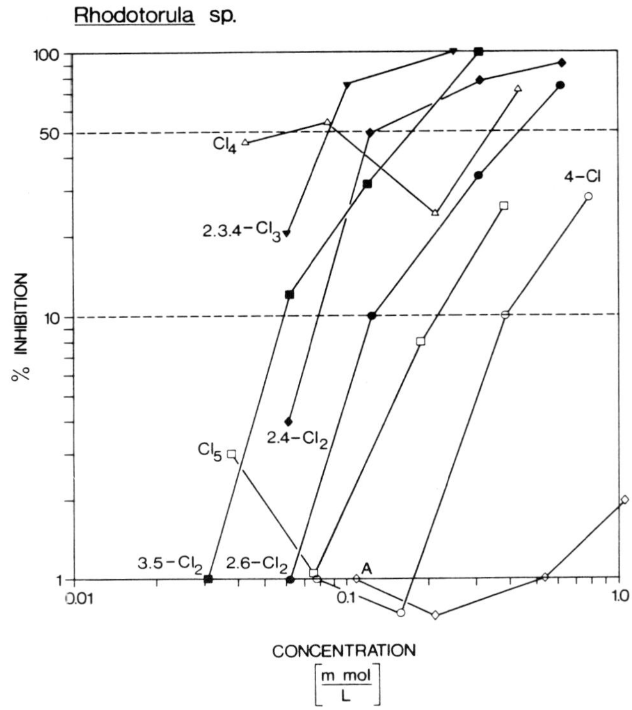

FIGURE 3: Growth inhibition of Rhodotorula sp. versus concentrations of aniline and chloroanilines

fermentative strains (FIGURES 1 and 2). In both cases, the observed toxicities increase in the order

$$A < 4\text{-Cl} < 2,6\text{-}Cl_2 < 2,4\text{-}Cl_2 < 3,5\text{-}Cl_2 < 2,3,4\text{-}Cl_3$$

However, *Rhodotorula rubra* appears to be more sensitive to the chloroanilines than *Rhod.* sp. as identical low concentrations of the three dichloroanilines effect growth inhibitions of between 12% and 45% in the former but only 1% to 4% in the latter. In contrast to the observations on the fermentative strains, the oxidative strains appear to show positive slopes of the dose/response curves. Although the trends are not very clearly seen, the tenfold increase in the concentration of Cl_5 results in an increase of the inhibitory effect from 3% to nearly 30% in *Rhod.* sp. and from 23% to 50% in *Rhod. rubra.* The

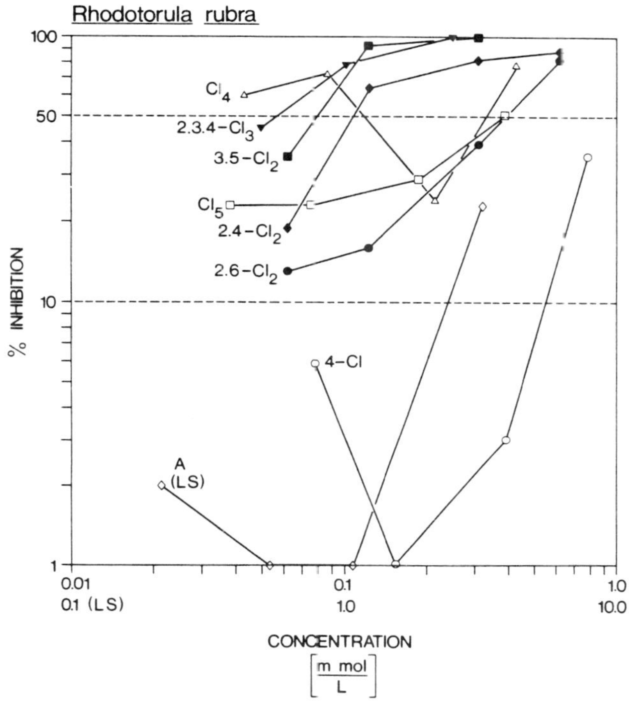

FIGURE 4: Growth inhibition of Rhodotorula rubra versus concentrations of aniline and chloroanilines

corresponding curves for Cl_4 are more difficult to interpret. As the solubility of Cl_5 in the growth medium should not be affected by the different yeast species, it appears possible that the Cl_5 is partially degraded to less chlorinated anilines by the oxidative strains, which, at higher Cl_5 concentrations accumulate and produce the observed increase in toxicity. An investigation of this hypothesis will be undertaken in the future.

TABLE 1 gives an overview of the observed toxicity data by way of chemical concentrations resulting in 10% (IC10) and 50% (IC50) growth inhibition of each yeast species. All concentrations are given as the logarithms of the inverse molar concentrations ($L \cdot mmol^{-1}$). These data have further been investigated for quantitative structure-activity correlations (QSAR) with the octanol/water

TABLE 1: Concentrations (log 1/IC50) of chloroanilines causing a 50% growth reduction of four strains of yeast

Compound	*Pichia* sp.	*Sacc.c.*	*Rhod.*sp.	*Rhod.rb.*	log P[a]
Aniline	-	-		~-0.60	0.90
4-Cl	0.21	-	~0.00	0.07	1.83
2,4-Cl_2	0.43	0.14	0.90	0.96	2.66
2,6-Cl_2	0.26	0.48	0.37	0.41	2.67
3,5-Cl_2	0.77	1.01	0.76	1.10	2.60
2,3,4-Cl_3	1.22	1.01	1.06	1.24	3.46
2,3,5,6-Cl_4	1.37	1.42	1.19	~1.52	4.24
2,3,4,5,6-Cl_5	-	-	~0.22	0.41	4.78

a From Ribo and Kaiser (1984).

partition coefficient (log P), also given in TABLE 1. The following QSAR equations were obtained:

$$\text{IC50 } \textit{Rhod. r.} = 1.04 + 0.65 \log P \tag{1}$$
$$n = 7 \text{ (excluding } Cl_5\text{)}; \; r^2 = 0.87$$

$$\text{IC50 } \textit{Pichia} \text{ sp.} = -0.71 + 0.51 \log P \tag{2}$$
$$n = 5 \text{ (excluding 2,6-}Cl_2 \text{ and } Cl_5\text{)}; \; r^2 = 0.89$$

Both equations exclude Cl_5 and equation 2 also 2,6-Cl_2 from the data sets. The comparatively low toxicity of the 2,6-dichloroaniline is not surprising in view of the intramolecular bonding of the amino-hydrogen and the two ortho chlorine substituents as well as their shielding of the NH_2 group. Similar observations have been made for the effects of ortho-chloro substituted phenols on bacteria (Liu *et al.* 1982, Ribo and Kaiser 1983) and other organisms (Ribo and Kaiser 1983, McLeese *et al.* 1979, Hattula *et al.* 1981). Indeed, also the correlation coefficient for equation 1 would improve from r^2 = 0.87 to r^2 = 0.89 if 2,6-Cl_2 is excluded from that set as well. However, this improvement is not significant. The reason for excluding the Cl_5 values were discussed above.

Of more interest appears to be a comparison of the various yeast toxicity data for the chloroanilines (TABLE 1) with those of the corresponding chlorophenols (Kwasniewska and Kaiser 1983) reported previously. TABLE 2 presents both data sets for the two yeast species used in both experiments, namely *Pichia* sp. and *Rhodotorula rubra*. The correlations between the respective phenol and aniline toxicities are:

Pichia sp: IC50 anilines = 0.18 + 0.61 IC phenols (3)
$n = 3;\ r^2 = 0.972$

Rhod. rubra: IC50 anilines = 0.05 + 0.65 IC50 phenols (4)
$n = 4;\ r^2 = 0.946$

TABLE 2: Acute toxic concentrations (log 1/IC50) of selected chlorophenols and anilines to the yeast strains *Pichia* sp. and *Rhodotorula rubra* and to *Photobacterium phosphoreum*. Values given are logarithms of the inverse concentrations [$L \cdot mmol^{-1}$]

Compound	*Pichia* sp.		*Rhod. rubra*		*Photob. ph.*	
	Phenol[a]	Aniline	Phenol[a]	Aniline	Phenol[b]	Aniline[c]
P,A	-0.98	-	-1.03	~-0.60	0.42	0.12
4-Cl	-0.05	0.21	0.31	0.07	1.19	1.40
2,4-Cl_2	0.58	0.43	0.99	0.96	1.47	1.54
2,4,5-Cl_2	1.66	-	1.99	-	2.19	-
2,3,4-Cl_3	-	1.22	-	1.24	-	1.92
2,3,4,5,6-Cl_5	1.40	-	2.08	-	2.71	1.35

a Data from Kwasniewska and Kaiser (1983).
b Data from Ribo and Kaiser (1983).
c Data from Ribo and Kaiser (1984).

Both equations 3 and 4 show good linear correlations of the chlorophenol with the chloroaniline toxicities. It is interesting to note that the slopes of the toxicity variations with increasing chlorine substitutions is much steeper for the phenols than for the anilines. This is in contrast for the corresponding bacteria toxicities in the Microtox test (Ribo and Kaiser 1983; 1984), where for the same limited set of compounds, the data are also shown in TABLE 2, the correlation equation 5 with identical slopes is found:

Photobact. phos.: EC50 anilines = -0.09 + 1.01 EC50 phenols (5)
$n = 4;\ r^2 = 0.90$

The interpretation of the relative toxicities of chlorophenols and anilines to the same organisms is not clear at this time. However, as additional data become available on these and other sets of compounds, the development of QSAR equations encompassing several classes of compounds may become a possibility.

SUMMARY

The toxicities of unsubstituted aniline and selected chloroanilines to the four yeast strains _Pichia_ sp., _Rhodotorula_ sp., _Rhodotorula rubra_, and _Saccharomyces cerevisiae_ (baker) have been determined. For each strain, the toxicity increased in the order aniline (A) < 4-Cl-A < 2,6-Cl_2-A < 2,4-Cl_2-A < 3,5-Cl_2-A < 2,3,4-Cl_3-A with normal dose/response curves. Both tetra-(2,3,5,6-Cl_4-A) and pentachloroaniline gave declining dose/response curves with the fermentative strains (_Pichia_ and _Sacch._), and first declining and then rising dose/response curves with the oxidative strains (_Rhod._). These effects are likely caused by limited solubilities and partial degradation of Cl_4-A and Cl_5-A. Linear correlation equations of the observed toxicities with the octanol/water partition were obtained, and a comparison of the chloroaniline data with those of the corresponding chlorophenols also resulted in linear correlations.

REFERENCES

Diachenko, G.W. 1979. Determination of several industrial aromatic amines in fish. **Environ. Sci. Technol. 13:** 329-333.

Hattula, M.L., Wasenius, V.M., Reunanen, H. and Arstila, A.V. 1981. Acute toxicity of some chlorinated phenols, catechols and cresols to trout. **Bull. Environ. Contam. Toxicol. 26:** 295-298.

Hodson, P.V., Dixon, D.G. and Kaiser, K.L.E. 1983. The measurement of median lethal doses (LD50's) as a rapid indication of contaminant toxicity to fish. **Environ. Toxicol. Chem,** submitted.

Jones, P.A. 1981. Chlorophenols and their impurities in the Canadian environment. **Environment Canada,** Environ. Prot. Serv., Economic and Technical Review, **Report EPS 3-EC-81-2,** 434 p.

Kaiser, K.L.E. 1983. Organic Contaminants in the Environment: Research Progress and Needs. **National Water Research Institute, Report 83-5,** 35 p.

Kaminsky, R., Kaiser, K.L.E. and Hites, R.A. 1983. Fates of organic compounds from Niagara Falls dumpsites in Lake Ontario. **J. Great Lakes Res. 9:** 183-189.

Kwasniewska, K. and Kaiser, K.L.E. 1983. Toxicities of selected phenols to fermentative and oxidative yeasts. **Bull. Environ. Contam. Toxicol. 31:** 188-194.

Liu, D., Thomson, K. and Kaiser, K.L.E. 1982. Quantitative structure - toxicity relationship of halogenated phenols on bacteria. **Bull. Environ. Contam. Toxicol. 29:** 130-136.

McLeese, D.W., Zitko, V. and Peterson, M.R. 1979. Structure-lethality relationships for phenols, anilies and other aromatic compounds in shrimp and clams. **Chemosphere 2:** 53-57.

Nelson, C.R. and Hites, R.A. 1980. Aromatic amines in and near the Buffalo River. **Environ. Sci. Technol. 14:** 1147-1150.

Oliver, B.G. and Nicol, K.D. 1982. Chlorobenzenes in sediments, water and selected fish from Lakes Superior, Huron, Erie and Ontario. **Environ. Sci. Technol. 16:** 532-536.

Ribo, J.M. and Kaiser, K.L.E. 1983. Effects of selected chemicals to photoluminescent bacteria and their correlations with acute and sublethal effects on other organisms. **Chemosphere 12:** 1421-1442.

Ribo, J.M. and Kaiser, K.L.E. 1984. Toxicities of chloroanilines to Photobacterium phosphoreum and their correlations with effects on other organisms and structural parameters. In **QSAR in Environmental Toxicology,** K.L.E. Kaiser (Ed.), D. Reidel Publ. Co., Dordrecht, Holland, pp. 319-336.

COMPARATIVE STRUCTURE-TOXICITY RELATIONSHIPS BETWEEN ACUTE AND CHRONIC EFFECTS TO AQUATIC ORGANISMS

Gerald A. LeBlanc[1]

EG&G Bionomics
790 Main Street
Wareham, MA 02571
U.S.A.

ABSTRACT

The effects of 2-chlorophenol, 2,4,6-trichlorophenol, 2,4-dimethylphenol, 1,2,4-trichlorobenzene, and butylbenzylphthalate on embryos and larvae of fathead minnows (Pimephales promelas) are presented and used to estimate their chronic toxicities. In conjunction with published toxicity data, structure-toxicity relationships are shown of acute (LC50) toxicity and chronic (MATC) toxicity endpoints with structure coefficients (SC's) for chlorophenols and chlorobenzenes. Results indicate similar relationships of the SC's with LC50 values and of the SC's with MATC values for chlorophenols. In contrast, chronic toxicity of chlorobenzenes increases significantly more than acute toxicity with increased chlorination. The chronic toxicity of 2,4-dimethylphenol is similar to that of chlorinated phenols. Acute toxicity of phthalate esters increases with increasing substituent carbon chain length, while chronic toxicity decreases. Substitution of a butyl group with a phenyl ring has little effect on acute toxicity but significantly increases chronic toxicity of the

[1] Present address: Biology Department, University of South Florida Tampa, Florida 33620, U.S.A.

K. L. E. Kaiser (ed.), QSAR in Environmental Toxicology, 235–260.

phthalate ester. The possible influence of metabolism and activation of parent compounds on structure-acute/chronic toxicity relationship differences is discussed.

INTRODUCTION

Our understanding of the role of physical properties and chemical structure in the toxicity of environmental pollutants has advanced considerably over the past decade commensurate with the toxicity data base increase. Structure-toxicity studies with fish and macroinvertebrates have been limited to acute toxicity (Veith *et al.* 1983; Kanazawa 1981; McLeese *et al.* 1979) since there is ample published data in this area. In addition, data can be generated in only 48 to 96 hours at relatively low cost with test materials of interest. Similar studies have also been performed with bacteria (Liu *et al.* 1982), algae (Wong *et al.* 1982), and ciliates (Schultz *et al.* 1983; Schultz and Cajina-Quezada 1982; Schultz *et al.* 1982). These investigators have used indices of toxicity such as population growth and primary productivity during exposure periods ranging from 30 mins to 60 hours. It is difficult to classify toxicity as being acute or chronic with such organisms due to their rapid rate of multiplication.

Relationships between chemical structure and chronic toxicity to fish have not been investigated. The paucity of such toxicity data and the long duration of such tests have precluded comprehensive evaluations. Toxicity tests using embryos and larvae of fish have been identified as accurate predictors of chronic toxicity for many groups of compounds (Macek and Sleight 1977; McKim 1977) and the U.S. EPA (1982) has accepted such tests in place of full chronic toxicity tests for hazard evaluations. Although these tests are not designed to detect chronic effects such as carcinogenicity and mutagenicity, they can be used to establish preliminary relationships between chemical structures and chronic toxicity.

The intent of this study was to compare structure-toxicity relationships between acute and chronic effects of several groups of compounds, and attempt to identify factors which may be responsible for differences in these relationships.

MATERIALS AND METHODS

Embryo-Larval Tests

The effects of 2-chlorophenol, 2,4,6-trichlorophenol, 2,4-dimethylphenol, 1,2,4-trichlorobenzene, and butylbenzylphthalate on embryos and larvae of fathead minnows (*Pimephales promelas*) were determined. The effects of the test materials were assessed during these tests on the hatching of embryos and survival and growth of larvae through 30 days post-hatch exposure. Tests were performed according to standard procedures (Benoit *et al.* 1982; U.S. EPA 1972) using Mount and Brungs (1967) type proportional diluters to deliver test solutions to the aquaria. Exposure concentrations were confirmed through weekly analyses of the concentration of test material in each aquarium. Analyses were performed using either a Hewlett-Packard model 5840A gas chromatograph equipped with a 15 mCi, Ni-63 electron capture detector or a Waters Associates model ALC/GPC-204 liquid chromatograph equipped with a model 440 UV absorbance detector. All test concentrations were replicated 1X. Significant ($P = 0.05$) effects of the test material on the parameters measured were determined using analysis of variance and Dunnett's procedure (Steel and Torrie 1960).

Results of these tests are presented as the **maximum acceptable toxicant concentration** (MATC). The MATC is the geometric mean of the lowest exposure concentration causing a significant effect on one of the parameters measured and of the highest concentration causing no measurable effect on the parameters. Therefore, the MATC is an estimate of the threshold effect concentration. Chronic toxicity was determined using the **acute-chronic toxicity ratio** (ACR). The ACR is derived by dividing a 96-hour LC50 value derived for the test material by it's MATC. Therefore, higher ACR's imply greater toxicity.

LC50 values were not available for fathead minnows and most of the chlorobenzenes used to quantitate structure-toxicity relationships during this study. Range-finding tests performed at this laboratory in preparation for the embryo-larvae tests, in conjunction with existing data, indicated that the acute response of minnows to chlorobenzenes is virtually the same as those of bluegill (*Lepomis macrochirus*) for which ample acute toxicity data with chlorobenzenes are available (Buccafusco *et al*. 1981). Therefore, ACR's for chlorobenzenes were estimated by dividing LC50 values determined with bluegill by MATC values determined with fathead or sheepshead minnows (*Cyprinodon variegatus*).

LC50 values were also not available for fathead minnows and most of the chlorophenols used in this study. Sufficient data were available to calculate ACR's using LC50 values derived with fathead minnows. A comparison of the acute sensitivity of fathead minnows and bluegill to chlorophenols shows that the responses of both organisms are well correlated (r^2 = 0.94; FIGURE 1) with the fathead minnows being slightly less sensitive. Since structure-acute toxicity relationships are available for bluegill (Buccafusco *et al*. 1981), bluegill MATC's were estimated by dividing the LC50 value by the appropriate ACR. This procedure has been demonstrated acceptable for a variety of industrial chemicals (Kenaga 1982).

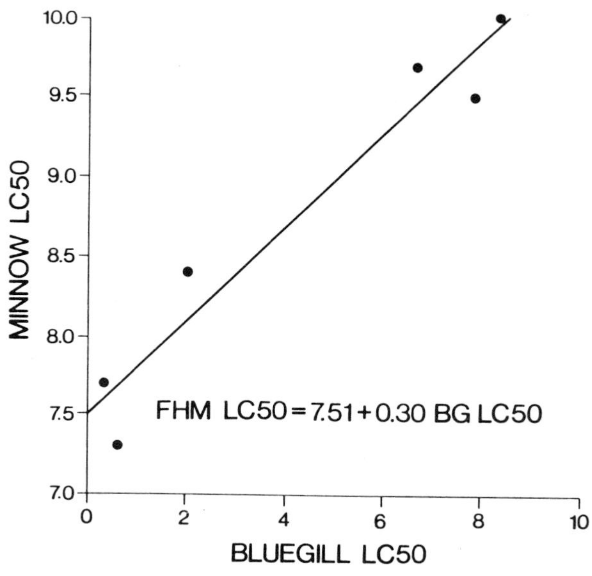

FIGURE 1: Plot of the acute toxicities of chlorophenols to bluegill (*Lepomis macrochirus*) versus those to fathead minnows (*Pimphales promelas*)

TABLE 1: Data used to examine structure-toxicity relationships

Compound	Species	LDC0 ($mg \cdot L^{-1}$)	MATC ($mg \cdot L^{-1}$)	ACR	Source
2-Chlorophenol	fhm[a]	12	5.7	2.1	Phipps *et al.* 1981; PS[b]
2,4,6-Trichlorophenol	fhm	9.2	1.4	6.6	Phipps *et al.* 1981; PS
Pentachlorophenol	shm[c]	0.442	0.064	6.9	Parrish *et al.* 1978
2-Chlorophenol	blg[d]	6.6	3.1[e]	2.1	Buccafusco *et al.* 1981; PS
2,4,6-Trichlorophenol	blg	0.32	0.048[e]	6.6	Buccafusco *et al.* 1981; PS
Pentachlorophenol	blg	0.048[f]	0.007[e]	6.9	PS
4-Chlorophenol	blg	3.8	-	-	Buccafusco *et al.* 1981
2,4-Dichlorophenol	blg	2.0	-	-	Buccafusco *et al.* 1981
2,4,5-Trichlorophenol	blg	0.45	-	-	Buccafusco *et al.* 1981
2,3,5,6-Tetrachlorophenol	blg	0.17	-	-	Buccafusco *et al.* 1981
2,3,4,6-Tetrachlorophenol	blg	0.14	-	-	Buccafusco *et al.* 1981
2,4-Dinitrophenol	blg	0.62	-	-	Buccafusco *et al.* 1981
2,4-Dimethylphenol	blg	7.8	-	-	Buccafusco *et al.* 1981
4-Nitrophenol	blg	8.3	-	-	Buccafusco *et al.* 1981
2,4-Dimethylphenol	fhm	17	-	-	Phipps *et al.* 1981
2,4-Dichlorophenol	fhm	8.2	-	-	Phipps *et al.* 1981
2-Chlorophenol	fhm	12	-	-	Phipps *et al.* 1981
2,4,6-Trichlorophenol	fhm	9.2	-	-	Phipps *et al.* 1981
2,4-Dinitrophenol	fhm	17	-	-	Phipps *et al.* 1981
4-Nitrophenol	fhm	60	-	-	Phipps *et al.* 1981
1-Chlorobenzene	blg	16	-	-	Buccafusco *et al.* 1981
1,2-Dichlorobenzene	blg-LC50 fhm-MATC	5.6	2.0 -	2.8	Buccafusco *et al.* 1981 U.S. EPA 1978
1,3-Dichlorobenzene	blg	5.0	-	-	Buccafusco *et al.* 1981
1,4-Dichlorobenzene	blg	4.3	-	-	Buccafusco *et al.* 1981
1,2,4-Trichlorobenzene	blg-LC50 fhm-MATC	3.4	0.29	11.7	Buccafusco *et al.* 1981; PS
1,2,3,5-Tetrachlorobenzene	blg	6.4	-	-	Buccafusco *et al.* 1981
1,2,4,5-Tetrachlorobenzene	blg-LC50 shm-MATC	1.6	0.13	12.3	Buccafusco *et al.* 1981 U.S. EPA 1978
Pentachlorobenzene	blg	0.25	-	-	Buccafusco *et al.* 1981
Butylbenzylphthalate	fhm	5.3	0.22	9.5	PS

a Fathead minnow.
b Present study.
c Sheepshead minnow.
d Bluegill.
e Calculated using the ACR derived with minnows.
f Determined from the structure-toxicity regression equation.

The LC50 and MATC values for chlorophenols and chlorobenzenes were correlated with the compounds' structure coefficients (SC's). The SC is a numerical representation of the chlorine content of the molecule and the position of the chlorine atoms. The SC is determined by dividing the total weight of the chlorine atoms on the molecule by the total weight of the carbon atoms and multiplying this value by 100. Within any groups of compounds containing the same number of chlorines the SC was increased by 15% for every position distance increase between the furthest two chlorines. For example, 1,2- and 1,3-dichlorobenzene both have formula SC's of 97 because the ratio of chlorine/carbon is the same. However, as the chlorines are one position distance further on 1,3-dichlorobenzene, it's SC is increased by 15% to 112. Both 1,2,3,5- and 1,2,4,5-tetrachlorobenzene have the same SC as the two furthest chlorines (1 and 5) are equally distant.

A summary of the data used to determine structure-toxicity relationships is presented in TABLE 1. Seventy-eight percent of the test results used in this evaluation were generated at EG & G Bionomics and were consistent in test procedures.

RESULTS

Embryo-Larval Tests

2-Chlorophenol: Fathead minnows were unaffected by 2-chlorophenol concentrations as high as 4.0 mg•L^{-1} (TABLE 2). A concentration of 8.1 mg•L^{-1} caused significant mortality of fathead minnows during the preliminary 96-hour test. Clearly, 2-chlorophenol is not chronically toxic to fathead minnows. Using 4.0 mg•L^{-1} and 8.1 mg•L^{-1} as the no-effect and effect concentrations, respectively, the MATC for this compound was estimated to be 5.7 mg•L^{-1} with an ACR of 2.1.

TABLE 2: The effects of 2-chlorophenol on the embryos and larvae of fathead minnows (*Pimephales promelas*)

Measured Concentration[a] (mg•L^{-1})	Hatch[a] (%)	30 Days Post Hatch		
		Survival[a] (%)	Length[a] (mm)	Weight[a] (mg)
4.0(0.2)	90(10)	84(16)	20(2)	64(11)
2.6(0.2)	94(5)	60(56)	20(2)	80(33)
1.7(0.1)	92(7)	98(0)	20(1)	62(1)
1.1(0.1)	92(5)	94(8)	20(2)	62(6)
0.78(0.06)	96(2)	92(4)	21(1)	66(1)
Solvent Control	92(0)	92(9)	21(2)	70(4)
Control	95(0)	96(6)	21(2)	67(3)

a Mean (standard deviation).

2,4,6-Trichlorophenol: Fathead minnow embryos were not affected by 2,4,6-trichlorophenol concentrations as high as 2.1 $mg \cdot L^{-1}$ (TABLE 3). Survival and growth of larvae exposed at 2.1 $mg \cdot L^{-1}$ for 30 days after hatch were significantly reduced, but were not significantly affected by concentrations as high as 0.97 $mg \cdot L^{-1}$. Based on a no-effect and effect concentration of 0.97 and 2.1 $mg \cdot L^{-1}$, respectively, the MATC for 2,4,6-trichlorophenol was estimated to be 1.4 $mg \cdot L^{-1}$. Based on an LC50 estimate of 9.2 $mg \cdot L^{-1}$, the ACR for this material was 6.6.

TABLE 3: The effects of 2,4,6-trichlorophenol on the embryos and larvae of fathead minnows (Pimephales promelas)

Measured Concentration[a] ($mg \cdot L^{-1}$)	Hatch[a] (%)	30 Days Post Hatch: Survival[a] (%)	Length[a] (mm)	Weight[a] (mg)
2.1(0.3)	99(1)	72(11)[b]	14(3)[b]	23(7)[b]
0.97(0.11)	98(4)	100(0)	21(1)	70(4)
0.53(0.04)	100(0)	100(0)	22(1)	80(1)
0.25(0.02)	98(0)	100(0)	22(1)	86(1)
0.13(0.01)	99(1)	96(6)	22(1)	90(4)
Solvent Control	98(2)	99(1)	22(1)	89(1)
Control	99(1)	99(1)	22(1)	88(4)

a Mean (standard deviation).

b Significantly (P = 0.05) different from controls.

2,4-Dimethylphenol: Exposure to 7.4 and 15 $mg \cdot L^{-1}$ 2,4-dimethylphenol significantly reduced the hatching of fathead minnow embryos (TABLE 4). No fathead minnow larvae survived exposure to 15 $mg \cdot L^{-1}$ and survival of larvae was significantly reduced at 7.4 $mg \cdot L^{-1}$. Larvae survival was not affected at 2,4-dimethylphenol concentrations as high as 3.2 $mg \cdot L^{-1}$. Larvae growth was significantly reduced at concentrations of 3.2 and 7.4 $mg \cdot L^{-1}$, but was not affected at 0.75 and 1.5 $mg \cdot L^{-1}$ 2,4-dimethylphenol. Based on no-effect and effect concentrations of 1.5 and 3.2, respectively, the MATC for 2,4-dimethylphenol was estimated to be 2.2 and with an LC50 estimate of 16.8 $mg \cdot L^{-1}$, the ACR for this compound was 7.6.

TABLE 4: The effects of 2,4-dimethylphenol on the embryos and larvae of fathead minnows (Pimephales promelas)

Measured Concentration[a] ($mg \cdot L^{-1}$)	Hatch[a] (%)	30 Days Post Hatch		
		Survival[a] (%)	Length[a] (mm)	Weight[a] (mg)
15(1)	56(31)[b]	0(0)[b]	-	-
7.4(0.9)	66(20)[b]	12(5)[b]	14(1)[b]	20(2)[b]
3.2(0.3)	84(6)	69(1)	19(2)[b]	53(3)[b]
1.5(0.3)	68(12)	62(2)	21(3)	78(8)
0.75(0.2)	98(1)	58(25)	22(2)	90(22)
Solvent Control	96(2)	86(2)	21(2)	80(6)
Control	94(6)	93(10)	21(2)	70(1)

a Mean (standard deviation).

b Significantly (P = 0.05) different from controls.

1,2,4-Trichlorobenzene: Fathead minnow embryos were not affected by 1,2,4-trichlorobenzene concentrations as high as 0.95 $mg \cdot L^{-1}$ (TABLE 5), however all larvae exposed at 0.95 $mg \cdot L^{-1}$ 1,2,4-trichlorobenzene died. In addition, survival of larvae was significantly reduced from 30 days exposure at 0.41 $mg \cdot L^{-1}$. Larvae growth was not affected at concentrations as high as 0.21 $mg \cdot L^{-1}$. Based on the no-effect and effect concentrations of 0.21 and 0.41 $mg \cdot L^{-1}$, respectively, the MATC for 1,2,4-trichlorobenzene was estimated to be 0.29 $mg \cdot L^{-1}$ and based on an LC50 estimate of 3.4 $mg \cdot L^{-1}$, the ACR for this material was 11.7.

Butylbenzlyphthalate: Hatching of embryos and survival of larvae was not adversely affected from exposure to butylbenzylphthalate concentrations as high as 0.36 $mg \cdot L^{-1}$ (TABLE 6). Growth of larvae exposed to 0.36 $mg \cdot L^{-1}$ for 30 days was significantly reduced as determined from length measurements. Normal growth was observed for larvae exposed to concentrations as high as 0.14 $mg \cdot L^{-1}$. Using the no-effect and effect concentrations of 0.14 and 0.36 $mg \cdot L^{-1}$, respectively, the MATC for butylbenzylphthalate was estimated to be 0.22 $mg \cdot L^{-1}$ and based on an LC50 estimate of 5.3 $mg \cdot L^{-1}$, the ACR was 9.5.

TABLE 5: The effects of 1,2,4-trichlorobenzene on the embryos and larvae of fathead minnows (*Pimephales promelas*)

Measured Concentration[a] ($mg \cdot L^{-1}$)	Hatch[a] (%)	30 Days Post Hatch		
		Survival[a] (%)	Length[a] (mm)	Weight[a] (mg)
0.95(0.27)	98(4)	0(0)[b]	-	-
0.41(0.34)	98(0)	31(16)[b]	21(4)	92(34)
0.21(0.09)	94(4)	70(17)	20(3)	64(20)
0.09(0.03)	94(9)	78(0)	19(3)	58(1)
0.04(0.01)	93(7)	91(1)	19(3)	52(3)
Solvent Control	92(2)	89(1)	19(2)	51(4)
Control	91(8)	82(0)	19(3)	55(8)

a Mean (standard deviation).

b Significantly (P = 0.05) different from controls.

TABLE 6: The effects of butylbenzylphthalate on the embryos and larvae of fathead minnows (*Pimephales promelas*)

Measured Concentration[a] ($mg \cdot L^{-1}$)	Hatch[a] (%)	30 Days Post Hatch		
		Survival[a] (%)	Length[a] (mm)	Weight[a] (mg)
0.36(0.16)	99(1)	84(8)	18(2)[b]	46(9)
0.14(0.03)	96(1)	78(19)	20(2)	56(6)
0.07(0.01)	99(1)	83(7)	20(2)	56(11)
0.03(0.01)	96(2)	88(9)	20(2)	50(1)
0.02(0.01)	98(0)	83(21)	20(2)	49(7)
Solvent Control	94(6)	84(12)	20(2)	58(9)
Control	96(2)	75(4)	20(2)	53(3)

a Mean (standard deviation).

b Significantly (P = 0.05) different from controls.

Structure-Acute Toxicity Relationships

Chlorinated phenols: A highly significant (P = 0.01) relationship exists between the log LC50 and the SC's of chlorinated phenols (FIGURE 2). The relationship corroborates the observation made by other investigators (Liu et al. 1982) that the toxicity of chlorinated phenols increases with increasing chlorine substitution. In addition, the distance between the furthest two chlorines on the carbon ring affects the acute toxicity of the substance. This relationship is concordant with that observed between the water solubility or log P and acute toxicity of chlorophenols (Veith et al. 1983; Zitko 1975) since the SC of chlorophenols increases with decreasing water solubility or increasing log P of the compounds (TABLE 7).

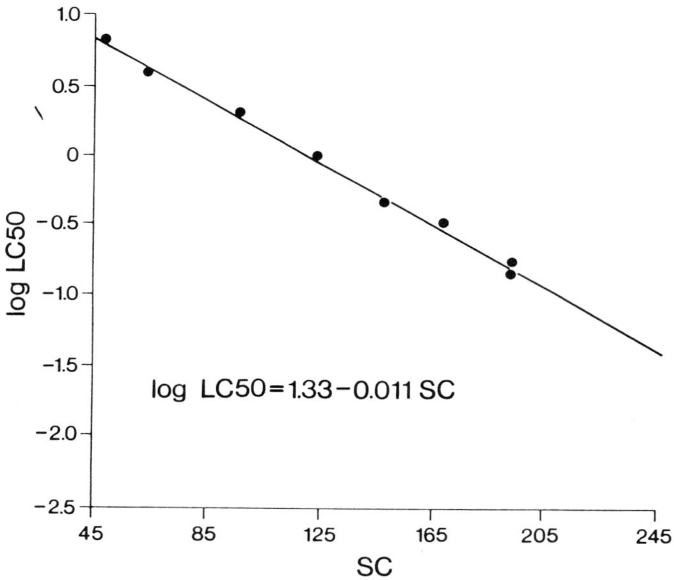

FIGURE 2: Relationship between the acute toxicities of chlorophenols to bluegill (*Lepomis macrochirus*) and the structure coefficients (SC)

Chlorinated benzenes: A highly significant (P = 0.01) relationship exists between the log LC50 and SC's of chlorinated benzenes

(FIGURE 3). One LC50 value (1,2,3,5-tetrachlorobenzene) appeared extraneous and was eliminated from the regression analysis on the presumption that it was inaccurate. The slope of the regression of chlorobenzene LC50 values and SC's was nearly identical to that of the chlorophenols. Structure-acute toxicity relationships, as they apply to chlorophenols, seem also pertinent to chlorobenzenes.

TABLE 7: Some physical and biological properties of chlorobenzenes, chlorophenols, and phthalate esters

Compound	log P[a]	SC	log BCF[a]	Water Solubility[a] (g/100 g)
Benzenes			-	0.0488
Monochloro-	2.83	49	-	0.0145
1,2-Dichloro-	-	97	-	0.0123
1,3-Dichloro-	3.44	112	-	0.0080
1,4-Dichloro-	3.37	126	-	-
1,2,3-Trichloro-	4.23	-	-	-
1,2,4-Trichloro-	4.23	146	-	-
1,2,3,5-Tetrachloro-	4.93	194	-	-
Pentachloro-	5.63	243	-	-
Hexachloro-	6.43	-	-	
Phenols				
2-Chloro-	-	49	2.33	2.85
4-Chloro-	-	64	-	2.71
2,4-Dichloro-	3.08	112	-	-
2,4,5-Trichloro-	-	146	-	0.2
2,4,6-Trichloro-	-	168	-	0.1
2,3,6,5-Tetrachloro-	-	194	-	0.1
Pentachloro-	-	243	3.00	0.0014
Phthalates				
Dimethyl-	-	-	-	0.43
Di-n-butyl-	-	-	-	0.04
Butylbenzyl-	-	-	-	0.00029

a Data from various sources.

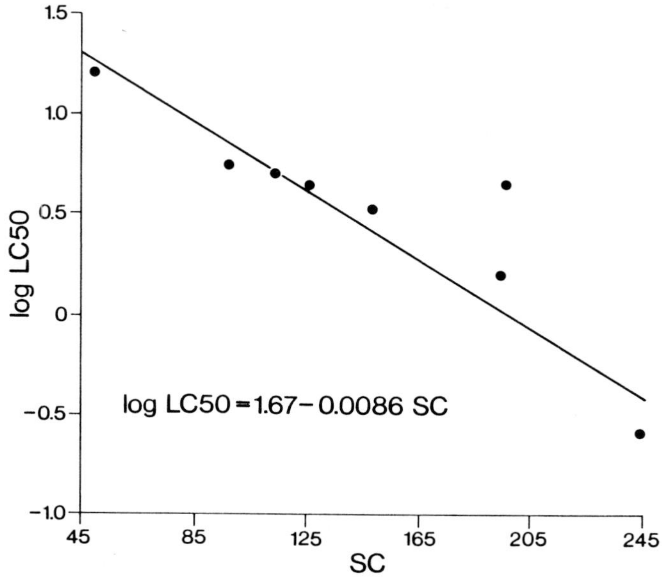

FIGURE 3: **Relationship between the acute toxicities of chlorobenzenes to bluegill (*Lepomis macrochirus*) and the structure coefficients (SC)**

Structure–Chronic Toxicity Relationships

Chlorinated phenols: A significant relationship exists between MATC's and SC's of chlorinated phenols (FIGURE 4). The slope of the regression line is nearly parallel to that of the LC50-SC regression line. The slope of the MATC regression line is slightly steeper than that of the LC50 indicating acute-chronic ratios (ACR) would increase slightly with increasing SC's. Using the regression equation 1:

$$ACR = -9.4588 + 7.0121 \log SC \qquad (1)$$

derived from the known ACR values for 2-chlorophenol, 2,4,6-trichlorophenol, and pentachlorophenol ($r^2 = 0.98$), the ACR's for several chlorophenols were calculated (FIGURE 5). These data indicate that the chronic toxicity (ACR) of chlorophenols increases by a factor of three between 2-chlorophenol and pentachlorophenol.

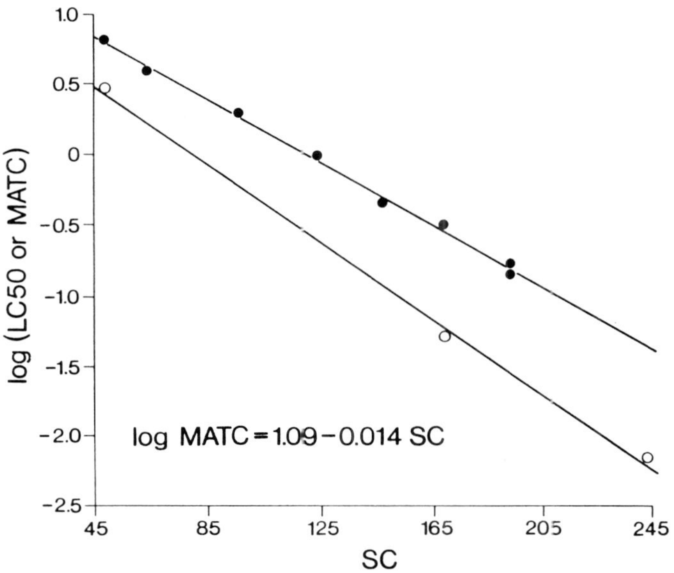

FIGURE 4: **Plot of the maximum acceptable toxicant concentrations (MATC; open circles) and the acute toxic concentrations (LC50; dark circles) of chlorophenols versus their structure coefficients (SC)**

Chlorinated benzenes: The correlation of chlorinated benzenes MATC's and SC's is presented in FIGURE 5. The slope of the regression line is considerably steeper than that of the LC50 regression line, indicating that ACR's of chlorobenzenes increase with increasing SC's to a greater extent than observed with chlorophenols. Using the regression equation 2:

$$ACR = -62.1675 + 33.0702 \log SC \quad (2)$$

derived from the known ACR values for 1,2-dichlorobenzene, 1,2,4-trichlorobenzene, and 1,2,4,5-tetrachlorobenzene ($r^2 = 0.88$), the ACR's for several chlorinated benzenes were calculated (FIGURE 7). A steady increase in ACR's is observed with increasing chlorine substitution. A fivefold increase in chronic toxicity is evident by increasing the chlorine substitution from 1,2-dichlorobenzene to pentachlorobenzene.

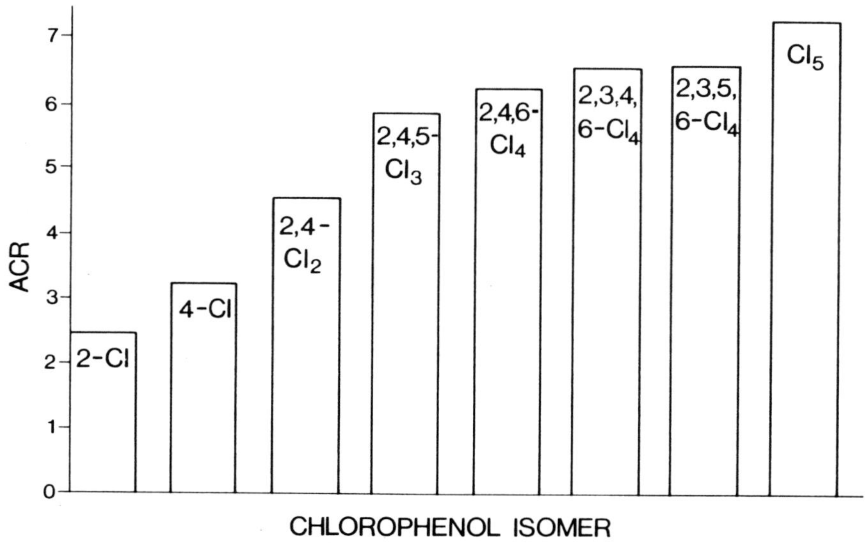

FIGURE 5: Bargraph of acute-chronic toxicity ratios (ACR) of chlorophenol isomers

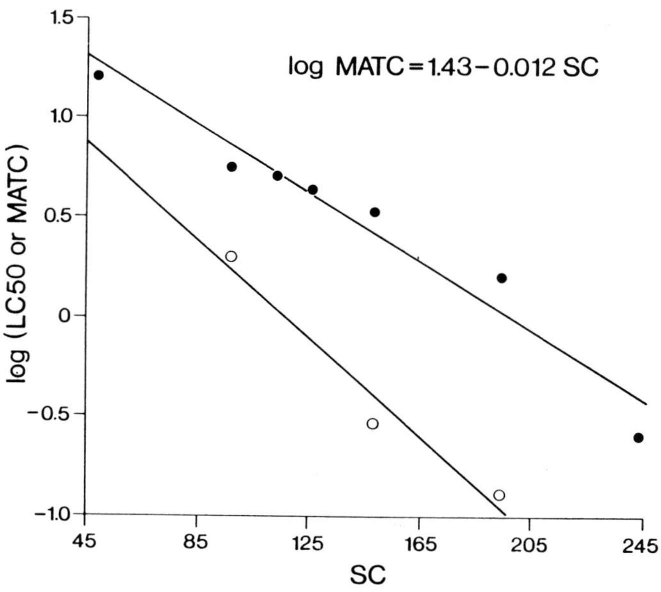

FIGURE 6: Plot of the maximum acceptable toxicant concentrations (MATC; open circles) and the acute toxic concentrations (LC50; dark circles) of chlorobenzenes versus their structure coefficients (SC)

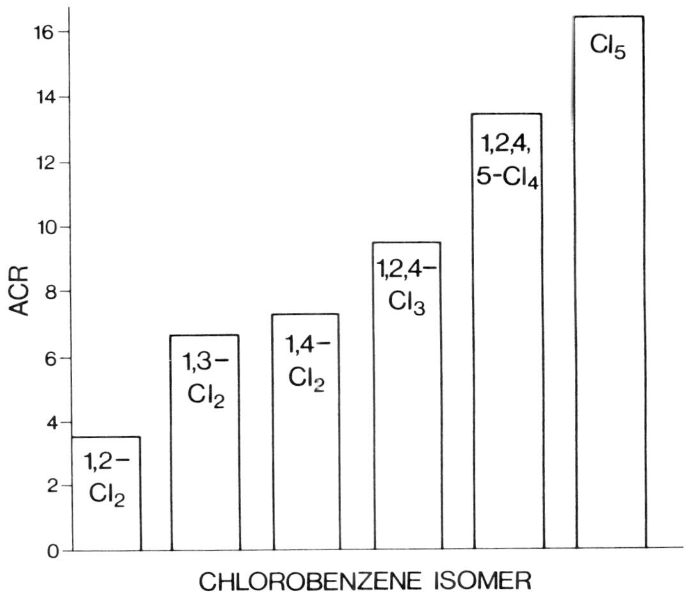

FIGURE 7: Bargraph of acute-chronic toxicity ratios (ACR) of chlorobenzene isomers

Methylated phenols: There are currently insufficient data to determine structure relationships of methylphenols with their acute or chronic toxicities to fish. Data by Phipps et al. (1981) suggest that the 2,4- and 2,6-dimethylphenols have acute toxicities which are relatively comparable to unsubstituted phenol. The ACR for 2,4-dimethylphenol of 7.6 determined during this study is within the range of ACR's estimated for chloro-substituted phenols.

Phthalate esters: The LC50 values and ACR's for selected phthalate esters are shown in FIGURE 8. Acute toxicity increased with increasing carbon chain lengths. In contrast, chronic toxicity decreases with increasing carbon chain lengths. The reduction in chronic toxicity appeared proportionate to the increase in acute toxicity. Substituting a phenyl ring for one butyl group (di-n-butyl changed to butylbenzyl) had little effect on acute toxicity; however, this substitution dramatically increased the phthalates' chronic toxicity.

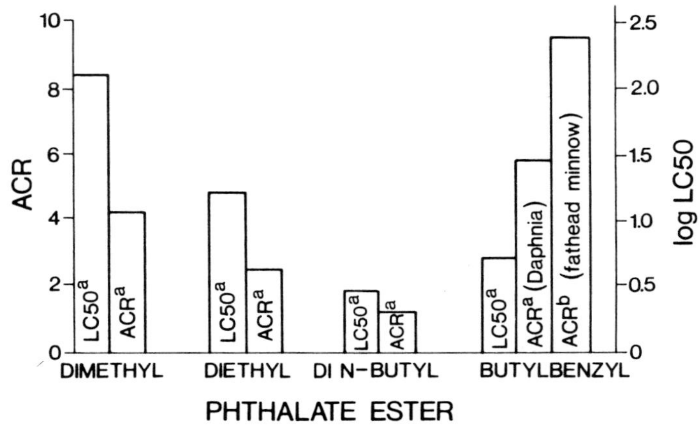

FIGURE 8: Bargraph of acute toxic concentrations (LC50) and acute-chronic toxicity ratios of phthalate esters. Data from Stubblefield *et al*. (1983) (a) and this study (b)

DISCUSSION

Relationships were determined, during this study, between the acute LC50 or chronic MATC of chlorinated phenols or chlorinated benzenes and the compound's SC. Structure coefficient's represent the number of chlorine atoms on the molecule and the distance between the furthest two chlorines in each molecule. The bio-activity of a variety of molecules is known to increase with increasing chlorination (Safe *et al*. 1981; Ecobichon and Comeau 1974). Increasing the number of chlorine atoms on a benzene or phenol ring decreases the compound's water solubility and increases its octanol/water partition coefficient (TABLE 7), which facilitates the chemical's entry into the site of action thus enhancing the potential for toxicity. Relationships between water solubility or octanol/water partition coefficient and toxicity have been demonstrated repeatedly (Veith *et al*. 1983; Kanazawa 1981; Zitko 1975).

The most toxic congeners of halogenated aromatic hydrocarbons such as dioxins and PCB's are those with chlorine substitutions in the lateral positions, i.e., opposite ends of the molecule (Safe 1983). It has been proposed that the polarizability of the lateral halogen substituents is an important factor in toxicity. The same principle may apply to the aromatic hydrocarbons studied here, where increased lateral positioning of the chlorines may increase the overall reactivity (toxicity) of the molecule. The observed increase in toxicity with increasing distance between the two furthest chlorine atoms during the present study was also observed with halogenated phenols and bacteria (Liu *et al*. 1982). Halogen substituents close to the hydroxyl group negate the normal toxic properties of the multi-halogenated molecule. This reduction in toxicity of 2,6-isomers was not observed in the present study, where such isomerization followed the same pattern in toxicity as the other isomers. Liu *et al*. (1982) observed that para-substituted phenols were the most toxic isomers. This observation also suggests that lateral positioning of the halogen substituents of aromatic compounds increases their toxicity.

The slopes of the regression lines of chlorophenol structure versus LC50 values and versus MATC values were nearly parallel. In addition, the two lines were relatively close together. This suggests that chlorinated phenols exhibit little chronic toxicity. The correlation between chlorophenol structure and MATC's consists of only three data points and is not considered definitive. Should further data indicate the MATC regression line is actually parallel to the LC50 regression line, then one could conclude that the mechanisms responsible for the chronic toxicity are the same as those for the acute toxicity. During metabolism, aromatic molecules are commonly converted to phenols. The phenol can then be conjugated with UDP glucuronic acid and readily excreted (Thorgeirsson and Nebert 1977). Therefore, phenols would have less propensity to elicit chronic toxicity simply because they are rapidly excreted.

The metabolic products of halogenated benzenes, through cytochrome P-450 mediated activity, appear to be responsible for in vitro liver necrosis (Reid and Krishna 1973). These authors demonstrated a correlation between the degree of binding of halogenated benzenes to liver receptor protein and necrosis of the liver. A correlation between the binding efficiency of poly-aromatic halogenated hydrocarbons to Ah receptor proteins and toxicity has also been documented (Safe *et al*. 1981; Poland *et al*. 1979; Poland and Glover 1977) as well as increased binding to Ah receptor proteins with increased chlorination (Ecobichon and Comeau 1974). In comparison with chlorophenols, chlorobenzenes have therefore a greater propensity to elicit chronic toxicity through metabolism to more reactive products; also, this propensity will increase with increasing chlorination through binding to receptor proteins. This conclusion is supported by the increasing ACR's as shown in FIGURE 7.

Schultz et al. (1982) and Schultz and Cajina-Quezada (1982) demonstrated that toxicity of aromatic compounds to a freshwater ciliate (Tetrahymena pyriformis) increased with successive methylation of the ring. Wong et al. (1982) reported that increased carbon chain length increased the toxicity of alkyl tin compounds to algae. Indeed, increased straight chain methylation increased the acute toxicity of phthalate esters to fish, however, the reverse was true for their chronic toxicity to Daphnia magna (FIGURE 8). The studies of Schultz et al. (1982), Schultz and Cajina-Quezada (1982) and Wong et al. (1982) were performed with organisms with which acute and chronic toxicity are difficult to differentiate due to the short life cycle of the organisms. However, the phthalate ester data suggest that structure-activity relationships with acute toxicity may not apply to chronic effects. The structure-activity relationship differences between acute and chronic effects is evident from the di-n-butyl and butylbenzyl phthalate results. Substitution of a straight chain butyl segment with a phenyl ring had little more effect on acute toxicity than might be predicted by increasing the number of atoms of the molecule (Schultz et al. 1982); however, this substitution significantly increased its chronic toxicity. The addition of the phenyl group doubled the aromatic content to the molecule. Increased toxicity with increased number of aromatic structures in a molecule was demonstrated by Schultz and Cajina-Quezada (1982). Poly-aromatic structures are classic substrates for Ah protein receptor binding even when not halogenated (Poland and Glover 1980). Furthermore, it is hypothesized that chemicals bound to the receptor proteins are carried into the nucleus where they can adversely affect genetic activity. Butylbenzylphthalate is a known carcinogen (NTP 1981), a probably genotoxic effect, while the straight chain substituted phthalate esters discussed are not.

Structure-activity relationships for phthalate esters using acute toxicity are parallel to hydrophobicity trends. That is, acute toxicity increases with decreasing water solubility. This relationship seems to exist with most groups of compounds which elicit acute toxicity. This unifying relationship in acute toxicity trends relating to solubility may have significance in the findings of Schanne et al. (1979). These investigators discovered that with a wide variety of direct-acting chemicals (i.e., no metabolism required for activation), acute cell death occurs after an influx of calcium into the cell, presumably due to a disruption of the cell membrane by the toxicant. If there is a unifying mechanism for the acute toxicity of toxicants, it is reasonable to assume that a unifying trend in structure-toxicity relationships relates to the materials propensity to enter the cell.

CONCLUSIONS

In summary, the data indicate that the acute toxicity of many groups of chemicals to aquatic organisms is directly related to water solubility and octanol/water partition coefficient. These properties are also important in structure-toxicity relationships regarding chronic toxicity. However, other important factors influence chronic toxicity (SCHEME 1). The effect of structure on metabolism must be considereed. Non-toxic levels of parent compound can be metabolized to active products which can interact with macromolecules eliciting effects on biochemical processes resulting in impaired growth or reproduction (i.e., butylbenzylphthalate), or cumulative lethality (i.e., tetrachlorobenzene). Such materials may also elicit genetic effects resulting in mutagenicity or carcinogenicity.

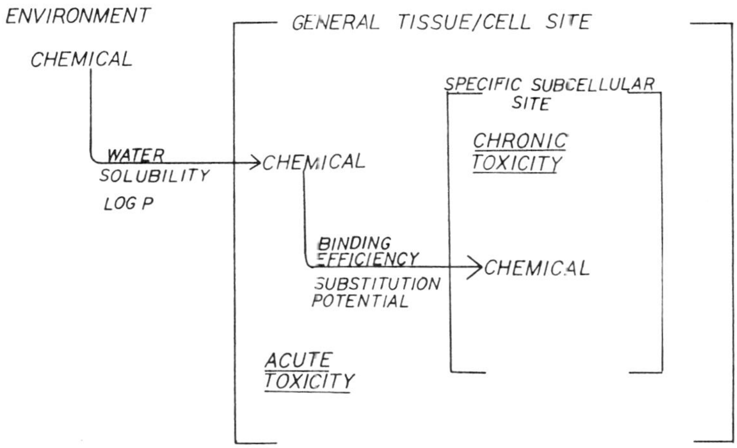

SCHEME 1: Proposed relationship between a chemical's physical/ biochemical properties and it's propensity to elicit acute or chronic toxicity

REFERENCES

Benoit, D.A., Puglisi, F.A. and Olson, D.L. 1982. A fathead minnow, *Pimephales promelas*, early life stage toxicity test method evaluation and exposure to four organic chemicals. **Environ. Pollut. 28:** 189-297.

Buccafusco, R.J., Ells, S.J. and LeBlanc, G.A. 1981. Acute toxicity of priority pollutants to bluegill (*Lepomis macrochirus*). **Bull. Environ. Contam. Toxicol. 26:** 446-452.

Ecobichon, D.J. and Comeau, A.M. 1974. Comparative effects of commercial Aroclors on rat liver enzyme activities. **Chem. Biol. Interact. 9:** 94-105.

Kanazawa, J. 1981. Measurement of the bioconcentration factors of pesticides by freshwater fish and their correlation with physico-chemical properties or acute toxicity. **Pestic. Sci. 12:** 417-424.

Kenaga, E.E. 1982. Predictability of chronic toxicity from acute toxicity of chemicals in fish and aquatic invertebrates. **Environ. Toxicol. Chem. 1:** 347-358.

Liu, D., Thomson, K. and Kaiser, K.L.E. 1982. Quantitative structure-toxicity relationship of halogenated phenols on bacteria. **Bull. Environ. Contam. Toxicol. 29:** 130-136.

Macek, K.J. and Sleight, B.H. III. 1977. Utility of toxicity tests with embryo and fry of fish in evaluating hazards associated with chronic toxicity of chemicals to fishes. In **Aquatic Toxicology Hazard Evaluation,** F.L. Mayer and J.L. Hamelink (Eds.), ASTM STP634, ASTM Philadelphia, pp. 137-146.

McKim, J.M. 1977. Evaluation of tests with early life stages of fish for predicting long-term toxicity. **J. Fish. Res. Board Can. 34:** 1148-1154.

McLeese, D.W., Zitko, V. and Peterson, M.R. 1979. Structure-lethality relationships for phenols, anilines and other aromatic compounds in shrimp and clams. **Chemosphere 2:** 53-57.

Mount, D.I. and Brungs, W.A. 1967. A simplified dosing apparatus for fish toxicological studies. **Water Research 1:** 21-29.

NTP Technical Bulletin. 1981. **NTP toxicology and carcinogenicity testing status of phthalates and related compounds.** PO Box 12233, Research Triangle Park, North Carolina 27709, Issue No. 4, April.

Parrish, P.R., Dyar, E.E., Enos, J.M. and Wilson, W.G. 1978. Chronic toxicity of chlordane, trifluralin, and pentachlorophenol to sheepshead minnow (*Cyprinodon variegatus*). **Ecological Research Series EPA-600/3-78-010.** U.S. EPA Gulf Breeze Florida, 53 p.

Phipps, G.L., Holcombe, G.W. and Fiandt, J.T. 1981. Acute toxicity of phenol and substituted phenols to the fathead minnow. **Bull. Environ. Contam. Toxicol. 26:** 585-593.

Poland, A. and Glover, E. 1977. Chlorinated biphenyl induction of aryl hydrocarbon hydroxylase activity: A study of structure-activity relationship. **Mol. Pharmacol. 13:** 924-938.

Poland, A., Greenlee, W.F. and Kende, A.S. 1979. Studies on the mechanism of action of the chlorinated dibenzo-p-dioxins and related compounds. **Ann. N.Y. Acad. Sci. 320:** 214-230.

Poland, A. and Glover, E. 1980. 2,3,7,8-tetrachlorodibenzo-p-dioxin: Segregation of toxicity with the *Ah* locus. **Mol. Pharmacol. 17:** 86-94.

Reid, W.D. and Krishna, G. 1973. Centrolobular hepatic necrosis related to covalent binding of metabolites of halogenated aromatic hydrocarbons. **Exp. Mol. Pathology. 18:** 80-99.

Safe, S. 1983. 2,3,7,8-TCDD - Biochemical effects. **Chemosphere 12:** 447-451.

Safe, S., Parkinson, A., Robertson, L., Safe, L. and Bandiera, S. 1981. Polychlorinated biphenyls: Effects of structure on biologic and toxic activity. In Workshop on the **Combined Effects of Xenobiotics,** H.F. Stich, H.W. Leung, J.R. Roberts (Eds.), Publication No. NRCC/CNRC 18978 of Environmental Secretariat, National Research Council of Canada, pp. 151-176.

Schanne, F.A.Y., Kane, A.B., Young, E.E. and Farber, J.L. 1979. Calcium dependence of toxic cell death: A final common pathway. **Science 206(9):** 700-702.

Schultz, T.W., Dumont, J.N., Sankey, F.D. and Schmoyer, R.L. Jr. 1983. Structure activity relationship of selected naphthalene derivatives. **Ecotox. Environ. Safety 7:** 191-203.

Schultz, T.W., Kier, L.B. and Hall, L.H. 1982. Structure-toxicity relationship of nitrogenous heterocyclic compounds. III. Relations using molecular connectivity. **Bull. Environ. Contam. Toxicol. 28:** 373-378.

Schultz, T.W. and Cajina-Quezada, M. 1982. Structure-toxicity relationships of selected nitrogenous heterocyclic compounds. II. Dinitrogen molecules. **Arch. Environ. Contam. Toxicol. 11:** 353-361.

Steel, R.G.D. and Torrie, J.H. 1960. **Principles and Procedures of Statistics.** McGraw Hill, New York, 481 p.

Stubblefield, W.A., Foster, R.B. and Howard, P.H. 1983. An environmental toxicological assessment of phthalate esters. **Society of Environmental Toxicology and Chemistry,** Abstracts from the Annual Meeting, November 1983, PO Box 352, Rockville, Maryland.

Thorgeirsson, S.S. and Nebert, D.W. 1977. The Ah locus and the metabolism of chemical carcinogens and other foreign compounds. **Adv. Cancer Research 25:** 149-193.

U.S. EPA. 1972. Proposed recommended bioassay procedures for egg and fry stages of freshwater fish. **U.S. EPA,** Duluth, Minnesota, 7 p.

U.S. EPA. 1982. Wildlife aquatic organism data requirements. U.S. Government Printing Office, Washington, D.C. **Federal Register 47:** 53212.

Veith, G.D., Call, D.J. and Brooke, L.T. 1983. Structure-toxicity relationships for the fathead minnow, Pimephales promelas: Narcotic industrial chemicals. **Can. J. Fish. Aquat. Sci. 40:** 743-748.

Wong, P.T.S., Chau, Y.K., Kramar, O. and Bengert, G.A. 1982. Structure-toxicity relationship of tin compounds on algae. **Can. J. Fish. Aquatic. Sci. 39:** 483-488.

Zitko, V. 1975. Structure-activity relationships in toxicology. In **Structure-Activity Correlations in Studies of Toxicity and Bioconcentration with Aquatic Organisms,** G.D. Veith and D.E. Konasewich (Eds.), International Joint Commission, Windsor, Ontario, pp. 7-24.

WORKSHOP DISCUSSION

McCARTY:

Is the toxicity of the phenols corrected for ionization? Consideration of that might change the relationship so to become similar to that of the chlorobenzenes.

LeBLANC:

No, I haven't done that. Just recently, a paper appeared in Chemosphere on Daphnia and the chronic toxicity of a variety of chlorobenzenes. I looked at these data similarly to the way I have done it here and the slopes were somewhere in between my results on chlorobenzenes and chlorophenols. The slope of the acute toxicities was the same. In the case of chlorobenzenes, if the toxicity relates to metabolites that may be why we are seeing a lesser slope with the Daphnia, they may not have the metabolic enzyme system to the extent of the fish.

HODSON:

In your models, you consider log P and solubility to be very important factors which determine the concentration of the chemical in the tissues, especially in acute toxicity tests. Do you think there is much influence of log P and solubility on compartmental transfer within the animal?

LeBLANC:

I guess it depends on the animal; certainly the potential is there. I am not so sure of its importance for acute toxicity. I think the model I showed last certainly indicates that real life is not black and white, there is a lot of gray between the parameters responsible for acute and chronic toxicity.

WELLS:

How sensitive is the developmental state in the embryos to some of these compounds because they can change significantly when the eggs change in the hardening process?

LeBLANC:

I think I said in the presentation that the eggs were less than 48 hours old. It would probably be more accurate in saying that they were between 24 and 48 hours old when we started the tests. They certainly were hardened. It doesn't take very long for the hardening to set in and I think it is demonstrated by most of the data here that the eggs were not sensitive and that it usually affects the survival and growth of the fish after they hatch. I think it probably relates to the low ability of these compounds to penetrate the egg.

SCHULTZ:

In regards to that question on the sensitivity of the eggs, if you take the chorion off fish eggs, chemically and mechanically, you do not have the problem of penetration and when you do that you find that gastrulation is the most sensitive stage in development. I don't know how this relates to other fish but we did some work with carp and frogs at Oak Ridge just because there is a lot of water around Oak Ridge and the carp and frogs were all coming up and spawning so we could get millions of eggs. We collected eggs and set them in a variety of compounds to see what happened; the problem is getting the compound into the fish eggs. There was no correlation between effects we could get with the easily penetrated frog eggs in comparison to carp eggs.

LeBLANC:

We have some recent comparative studies between fathead minnows and rainbow trout; looking at their early life stages. The fathead minnow were treated basically the same as here but with the rainbow trout, we fertilized the eggs in the laboratory and started the exposure shortly afterwards. The rainbow trout test was run at 12° and the duration of the egg development stage is much longer with some significant differences in response whereas, again, with the fatheads we generally didn't see the effects with the eggs. It is probably a function of the duration of exposure and the fact that the exposure started much earlier with the rainbow trout eggs.

MACKAY:

How did you arrive at your 15% adjustment?

LeBLANC:

By trial and error. I was looking for chlorphenol data and just looking at the molecular ratio of the chlorinated isomers. There were obvious points that were away from the line and it appeared to be a consistent trend. It related to the distance of the two furthest chlorines on the molecule. Through trial and error the 15% correction factor applied very well to the chlorophenols and then I went to the chlorobenzenes and used it and it applied pretty well. I looked for a better correlation factor with the chlorobenzenes and there really wasn't any that applied as well as the chlorophenols. In cases of, let's say tetrachloro isomers where there are four chlorines on the molecule, that correction factor only applies to the position of the two chlorines that are the furthest apart, 1 and 4. So, if you had 1,2,4,5 and 1,2,3,5 the structure coefficient would be the same.

MACKAY:

So, 1,3 and 1,4 dichlorobenzenes would have the same value?

LeBLANC:

No, 1,3 and 1,4 would differ by 15%

WELLS:

When you decided to use LC50 as your method of describing the endpoint, did you look at the toxicity curves for all the compounds?

LeBLANC:

No, I strictly chose the LC50 as endpoint.

HODSON:

It would be interesting to look at your curves and to study if all those LC50's are exactly comparable, i.e., if the same type of acute toxicity is occurring.

LeBLANC:

You are right and it may make the values less relevant in terms of acute lethality occurring but my intent was to have a comparable endpoint for all these data and I really had to choose a consistent point and time.

Comment:

I think that what Peter Hodson is saying is that by choosing the time endpoint you are not necessarily choosing the common endpoint and that is a problem with all short-term tests.

Comment:

That applies to a lot of papers I have heard in the last two days. You really have to be careful.

ZITKO:

Do you or anyone else who listed chlorophenols check them for purity because when we talk about all the dioxins and all these things which could be in the chlorophenols, they could affect the toxicities?

LeBLANC:

I certainly did not, I asked my chemist how pure the phenol was and he gave me a number, but we didn't do the analysis ourselves.

CHU:

We did some studies on the purity of the tetrachlorophenols, all three isomers. The TCDD concentrations were below one part per million, comparatively normal. We concluded that the impurities were not a significant factor in contributing to the toxicity of chlorophenols in the mammilian systems.

RELATIONSHIPS BETWEEN PHYSICAL-CHEMICAL AND ENVIRONMENTAL PARTITIONING COEFFICIENTS

Donald Mackay and Wan Ying Shiu

Department of Chemical Engineering
and Applied Chemistry
University of Toronto
Toronto, Ontario, M5S 1A4
CANADA

ABSTRACT

A review is presented of the relationships between aqueous solubility, melting point, vapour pressure, boiling point, Henry's Law constant, sorption partition coefficient, bioconcentration factor, octanol/water partition coefficient, and molecular structure. It is suggested that progress towards establishing quantitative structure-activity relationships can best be made by considering four categories of properties and their linking equations, namely molecular descriptors, fundamental thermodynamic functions, physical-chemical properties, and environmental partitioning quantities.

INTRODUCTION

In a series of recent papers we have advocated the use of the fugacity concept for modelling the behaviour of chemicals in evaluative environments (Mackay et al. 1983a; Mackay and Paterson 1981, 1982; Mackay 1979) and in real lake and river environments (Mackay et al. 1983b,c). Other approaches involving the use of partition coefficients rather than fugacity can achieve the same results,

K. L. E. Kaiser (ed.), QSAR in Environmental Toxicology, 261–278.

albeit we believe, at the expense of greater mathematical complexity. The recent text by Dickson et al. (1982) reviews many of these modelling approaches. Regardless of the nature of the equations, there is a need for fundamental data on the chemical's properties in the various media which comprise the environment. These include the following:

A) Equilibrium phase partitioning data either in the form of partition coefficients or fugacity capacities.
B) Reaction rate constants for processes such as hydrolysis, photolysis, biolysis and oxidation.
C) Interphase transfer rate data for processes such as volatilization, bio-uptake and sedimentation.
D) Toxicological information.

In this paper, we review some work by our group and others on the predictability of data for the first of the above classes restricting the discussion to relatively hydrophobic organic chemicals. We further explore the relationships between the various partition properties and suggest that structure-activity relationships can play a valuable role in correlating data, checking data consistency and in predicting aspects of the environmental behaviour of new chemicals.

Elsewhere in this volume and in the literature, Bobra et al. (1983) have discussed some recent efforts in our group to establish toxicological structure-activity relationships in which we believe that aqueous solubility is a primary determinant of the partitioning characterisitcs of hydrophobic organics into membranes and hence of the acute, but non-specific, toxicity exerted by the chemical.

Regarding interphase transport, our work on volatilization has been compiled in a recent report (Mackay et al. 1982a). Recently, we have examined the issue of the rate of uptake of organic chemicals by fish and have proposed a pharmacological fugacity model which is, in effect, a correlation of k_2, the "depuration rate constant", with solubility and octanol/water partition coefficient (Mackay and Hughes 1983).

PARAMETERS AND PRINCIPLES

Equilibrium Partitioning Data

The usual approach for quantifying partitioning tendencies is to invoke a partition coefficient (K) or ratio of concentrations at equilibrium. This ratio may be dimensionless, but often it is not. Examples are the Henry's law constant (H) for air/water, sorption coefficients (K_p), organic carbon sorption coefficients (K_{oc}), and bioconcentration factors (K_B). It is not always realized that aqueous solubility (C^S) and vapour pressure (P^S) are in reality, partition coefficients between a pure phase and water, and a pure phase and a vapour phase, respectively.

The other approach is to use an intermediate criterion of equilibrium such as chemical potential, fugacity or activity and relate the two (or more) phase concentrations independently to this criterion. We have used the concept of fugacity as an equilibrium criterion and fugacity capacity (Z) as the link between concentration and fugacity. In effect, the partition coefficient is the ratio of Z for the solute in the two phases. Algebraically,

$$K_{12} = C_1/C_2$$

but

$$C_1 = Z_1 f, \quad c_2 = Z_2 f$$

thus

$$C_1/C_2 = Z_1 f/Z_2 f = Z_1/Z_2 = K_{12}$$

This is illustrated in FIGURE 1.

The aim is thus to develop methods of predicting the values of Z or K from molecular structures using suitable molecular descriptors. In principle, it should be easier to predict Z since it contains information about only one phase, the phase properties being separated. When using K_{12} it may not always be clear how the properties of phase 1 and phase 2 contribute to affect the overall value.

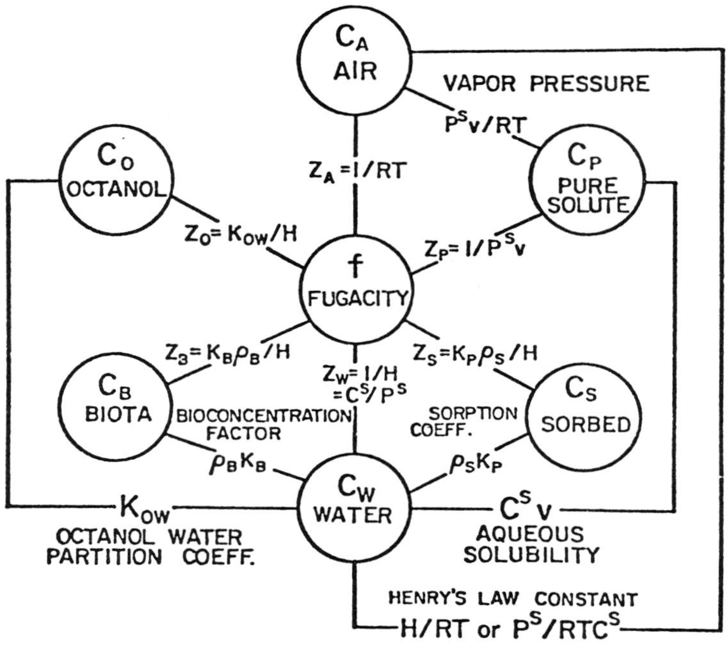

FIGURE 1: Relationships between partitioning coefficients and fugacity capacities

Consideration of the fundamental thermodynamic relationships which apply to environmental systems suggests that we can categorize the various properties as shown in FIGURE 2.

Types of Descriptors

Most basic are **molecular descriptors** (MD) i.e., size, shape and the nature of the intermolecular forces which exist between the molecule and its neighbours which comprise the "medium" of air, water, lipid, etc. These interactions are exceedingly complex and statistical in nature and there seems little prospect that their magnitude can be calculated *a priori* in any but the simplest cases. For molecules of similar interactive nature, descriptors such as molecular volume, area and shape can be very valuable determinants of properties. For example, solubility can be correlated with total surface area (TSA) (Yalkowsky and Valvani 1979) or with molar volume (Mackay and Shiu 1977).

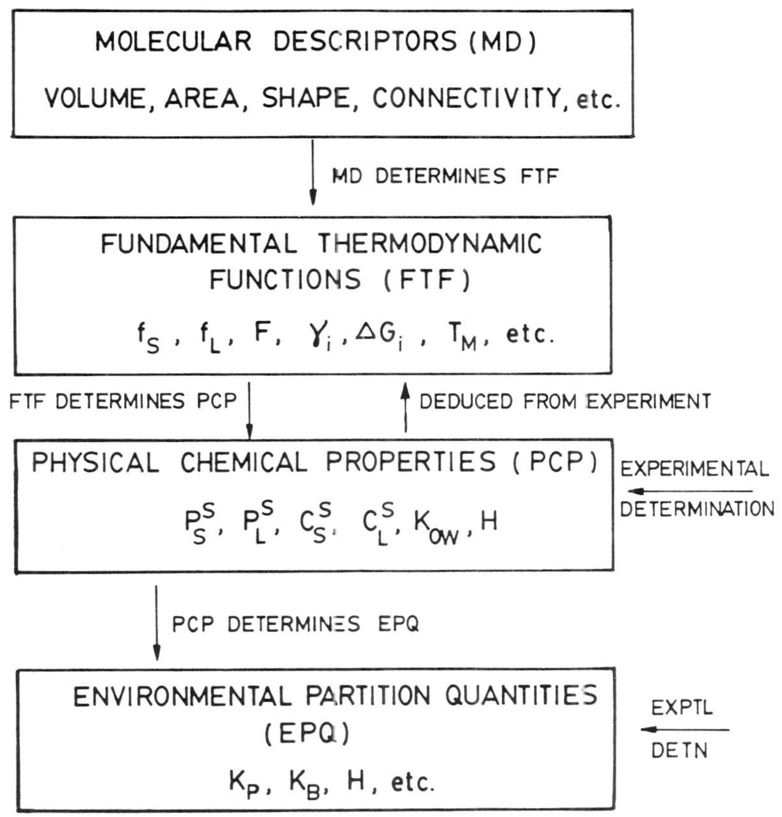

FIGURE 2: Diagram of the four categories of properties and their interdependencies

A second group of descriptors consists of the **fundamental thermodynamic functions** (FTF's) which control equilibrium and are usually not measurable directly, but their values can be estimated from other measurements. These are the following:

i) The fugacity of the pure liquid solute.
ii) The fugacity of the pure solid solute (if relevant).
iii) The excess Gibbs Free Energy or infinite dilution activity coefficients of the solute in various media such as water, lipid phases, octanol or sorbed phases.

These are all functions of temperature and to a negligible extent, pressure, at least under environmental conditions. The temperature dependencies can be characterized by enthalpies appropriate to the various phase transitions. For liquid solutes, in which the environmental temperature exceeds the melting point, the solid fugacity can be ignored. For solid solutes, both are of interest as is their ratio, F (the fugacity ratio).

A third group of descriptors consists of the measurable **physical-chemical parameters** (PCP's) such as vapour pressure, aqueous solubility, octanol/water partition coefficient, Henry's law constant, melting and boiling points, etc. Values of these quantities are controlled by the fundamental thermodynamic functions. In some cases several different measurements can be used to estimate values of the same basic function - which raises the problem that the separate estimates may be conflicting. Most schemes for hazard assessment of chemicals involve measurement of these properties by standard laboratory procedures at selected temperatures.

The fourth and final group is the **environmental partitioning quantities** (EPQ's) which are subject to less precise definition because of media variability, but they are the quantitites needed for environmental modelling purposes. These may be expressed as Z or K values. They may be measured in the real environment, in microcosms, in flasks or in aquaria.

The aim is to establish the linking equation between these four groups so that ultimately the environmental partitioning quantities can be predicted from molecular descriptors. We suggest that this task is best accomplished by approaching it stepwise namely in the sequence MD-FTF-PCP-EPQ.

LINKING EQUATIONS

In a recent report (Mackay _et al_. 1982a) we have set out to justify the relationships between the various quantities. Here, we only present the results.

Molecular Descriptors – Fundamental Thermodynamic Functions

In general it appears that relationships of the form below often apply to homologous series.

$$\ln(\mathrm{FTF}) = A_1 + A_2 \ln(\text{Molecular Size}) = A_1 + A_2 \ln(\mathrm{MD})$$

The "size" term can be volume or area and may be corrected for branching. Fragment constants can be used to characterize increases in molecular size due to addition of, say a chlorine atom or a methyl group. The relationship is only approximate and there are dangers in extrapolation. The (FTF) term may be γ_i or f_L. The term f_S contains a melting point correction while γ_o is apparently relatively constant, thus neither can be expressed by this equation.

Fundamental Thermodynamic Functions – Physical-Chemical Parmeters

The following relationships apply, the symbols being defined in the appended nomenclature section:

$$P_L^S \approx f_L, \quad P_S^S \approx f_S \text{ (vapour pressure and fugacity are equal)}$$

$$C_L^S = 1/v_W\gamma_W, \quad C_S^S = 1/Fv_W\gamma_W$$

$$F = C_S^S/C_L^S = P_S^S/P_L^S$$

$$F = f_S/f_L \approx \exp(-A_3(T_M/T - 1)), \quad A_3 \approx 6.8$$

also,

$$K_{ow} \approx \gamma_W v_W/\gamma_o v_o$$

$$H \approx P_S^S/C_S^S \approx P_L^S/C_L^S = \gamma_W v_W f_L$$

In addition P_S^S and P_L^S can be estimated from boiling point (and if necessary melting point) as discussed by Mackay et al. (1982b).

The important conclusion is that if accurate values are available for the quantities f_L, f_S, γ_W and γ_O, then all the physical-chemical, measurable quantities can be calculated. In practice, PCP measurements are used to estimate the FTF's and there may be conflicting determinations. For example, if P_L^S, C_L^S and H are measured as 100 Pa, 20 mol•m^{-3} and 6 Pa m^3•mol^{-1}, there is obviously a conflict since H must be P_L^S/C_L^S. Often there are several different determinations of each quantity, each with a different perceived accuracy. In some current work we are attempting to apply the maximum likelihood principle to this problem to establish a method by which all available data (at various temperatures) may be combined in a statistically rigorous manner to select the "best" or "most likely" set of FTF's.

It is very important to appreciate the role of the fugacity ratio F, especially as it is a determinant of the solubility of solids. FIGURE 3 shows a common vapour pressure (or fugacity) temperature diagram. This characterizes the partitioning of the chemical between its pure phase (solid or liquid) and vapour. Similar diagrams could be constructed for partitioning into water and other phases.

Above the melting point (or triple point) only the liquid is stable, thus vapour pressure and solubilities always refer to the liquid state, unless accidentally extrapolated. The melting point is determined by the entropic and enthalpic effects which occur when the liquid freezes, or the solid melts. More symmetrical, "ordered" molecules generally form, or remain as, solids more readily and have higher melting points. For example, the linear anthracene melts at 216°C whereas the non-linear phenanthrene melts at 101°C.

Now the solid and liquid vapour pressures (or solubilities) differ by the factor F which depends on the difference in slope of the two lines (effectively the enthalpy of fusion ΔH_F and the departure from the melting point. Indeed,

$$\ln F \approx -\Delta H_F(1/T - 1/T_M)/R \approx (-\Delta H/RT_M)(T_M/T - 1)$$

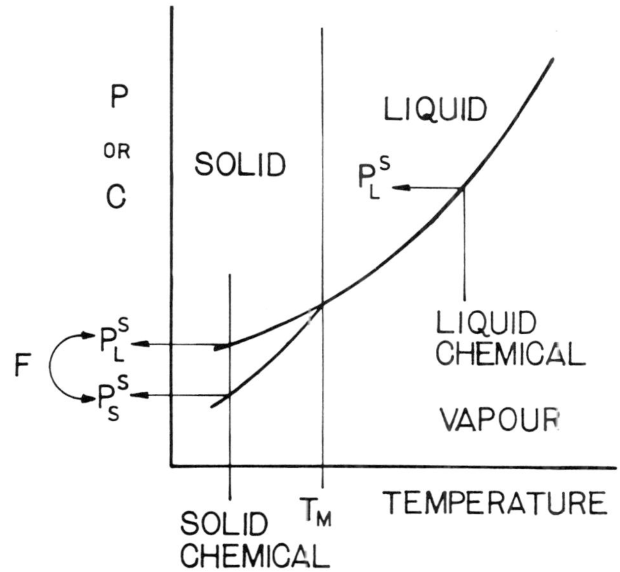

FIGURE 3: Phase diagram illustrating the fugacity ratio F as P_S^S/P_L^S

Solids thus have lower solubilities than corresponding liquids because they are able to form a pure solid phase at a lower concentration. Likewise high melting point solids (e.g., anthracene 0.042 $mg \cdot L^{-1}$) have lower solubilities than low melting point solids (e.g., phenanthrene 1.2 $mg \cdot L^{-1}$).

Particularly important is the relationship between solubility (C_L^S or C_S^S) and octanol/water partition coefficient (K_{ow}). Rearranging the equations suggests that (Mackay _et al_. 1980),

$$C_L^S \approx A_4/K_{ow}$$

where $A_4 \approx$ 1700 to 2300 $mol \cdot m^{-3}$, and

$$C_S^S \approx FA_4/K_{ow}$$

These equations can also be written (for liquids) in the equivalent form:

$$\log C_L^S \approx A_5 - A_6 \log K_{ow}$$

in which we would expect A_6 to be close to unity. Correlations of this type have been devised by Chiou et al. (1977, 1982) and Yalkowsky and Valvani (1980). In principle, we believe that A_6 should be forced to be unity, thus eliminating one constant; and if a second constant is desired it can be used to "correct" or modify A_4. Since C_S^S and C_L^S are related through F, and F can be related to melting point, a different equation can be written for solids of the form,

$$\log C_S^S = A_7 + A_8 T_M - A_6 \log K_{ow}$$

It is critically important to include F or the melting point correction for solids. The simple $C_L^S - K_{ow}$ equation applies only to liquids.

Inspection of the properties of homologous series show that just as the FTF values vary systematically with molecular structure, so too do their dependent properties such as C_L^S, P_L^S, H, and of course, K_{ow}. Interestingly, substitution of a chlorine or a methyl for an aromatic hydrogen tends to reduce C_L^S and P_L^S by a factor of 3 (approximately) thus K_{ow} increases by this factor or log K_{ow} increases by log 3 or 0.5, but H remains relatively constant. For a series of congeners such as the chlorobenzenes or chlorobiphenyls it is possible to infer values of the various properties from congeners to congeners and build up a "consistent" set of FTF, which apply to the entire series. This is particularly valuable when the number of congeners is large, as in the case of the chlorinated biphenyls, dibenzodioxins, dibenzofurans and naphthalenes. Erroneous data are fairly easily identified and considerable economies of effort are possible. We are currently attempting to formalize this procedure using the PCB's as a case study and have published a first attempt for the lower congeners (Mackay et al. 1982c).

Physical-Chemical Parameters - Environmental Partitioning Quantities

The Henry's law constant H, or in dimensionless form H/RT or K_{AW}, is both a physical-chemical and environmental property.

The organic carbon partition coefficient K_{oc}, which is often defined as K_P/ϕ where ϕ is the mass fraction organic carbon, has been shown to be related to K_{ow} by Karickhoff _et al._ (1979) through the simple relationship:

$$K_{oc} \approx 0.41\ K_{ow} \approx 0.41\ (1795/C_L^S) \approx 800/C_L^S$$

The last two terms involv e substitution of the $C_L^S - K_{ow}$ equation suggested by Mackay _et al._ (1980). Interestingly, this equation can be interpreted that the "effective solubility" of a hydrophobic organic solute in organic carbon is approximtely 800 $mol{\cdot}m^{-3}$.

We have recently suggested (Mackay 1982) a similar relationship for bioconcentration factor K_B for fish, namely,

$$K_B \approx 0.048\ K_{ow} \approx 92/C_L^S$$

Thus the "effective solubility" of a hydrophobic organic solute in a fish is 92 $mol{\cdot}m^{-3}$.

CONCLUSIONS

These relationships can also be used to estimate the EPQ's, (i.e., the fugacity capacities (Z)), or partition coefficients from the PCP's or the FTF's, or in some cases all the way from MD (molecular descriptor). It is thus apparent that the "journey" from molecular descriptor to partition coefficient or Z is a long one and it is presently difficult or even impossible to accomplish it in one step with satisfactory accuracy. By breaking the "journey" into a series of shorter steps and devoting efforts into making each smaller step more accurate it should be possible to facilitate the entire "journey" and bring closer the ultimate goal of calculating the environmental partitioning behaviour of chemicals more easily, accurately and economically with only a knowledge of molecular structure.

Finally, if this task can be accomplished it provides a sound foundation for tackling the more complex problem of relating toxicity to molecular descriptors through the intermediate quantities such as octanol/water partition coefficient. It appears from the trends described elsewhere in this volume that many manifestations of toxicity are at least in part controlled by water/organism partitioning, which should be predictable in a manner similar to bioconcentration factor. By quantifying this process it should become possible to identify and focus more clearly the subsequent and more complex biochemical phenonomena which ultimately result in toxicity.

GLOSSARY

Symbol	Meaning
A_i	A series of constants
C	Solubility ($mol \cdot m^{-3}$)
f	Fugacity (Pa)
F	Fugacity ratio (dimensionless)
H	Henry's law constant ($Pa \cdot m^3 \cdot mol^{-1}$)
ΔH_F	Enthalpy of fusion ($J \cdot mol^{-1}$)
K_{ow}	Octanol/water partition coefficient
K_{aw}	Air/water partition coefficient
K_B	Bioconcentration factor
K_P	Sorption partition coefficient
K_{oc}	Organic carbon partition coefficient
P^S	Vapour pressure
R	Gas constant ($8.314\ J \cdot mol^{-1} \cdot K^{-1}$)
T	Absolute temperature (K)
v	Molar volume ($m^3 \cdot mol^{-1}$)
Z	Fugacity capacity
γ	Activity coefficient
ϕ	Mass fraction organic carbon

Subscript	Meaning
L	Liquid
S	Solid
M	Melting
W	Water
O	Octanol
A	Air

REFERENCES

Bobra, A.M, Shiu, W.Y. and Mackay, D. 1983. Toxicity of hydrocarbons and crude oils. Proceedings of the **6th Arctic Marine Oil Spill Program Technical Seminar,** Edmonton, Alberta, pp. 52-58.

Chiou, C.T., Freed, V.H., Schmedding, D.W. and Kohnert, R.L. 1977. Partition coefficient and bioaccumualtion of selected organic chemicals. **Environ. Sci. Technol. 11:** 475-478.

Chiou, C.T. Schmedding, D.W. and Manes, M. 1982. Partitioning of organic compounds in octanol/water systems. **Environ. Sci. Technol. 16:** 4-10

Dickson, K.L., Maki, A.W. and Cairns, J., Eds. 1982. **Modelling the fate of chemicals in the aquatic environment.** Ann Arbor Science, Ann Arbor, Michigan.

Karickhoff, S.W., Brown, D.S. and Scott, T.A. 1979. Sorption of hydrophobic pollutants on natural sediments. **Water Res. 13:** 241-248.

Mackay, D. and Shiu, W.Y. 1977. Aqueous solubility of polynuclear aromatic hydrocarbons. **J. Chem. Eng. Data 22:** 399-402.

Mackay, D. 1979. Finding fugacity feasible. **Environ. Sci. Technol. 13:** 1219-1223.

Mackay, D., Bobra, A., Shiu, W.Y. and Yalkowsky, S.H. 1980. Relationship between aqueous solubility and octanol/water partition coefficients. **Chemosphere 9:** 701-711.

Mackay, D. and Paterson, S. 1981. Calculating fugacity. **Environ. Sci. Technol. 15:** 1006-1014.

Mackay, D. and Paterson, S. 1982. Fugacity revisited. **Environ. Sci. Technol. 16:** 654A-660A.

Mackay, D., Shiu, W.Y., Bobra, A., Billington, J., Chau, E., Yuen, A., Ng, C. and Szeto, F. 1982a. Volatilization of organic pollutants from water. **U.S. Environmental Protection Agency,** EPA-600/S3-82-019.

Mackay, D., Bobra, A., Chau, D.W. and Shiu, W.Y. 1982b. Vapor pressure correlations for low-volatility environmental chemicals. **Environ. Sci. Technol. 16:** 645-649.

Mackay, D., Shiu, W.Y., Billington, J. and Huang, G.L. 1982c. Physical-chemical properties of polychlorinated biphenyls. In **Physical Behaviour of PCB's,** D. Mackay, S. Paterson, S.J. Eisenreich and M.S. Simmons, (Eds.), Ann Arbor Science, pp. 59-69.

Mackay, D. 1982. Correlation of bioconcentration factors. **Environ. Sci. Technol. 16:** 274-278.

Mackay, D., Paterson, S. and Joy, M. 1982a. Application of fugacity models to the estimation of chemical distribution and persistence in the environment. In **Fate of Chemicals in the Environment,** R.L. Swann and A. Eschenroeder (Eds.), ACS Symposium Series 225, Washington, D.C., 1983, pp. 175-196.

Mackay, D. Joy, M. and Paterson, S. 1983b. A quantitative water, air, sediment interaction (QWASI) fugacity model for describing the fate of chemicals in lakes. **Chemosphere 12:** 981-997.

Mackay, D., Paterson, S. and Joy, M. 1983c. A quantitative water, air, sediment interaction (QWASI) fugacity model for describing the fate of chemicals in rivers. **Chemosphere 12:** 981-997.

Mackay, D. and Hughes, A.I. 1983. A three parameter equation describing the uptake of organic compounds by fish. **Environ. Sci. Technol.,** in press.

Yalkowsky, S.H. and Valvani, S.C. 1979. Solubility and partioning 2. Relationship between aqueous solubilities, partition coefficients and molecular surface areas of rigid aromatic compounds. **J. Chem. Eng. Data 24:** 127-129.

Yalkowsky, S.H. and Valvani, S.C. 1980. Solubility and partitioning I. Solubility of nonelectrolytes in water. **J. Pharm. Sci. 69:** 912-922.

WORKSHOP DISCUSSION

SCHULTZ:

In the second-to-last graph, you mentioned water exposure versus the uptake time. Are they functions of the chemical that you are testing and also of the system you are testing it in? In other words, would the uptake time be different for Daphnia, for a 2 g trout, a 20 g trout, or the fingerlings, etc?

MACKAY:

I believe that if you measure the concentration in the organism, as a function of type, when it is in a solution of constant concentration and no volatilisation, adsorption or metabolism occurs, the concentration will tend to rise and will reach an equilibrium value in time. Now, if the exposure time of the organism is relatively long, the uptake period can be neglected as the organism is experiencing this concentration for most of the time. If the exposure time is much shorter, then the exposure may be reduced because of the kinetic period. I think the steepness of the curve or the time of approach to equilibrium depends upon the chemical and particularly on its K_{ow}. It also depends upon the organism's size and the transport resistance between outside and inside the organism, through the gills, through food or through an injection. The beauty of the injection is that the chemical is conveyed directly to the site.

Question:

Does it not also depend upon the concentration that you are using because if you start using low concentrations does that not affect the driving force into the animal and affect the uptake time?

MACKAY:

I would say not. Concentration, in my opinion, does affect the driving force into the animal but not the rate of approach to the higher equilibrium. Lower concentrations will take an equal time but they have not got to go so far so the uptake rate is reduced because the water concentration is reduced but also the final destination is closer, so these cancel each other; Barry Oliver, however, doesn't agree with that.

Comment:

Your uptake time as it is defined here, is independent of all the processes conveying materials into the animal. You can calculate the uptake time as, say three or four times the biological halftime which is a clearance function. It is a function of the rate of loss, not the rate of uptake. The level within the organism is determined by the uptake process. If the interaction results in a loss, the time to reach that state is a function of clearance.

MACKAY:

I don't agree with you but I don't believe that it is a point worth arguing over. It is a matter of semantics what we mean by uptake and rate constants. You are right, insofar as when you shut off the exposure, you can work out a K_2 from the drop in concentration.

Comment:

Working from the K_2 you can calculate your time t_u.

MACKAY:

Yes, I agree.

Comment:

You can't calculate time t_u from your K_2 uptake.

MACKAY:

Right. I see your point. We are talking about semantics.

HALLETT:

What happens when you take the extremes, like TCDD compared to phenol or chlorobenzene, or in a more complex organism? The solubility of TCDD is fairly low and it would be interesting to see tests with a compound of extreme properties.

MACKAY:

I would argue that if someone is going to do TCDD toxicity tests they must be sure what some of its properties are. There is a controversy about its solubility. There is a danger that if the concentration has toxic effects on the fish then the whole model breaks down.

HALLETT:

This is something I think we should consider with molecules like that, its basic chemical properties and so forth.

DUNN:

You say that concentration in the organism is to a first approximation, a function of the partition coefficient, when does this break down and what are the exceptions?

MACKAY:

Well, what we find, is that solubility times K_{ow} is about 2,000 for a number of hydrophobic organics. There is another number for nitro compounds, another for phenols, another for alcohols, etc. Another area where it tends to break down is in the extreme regions where you have very high K_{ow}'s and very low solubility. We really don't know enough about them. Strange effects can occur with very high solubilities of water in the substance, especially alcohols. This tends to alter the thermodynamics. There are fuzzy regions at high and low solubilities and we must be very careful about extrapolating into these.

KAISER:

One could argue here that the problem is not with the octanol but with the water; we should have water with different properties.

LANDE:

A comment on the limits of the water partitioning, that is where the solute is either very hydrophilic or lipophobic. The octanol/water partition coefficient correlation with solubility is based on the assumption that the activities of the solutes in octanol are always constant. As the solutes become lipophobic, their activities will not remain constant and the correlation is no longer valid. It can be corrected by involving another term, using Hildebrand's solubility parameter as discussed by Amidon and Williams (Int. J. Pharm. 11: 249-256 [1982]). As far as the high solubility region, it is probably influenced by interactions with portions of the molecule that are hydrophilic, because of, for example, hydrogen bonding and so forth.

MACKAY:

You have to bear in mind that this region of low K_{ow} is of less interest to us because they are hydrophilic substances and do not tend to bioconcentrate.

Comment:

If we look at the log K_{ow} of 6 and we use it to predict LC50, which you really don't agree with, how would you explain Veith's work on toxicity versus K_{ow} which shows the same slope for different species, fathead minnows and all the different exposure periods?

MACKAY:

I believe that Veith's data, in fact, suggest that you are running into kinetic limitations at a K_{ow} of 5 or 6.

VALIDATION OF FISH TOXICITY QSARs FOR CERTAIN NON-REACTIVE NON-ELECTROLYTE ORGANIC COMPOUNDS

Larry D. Newsome, Robert L. Lipnick* and David E. Johnson

Office of Toxic Substances
U.S. Environmental Protection Agency
Washington, D.C. 20460
U.S.A.

ABSTRACT

A comparison was made of the predicted fish LC50 values for five classes of non-reactive non-electrolyte herbicides with measured data. The predictions were based upon the log P parameter and the QSAR equation of Könemann. The required log P values were obtained from experimental measurements, or else, calculated directly from chemical structures using the CLOGP computer program. The analysis was performed on a total of 37 chemicals and 49 fish LC50 tests. Several herbicides were considerably less toxic to fish than predicted by the QSAR equation. These results were interpreted in terms of limiting water solubility, even in the presence of organic carrier solvents employed in the fish toxicity testing.

INTRODUCTION

The Environmental Effects Branch of the Office of Toxic Substances is responsible for providing initial hazard assessments for

* Address correspondence to this author.

K. L. E. Kaiser (ed.), QSAR in Environmental Toxicology, 279–299.

new chemicals prior to their manufacture under Section 5 of the Toxic Substances Control Act (TSCA). In the absence of submitted test data, toxicity to fish is estimated based upon structure activity relationships (SARs) or quantitative structure activity relationships (QSASs).

It has been shown that the acute toxicity to several species of freshwater fishes of certain non-reactive, non-electrolyte organic chemicals is well correlated by QSAR equations based solely upon the log P parameter (Lipnick and Dunn 1983; Veith et al. 1983; Könemann 1981). Since these equations exhibit little variation in their slopes and intercepts, acute toxicity data for a variety of freshwater fishes may be treated in a combined fashion, as a first approximation, for QSAR analysis (Hermans 1983). Toxicity of this type, referred to as narcotic action, non-specific action, or physical toxicity, represents the adverse effect induced in organisms due to the presence of a minimum molar concentration of the toxicant in the biophase, and corresponds mechanistically to the same mode of action as induced by gaseous anesthetics and hypnotic drugs (Franks and Lieb 1982). Since theory predicts that such action represents baseline toxicity (Albert 1979), non-electrolytes are expected to be at least as toxic as predicted by such QSAR equations. Those non-electrolytes which contain specific pharmacophore groups or electrophilic centers may exhibit enhanced molar toxicity due to reversible interactions or the formation of permanent covalent bonds with critical biological macromolecules. For such narcotic type chemicals, a cutoff in toxicity, or no observed acute lethality, is observed beyond the point at which the LC50 exceeds the water solubility of the test solute. This phenomenon, the "Ferguson cutoff" (Ferguson 1939) takes place at log P $\approx$ 6 for liquids. In the case of solids, this cutoff occurs at lower log P values which diminish with increasing melting point (Banerjee et al. 1980). The appearance of this cutoff is expected to be extended partially by the addition of organic carrier solvents commonly employed by aquatic toxicologists in performing such tests on highly insoluble substances.

The study which we report herein was undertaken to explore the predictive limitations of such baseline fish toxicity QSAR equations for a set of more complex non-electrolytes containing multiple functional groups.

DATA AND METHODS

Fish toxicity and physicochemical property data on five classes of non-electrolyte herbicides were obtained from a recent monograph by Hisada and Imamura (1982). However, the reliability of these data could not be evaluated since no references to primary sources were provided by the authors. These classes consist of phenoxys, amides, toluidines, ureas, and diphenyl ethers. Those herbicides belonging to the toluidine class are very weak electrolytes. An analogous substance, 2,4-dinitro-N,N-diethylaniline, has a reported measured pKa of 0.21 (Perrin 1972), which corresponds to less than 10^{-4}% ionization in aqueous solution at pH 7 (Albert 1979).

While these herbicides may contain pharmacophores designed to exhibit specific toxic modes of action in plants, this same specificity may not be expressed in fishes in which there exist large biochemical differences. Chemicals within these groups were selected based upon availability of fish toxicity data, absence of chemically reactive functional groups, and the ability to assign a log P value. The chemicals selected are shown in FIGURES 1 to 5, and TABLE 1.

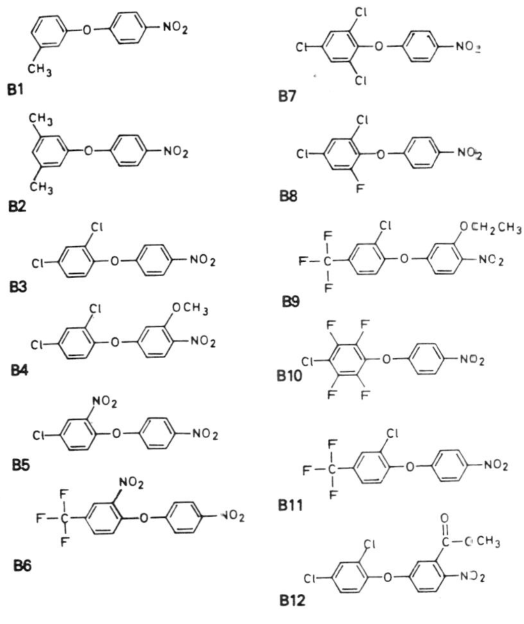

FIGURE 1: Structures and symbols of biphenyl ethers (B) investigated

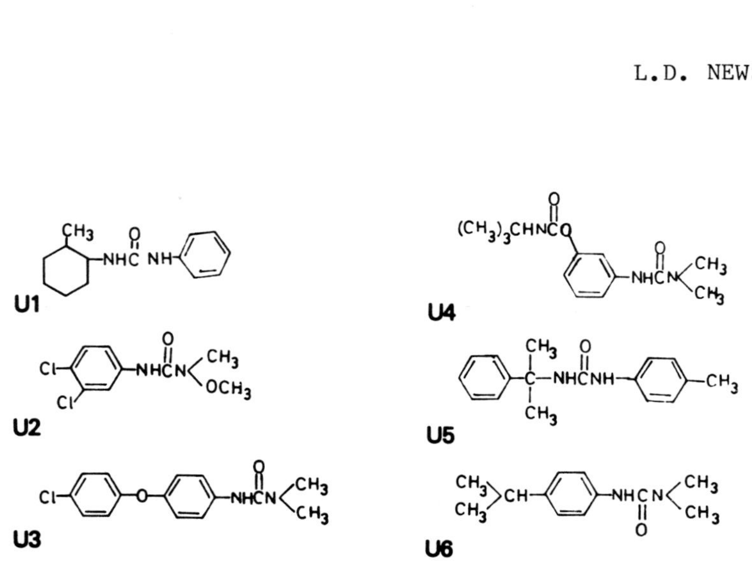

FIGURE 2: Structures and symbols of urea derivatives (U) investigated

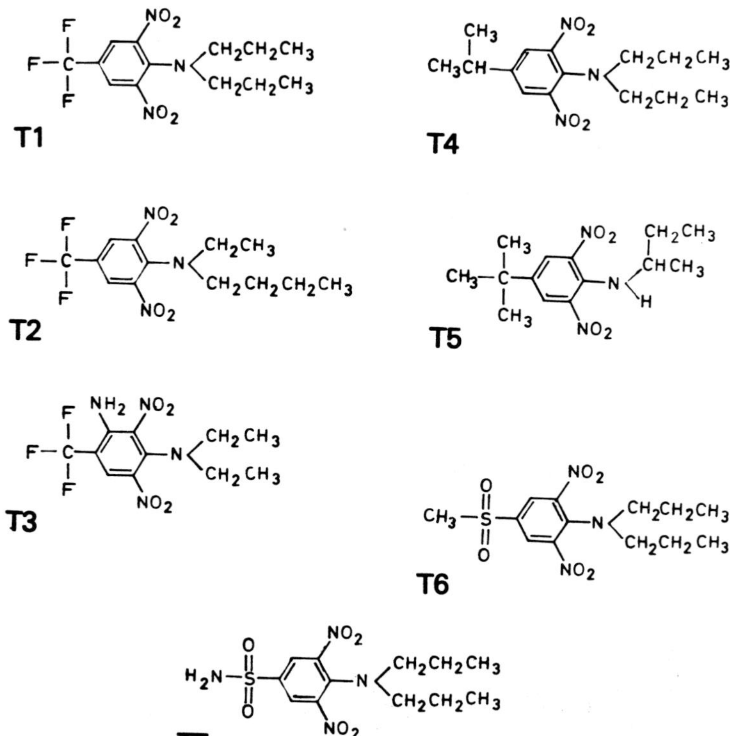

FIGURE 3: Structures and symbols of toluidine derivatives (T) investigated

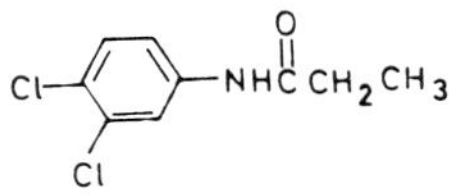

A1

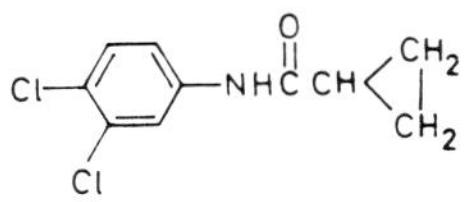

A2

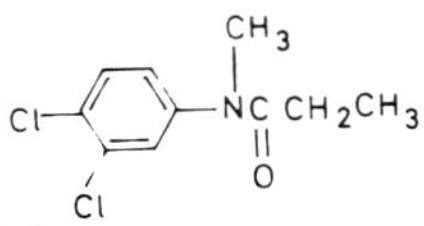

A3

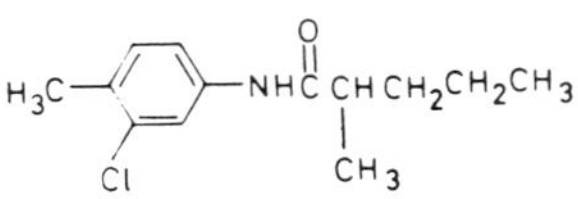

A4

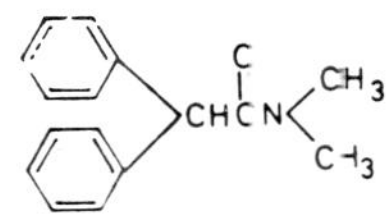

A5

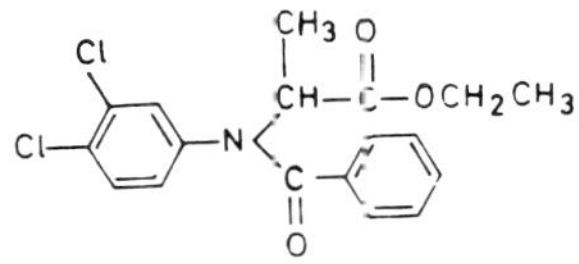

A6

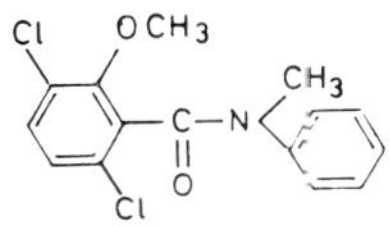

A7

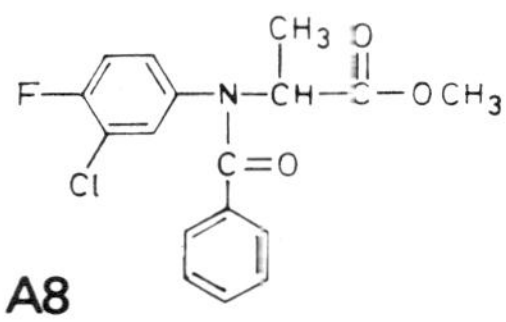

A8

FIGURE 4: Structures and symbols of amides (A) investigated

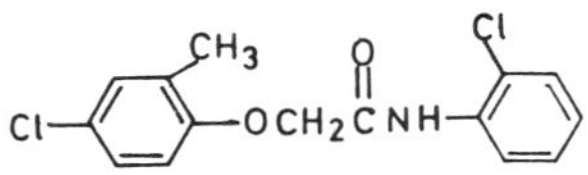

P2

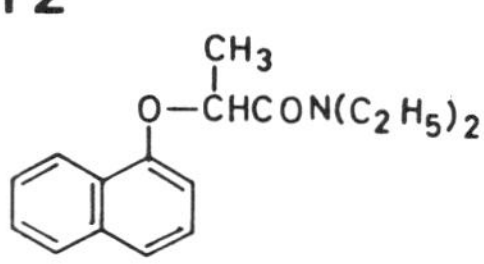

P3

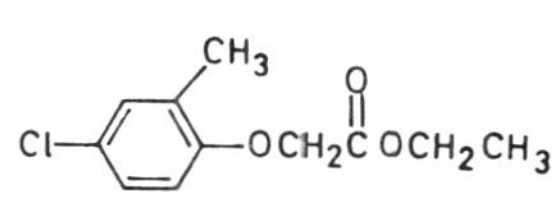

P4

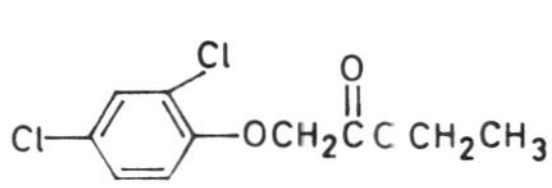

P5

FIGURE 5: Structures and symbols of phenoxy derivatives (P) investigated

TABLE 1: Physicochemical properties and toxicity data for herbicide study[a]

Study Number	Chemical Name and Formula	CAS RN (Hodagaya Number)	MW	Calculated[b] log P	MP°C	Measured[c] (Calculated) Water Sol. $mg \cdot L^{-1}$	Measured[d,f] LC50 $mg \cdot L^{-1}$ (Fish Species)	Calculated[g] LC50 $mg \cdot L^{-1}$	Measured log(1/LC50) $micromole \cdot L^{-1}$
B-1	3-methylphenyl-4-nitrophenyl ether (TOPE) $C_{13}H_{11}NO_3$	2303255 11-1	229.24	4.63#	63	5 2.5	1.2 Carp	1.58	-0.719
B-2	3,5-dimethylphenyl-4-nitrophenyl ether (farmaid) $C_{14}H_{13}NO_3$	1630177 11-2	243.26	5.30#	78-80	1.0 0.26	5 Carp	0.44	-1.313
B-3	2,4-dichlorophenyl-4-nitrophenyl ether (nitrofen) $C_{12}H_7Cl_2NO_3$	1836755 11-3	284.10	5.40# 5.37+	70-1	0.7-1.2 0.35	2.5 Carp	0.44	-0.944
B-4	2,4-dichlorophenyl-4-nitro-3-methoxyphenyl ether (chlormethoxynil) $C_{13}H_9Cl_2NO_4$	32861851 11-4	314.13	*5.72#	113-4	0.3 0.03	237 Carp	0.24	-2.878
B-5	2-nitro-4-chlorophenyl 4-nitrophenyl ether (KK-60) $C_{12}H_7ClN_2O_5$	20115348 11-7	294.65	4.43#	97-8	<1 1.45	2.2 Carp	3.03	-0.873
B-6	4-nitrophenyl-2-nitro-4-trifluoromethylphenyl ether (fluorodifen) $C_{13}H_7F_3N_2O_5$	15457053 11-9	328.21	4.61#	94- 94.5	2 1.17	0.8 Carp	2.35	-0.387
B-6[a]							0.18 Trout		0.261
B-7	2,4,6-trichlorophenyl 4-nitrophenyl ether (chlornitrofen) $C_{12}H_6Cl_3NO_3$	1836777 11-10	318.55	6.11#	107	-- 0.01	>40 Carp	0.11	-2.10
B-8	2,4-dichloro-6-fluoro-phenyl-4-nitrophenyl ether (fluoronitrofen) $C_{12}H_6Cl_2FNO_3$	13738631 11-11	302.09	5.54#	67.1- 67.9	0.66 0.27	0.5 Carp	0.33	-0.219

TABLE 1: Cont'd

Study Number	Chemical Name and Formula	CAS RN (Hodagaya Number)	MW	Calculated[b] log P	MP°C	Measured[c] (Calculated) Water Sol. $mg \cdot L^{-1}$	Measured[d,f] LC50 $mg \cdot L^{-1}$ (Fish Species)	Calculated[g] LC50 $mg \cdot L^{-1}$	Measured log(1/LC50) $micromole \cdot L^{-1}$
B-9	2-chloro-4-trifluoromethylphenyl 3-ethoxy-	42874033 11-12	361.71	*6.04#	84-5	<0.1 0.05	0.2 Bluegill	0.15	-0.257
B-9[a]	4-nitrophenyl ether (oxyfluorofen) $C_{15}H_{11}ClF_3NO_4$						0.26 Trout		0.143
B-10	4-chloro-2,3,5,6-tetrafluorophenyl 4-nitrophenyl ether (HW-24) $C_{12}H_4ClF_4NO_3$	52418083 11-13	321.61	5.26#	80	-- 0.36	10-20 Killifish	0.62	-1.493 to -1.794
B-11	2-chloro-4-trifluoromethylphenyl-4-nitrophenyl ether (nitrofluorfen) $C_{13}H_7ClF_3NO_3$	42874011 11-14	317.66	5.58#	70-1	0.3 0.23	0.8 Carp	0.32	-0.401
B-12	2,4-dichlorophenyl-3-methoxycarbonyl-4-nitrophenyl ether (bifenox) $C_{14}H_9Cl_2NO_5$	42576023 11-15	342.14	5.58#	84-6	0.35 0.14	3 Carp	0.35	-0.943
U-1	1-(2-methylcyclohexyl)-3-phenylurea (siduron)	1982496 5-11	232.33	3.81# 3.80[+]	133-8	18 1.44	50 Trout	8.44	-2.333
U-1[a]	$C_{14}H_{20}N_2O$						50 Salmon		-2.333
U-2	3-(3,4-dichlorophenyl)-1-methoxy-1-methylurea (linuron) $C_9H_{10}Cl_2N_2O_2$	330552 5-21	249.10	2.96[+]	93-4	75 64.0	10 Carp	48.8	-1.600
U-3	3-[4-(4-chlorophenoxy) phenyl]-1,1-dimethylurea (chloroxuron) $C_{15}H_{15}ClN_2O_2$	1982474 5-32	290.75	3.96# 3.94[+]	151-2	3.7 0.62	15 Killifish	7.98	-1.713

TABLE 1: Cont'd

Study Number	Chemical Name and Formula	CAS RN (Hodagaya Number)	MW	Calculated[b] log P	MP°C	Measured[c] (Calculated) Water Sol. $mg{\cdot}L^{-1}$	Measured[d,f] LC50 $mg{\cdot}L^{-1}$ (Fish Species)	Calculated[g] LC50 $mg{\cdot}L^{-1}$	Measured log(1/LC50) $micromole{\cdot}L^{-1}$
U-4	3-[3-(N-tert-butyl-carbamoyloxy) phenyl]-	4849325 5-34	279.34	1.51# 1.73∞ 1.66+	176-6.5	325 69.0	>75 Bluegill	644	-2.429
U-4[a]	1,1-dimethylurea (karbutilate) $C_{14}H_{21}N_3O_3$						>135 Trout		-2.684
U-5	1-(α,α-dimethyl-benzyl)-3-(4-methyl-phenyl) urea (dymron) $C_{17}H_{20}N_2O$	42609529 5-37	268.36	4.05#	203	1.7 0.06	300 Carp	5.9	-3.048
U-6	3-(4-isopropylphenyl)-1,1-dimethylurea (isoproturon)	34123596 5-41	206.29	2.77#	158-9	55 6.96	31 Trout	59.1	-2.177
U-6[a]	$C_{12}H_{18}N_2O$						96h 100 Carp		-2.686
U-6[b]							90 Guppy		-2.640
T-1	α,α,α-trifluoro-2,6-dinitro-N,N-dipropyl-p-toluidine (trifluralin) $C_{13}H_{16}F_3N_3O_4$	1582098 4-1	335.29	*4.82# 5.05+ 5.34∞	48.5-9	1 1.03	96h 0.089 Bluegill	0.55	0.571
T-2	α,α,α-trifluoro-2,6-dinitro-N-butyl-N-ethyl-p-toluidine (benfluralin) $C_{13}H_{16}F_3N_3O_4$	1861401 4-2	335.29	*4.82# 4.54+	65-6.5	1 4.29	96h 0.37 Bluegill	2.76	-0.043
T-3	N^3,N^3-diethyl-2,4-dinitro-6-trifluoro-	29091052 4-9	322.25	*3.45# 3.86+	98-9	1 6.65	EC[e] 6.7 Trout	10.4	-1.318
T-3[a]	methyl-m-phenylene-diamine (dinitramin)						11.0 Bluegill		-1.533
T-3[b]	$C_{11}H_{13}F_3N_4O_4$						3.7 Catfish		-1.060

TABLE 1: Cont'd

Study Number	Chemical Name and Formula	CAS RN (Hodagaya Number)	MW	Calculated[b] log P	MP°C	Measured[c] (Calculated) Water Sol. mg•L^{-1}	Measured[d,f] LC50 mg•L^{-1} (Fish Species)	Calculated[g] LC50 mg•L^{-1}	Measured log(1/LC50) micromole•L^{-1}
T-4	4-isopropyl-2,6-dinitro-N,N-dipropylaniline (isopropaline) $C_{15}H_{23}N_3O_4$	33820530 4-12	309.37	*5.54#	--	0.5 --	96h>0.15 Goldfish	0.34	0.314
T-4[a]							96h>0.1 Fathead Minnow		0.490
T-5	2,6-dinitro-N-sec-butyl-4-tert-butylaniline (butralin) $C_{14}H_{21}N_3O_4$	33629479 4-13	295.34	5.46# 5.46+ 5.96+	59-60	188 0.44	4.2 Bluegill	0.38	-1.153
T-5[a]							3.4 Trout		-1.061
T-6	4-(methylsulphonyl)-2,6-dinitro-N,N-dipropylaniline (nitralin) $C_{13}H_{19}N_3O_6S$	4726141 4-14	345.37	*2.30#	151-2	0.6 51.68	96h 46-68 Fish	254	-2.124 to -2.294
T-7	3,5-dinitro-N^4,N^4-dipropylsulphanilamide (oryzalin) $C_{12}H_{18}N_4O_6S$	19044883 4-17	346.36	*2.66#	141-2	2.4 30.0	96h>1.4 Goldfish	123	-0.607
A-1	3',4'-dichloropropionanilide (propanil) $C_9H_9Cl_2NO$	709988 3-4	218.08	3.14# 3.12+	85-9	225 50.6	13 Carp	31.08	-1.775
A-1[a]							11 Killifish		-1.703
A-1[b]							14 Goldfish		-1.808
A-2	3',4'-dichlorocyclopropanecarboxanilide (cypromid) $C_{10}H_9Cl_2NO$	2759719 3-6	230.10	3.06#	131	0.01% 10.5	15 Bluegill	36.9	-1.814
A-3	N-methyl-3',4'-dichloropropionanilide $C_{10}H_{11}Cl_2NO$	15533810 3-9	232.11	3.22#	39	-- 248	6.2 Killifish	27.0	-1.427

TABLE 1: Cont'd

Study Number	Chemical Name and Formula	CAS RN (Hodagaya Number)	MW	Calculated[b] log P	MP°C	Measured[c] (Calculated) Water Sol. $mg \cdot L^{-1}$	Measured[d,f] LC50 $mg \cdot L^{-1}$ (Fish Species)	Calculated[g] LC50 $mg \cdot L^{-1}$	Measured log(1/LC50) $micromole \cdot L^{-1}$
A-4	N-(3-chloro-4-methyl-phenyl)-2-methyl-pentanamide (solan) $C_{13}H_{18}ClNO$	2307688 3-12	239.75	4.49# 4.56+	82-6	8-9 1.50	1.8 Carp	1.89	-0.876
A-5	N,N-dimethyl-2,2-diphenyl-acetamide (diphenamide) $C_{16}H_{17}NO$	957517 3-25	239.32	2.23# 2.17+	134-5	260 96.8	40 Carp	229	-2.223
A-6	ethyl-N-benzoyl-N-(3,4-dichlorophenyl)-2-aminopropionate (benzoylpropethyl) $C_{18}H_{17}Cl_2NO_3$	33878501 3-39	366.25	*5.19# 5.26+	70-1	20 0.60	3-10 Guppy	0.71	-0.913 to -1.436
A-7	N-methyl-2-methoxy-3,6-dichlorobenz-anilide $C_{15}H_{13}Cl_2NO_2$	4659023 3-41	310.18	3.90#	--	-- --	18 Killifish	9.22	-1.764
A-8	methyl-N-benzoyl-N-(3-chloro-4-fluoro-phenyl)-2-amino-propionate (flamprop-methyl) $C_{17}H_{15}ClFNO_3$	57973667 3-43	335.77	*4.08#	81-2	35 7.58	96h 4.7 Trout	6.96	-1.146
P-2	N-(2-chlorophenyl)-2-(2-methyl-4-chloro-phenoxy) acetamide (mapica) $C_{15}H_{13}Cl_2NO_2$	2453965 1-28	310.18	*4.98# 4.98+	111-3	3.0 0.21	0.42 Carp	1.06	-0.132
P-3 P-3[a]	2-(1-naphthoxy)-N,N-diethylpropionamide (napropamid) $C_{17}H_{21}NO_2$	15299997 1-31	271.36	*4.31#	74.8- 75.5	73.0 4.30	3.2 Carp	3.54	-1.072

TABLE 1: Cont'd

Study Number	Chemical Name and Formula	CAS RN (Hodagaya Number)	MW	Calculated[b] log P	MP°C	Measured[c] (Calculated) Water Sol. $mg \cdot L^{-1}$	Measured[d] LC50 $mg \cdot L^{-1}$ (Fish Species)[f]	Calculated[g] LC50 $mg \cdot L^{-1}$	Measured log(1/LC50) $micromole \cdot L^{-1}$
P-4	(2-methyl-4-chloro-phenoxy) acetic acid, ethyl ester (MCPA) $C_{11}H_{13}ClO_3$	2698386 1-3	228.68	*3.80#	--	-- --	14 Carp	8.31	-1.787
P-5	(2,4-dichlorophenoxy) acetic acid, ethyl ester (2,4-D) $C_{10}H_{10}Cl_2O_3$	533233 1-1	249.10	*3.85#	--	-- --	1 Carp	8.19	-0.604

a Chemical names, melting points (MP), Chemical Abstracts Service registry number (CAS RN), water solubilities, and fish toxicity data, from Hisada and Imamura (1982). **The CAS RN were obtained from the Eigth and Ninth Decennial Indices of Chemical Abstracts.** Unless otherwise specified, the substance tested was the technical and not formulated product. The common names are given in parentheses.

b Using the CLOGP computer program, calculated log P values were obtained via timesharing from the following sources:

\# SFILES interpreter at EPA National Computer Center, Research Triangle Park, N.C.

\+ ISHOW Data Base (Magnuson et al. 1981); calculated by A. Leo, Pomona College.

* An additional hand correction was applied, based upon the fragment constant methodology (Hansch and Leo 1979). These corrections consisted of the following, in which the CLOGP value (in parentheses) is compared with the corrected value:

(i) -H- polar proximity correction $Fpl = -0.42 (F_1 + F_2)$

A-8: 4.08 (3.80) P-4: 3.80 (3.63)
A-6: 5.28 (4.91) P-3: 4.31 (4.02)
P-5: 3.85 (3.68) P-2: 4.98 (4.81)

(ii) - No value for fx^1-O- The corrected log P values are:

B-4: 5.72 (5.94)
B-9: 6.04 (6.66)

(iii) $-F_{byn}$ for t-amines not computed. The rule recommended by the CLOGP directive (if no polar group is present, add (n-1)x(-0.08)+0.13, where n=total of bonds), was applied.

T-1: 4.82 (5.17) T-4: 5.54 (5.89)
T-2: 4.82 (5.17) T-6: 2.30 (2.65)
T-3: 3.45 (3.64) T-7: 2.66 (3.01)

∞ Measured log P value from ISHOW.

c Water solubility was estimated using equation 2.

d Unless otherwise specified, measured fish toxicity is a 48-hour TLm (median tolerated limit) which is equivalent to the LC50.

e Emulsifiable concentrate (EC) tested.

f Scientific species name not provided in reference (Hisada and Imamura 1982).

g Calculated LC50 derived using equation 1.

RESULTS AND DISCUSSION

All study chemicals were searched initially in the ISHOW data base (Magnuson and Harriss 1977) for measured or calculated log P values. In addition, values were calculated based upon the fragment constant methodology (Hansch and Leo 1979) using CLOGP (Chou and Jurs 1979, 1980). A version of CLOGP installed on a Vax 11/780 computer at the EPA National Computer Centre was used via timesharing for this study. Chemical structures were input via a driver program (SFILES) that permits cursor drawing on a Tektronix terminal. For most chemicals there was excellent agreement between the ISHOW values and those calculated using CLOGP. Preference for use in this study was given to measured or previously estimated values of log P. For each herbicide, the preferred log P value was used to predict fish toxicity, based upon equation 1. This QSAR was derived by Könemann (1981) from test data for 50 narcotic chemicals tested on the guppy:

$$\log 1/LC50 = -4.87 + 0.87 \log P \qquad (1)$$
$$n = 50;\ r^2 = 0.98;\ s = 0.237$$

where LC50 values are in micromol$\cdot L^{-1}$ and P is the octanol/water partition coefficient.

This correlation is shown graphically in the form of a Hansch type plot in FIGURE 6. The dependence of predicted toxicity with respect to log P is represented by a solid line. Overall there is reasonable agreement between the measured LC50 values and those predicted by equation 1. Several herbicides, especially Dymron (U5) and Chloromethoxynil (B4) exhibit significantly lower toxicity than predicted for baseline narcosis. Since such behaviour is contrary to theoretical expectations, an experimental artifact was suspected. A similar finding in another study was interpreted in terms of reduced concentration due to loss through volatilization in static fish toxicity experiments (Lipnick and Dunn 1983). In this study, a problem with inadequate water solubility even in the presence of an organic carrier solvent was suspected. In order to test this hypothesis, the relationship between measured water solubility and observed fish toxicity was examined (FIGURE 7). The solid line represents a region in which toxicity equals water solubility. Those chemicals falling above this line were toxic at concentration in excess of their water solubilities, particularly Dymron (U5) and Chloromethoxynil (B4), which exhibit much lower toxicity than predicted.

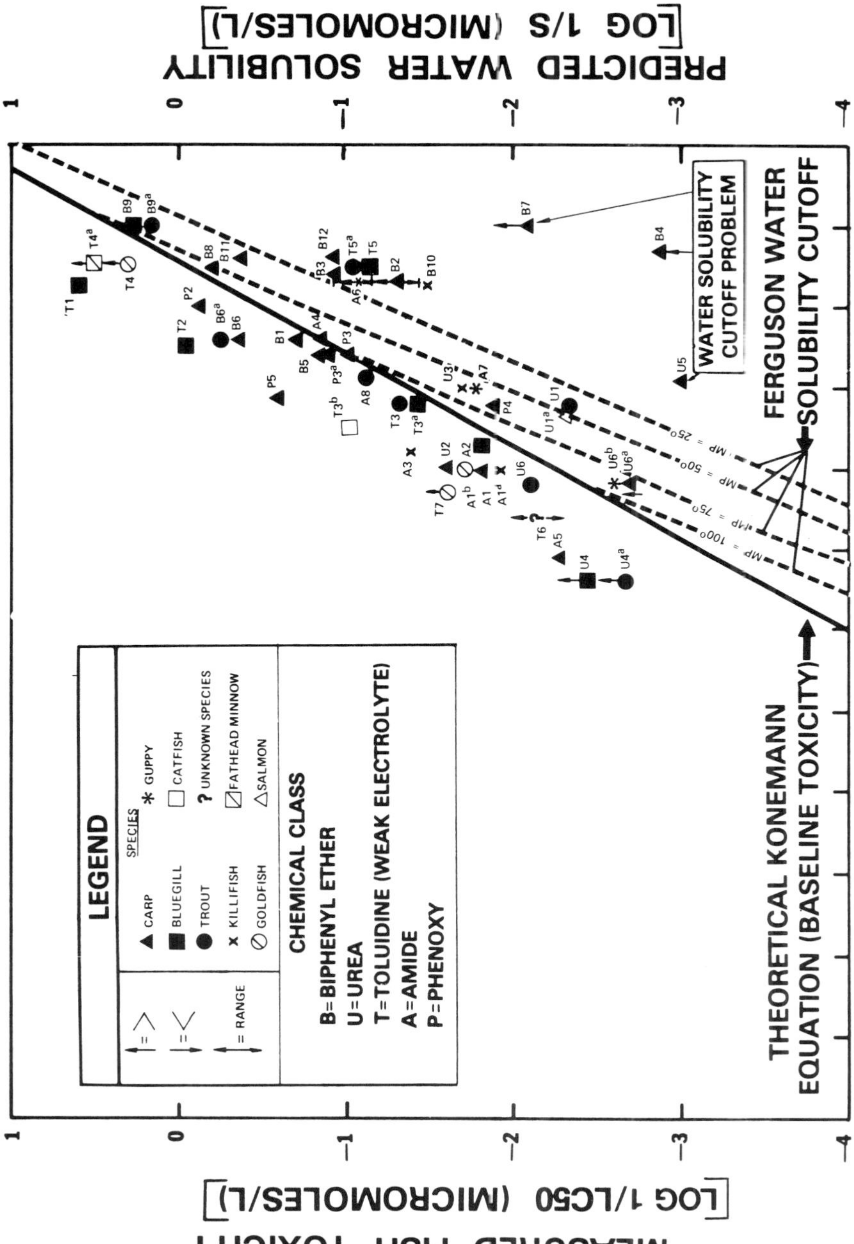

FIGURE 6: Plot of measured toxicities to several fish species and computed aqueous solubilities of the non-reactive non-electrolyte herbicides listed in TABLE 1 versus their octanol/water partition coefficients (log P)

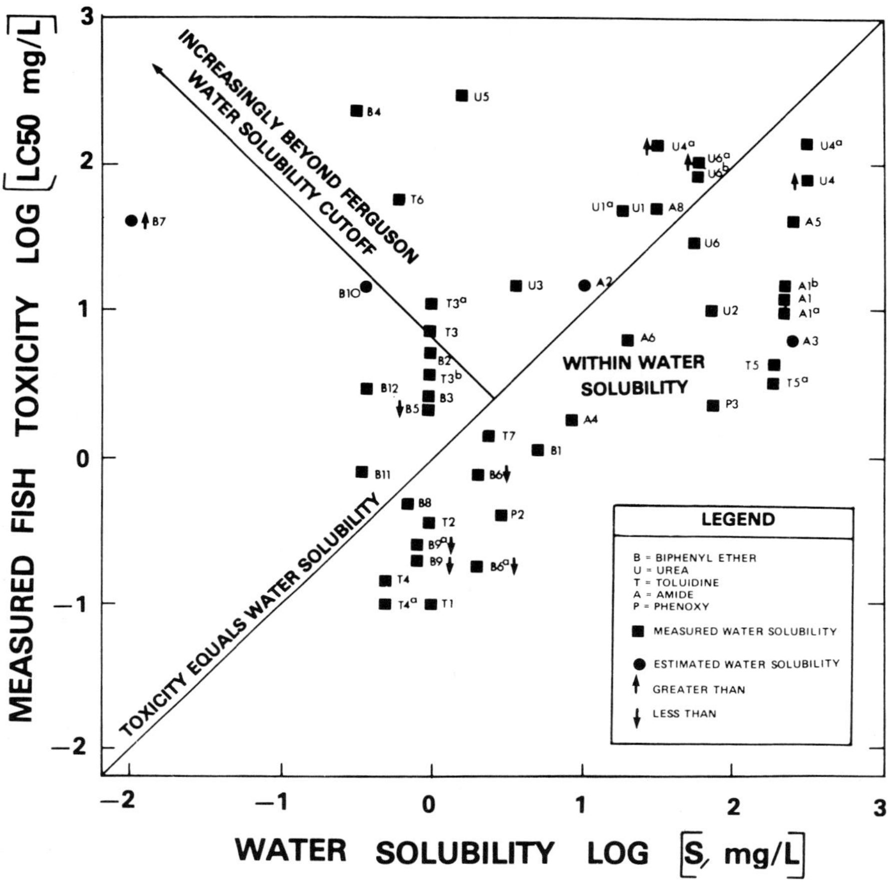

FIGURE 7: Plot of measured fish toxicities versus measured aqueous solubilities of the non-reactive non-electrolytes listed in TABLE 1

Since measured water solubility and melting point data were obtained for the majority of the study chemicals, it was possible to test the reliability of using equation 2 to estimate water solubility (Banerjee et al. 1980). Equation 2 was derived from measured data for 27 organic compounds which are members of the 65 EPA Consent Decree Chemicals.

$$\log P = 6.5 - 0.89 \log S - 0.015 \, MP \qquad (2)$$
$$n = 27; \; r^2 = 0.92$$

where MP is the melting point in °C (for liquid solutes, a nominal value of 25° is used), and S is the water solubility, in micromol•L^{-1}. While the estimated and measured water solubilities (TABLE 2 and FIGURE 8) compare favorably, there are significant outliers. These may represent problems related to log P estimation, limitations of the regression equation, artifacts of the experimental data, or a combination of these factors.

TABLE 2: Measured and estimated solubilities (S) and measured fish toxicities of compounds studied

Study Number	Measured[a,b] log S (mg•L^{-1})	Estimated[b,c] log S (mg•L^{-1})	Measured[d] log LC50 (mg•L^{-1})
B-1	+0.70	+0.40	+0.08
B-2	0.00	-0.59	+0.70
B-3	-0.02	-0.46	+0.40
B-4	-0.52	-1.52	+2.38
B-5	<+0.00	+0.16	+0.34
B-6	<+0.30	+0.07	-0.10
B-6[a]	<+0.30	+0.07	-0.75
B-7	----	-2.00	> 1.60
B-8	-0.18	-0.57	-0.30
B-9	<-0.10	-1.30	-0.70
B-9[a]	<-0.10	-1.30	-0.59
B-10	----	-0.44	+0.17
B-11	-0.52	-0.64	-0.10
B-12	-0.46	-0.86	+0.48
U-1	+1.26	+0.16	+1.70
U-1[a]	+1.26	+0.16	+1.70
U-2	+1.88	+1.81	+1.00
U-3	+0.57	-0.20	+1.18
U-4	+2.51	+1.84	> +1.88
U-4[a]	+2.51	+1.84	> +2.13
U-5	+0.23	-1.22	+2.48
U-6	+1.74	+0.84	+1.49
U-6[a]	+1.74	+0.84	> +2.00
U-6[b]	+1.74	+0.84	+1.95
T-1	0.00	+0.01	96h -1.05
T-2	0.00	+0.63	96h -0.43

TABLE 2: Cont'd

Study Number	Measured[a,b] log S ($mg \cdot L^{-1}$)	Estimated[b,c] log S ($mg \cdot L^{-1}$)	Measured[d] log LC50 ($mg \cdot L^{-1}$)
T-3	0.00	+0.82	+0.83
T-3[a]	0.00	+0.82	+1.04
T-3[b]	0.00	+0.82	+0.57
T-4	-0.30	----	96h>-0.82
T-4[a]	-0.30	----	96h>-1.00
T-5	+2.27	-0.36	+0.62
T-5[a]	+2.27	+0.46	+0.53
T-6	-0.22	+1.72	96h +1.76
T-7	+0.38	+1.48	96h> 0.15
A-1	+2.35	+1.70	+1.11
A-1[a]	+2.35	+1.70	+1.04
A-1[b]	+2.35	+1.70	+1.15
A-2	----	+1.02	+1.18
A-3	----	+2.39	+0.80
A-4	+0.92	+0.18	+0.26
A-5	+2.41	+1.99	+1.60
A-6	+1.30	-0.22	+0.81
A-7	----	----	+1.26
A-8	+1.54	+0.88	96h +0.67
P-2	+0.477	-0.68	+0.38
P-3	+1.863	+0.63	+0.48
P-3[a]	+1.863	+0.63	+0.30
P-4	-----	----	+1.15
P-5	-----	----	0.00

a Log of measured water solubility from TABLE 1.

b A dashed line indicates the appropriate data were not available for this chemical.

c Log of estimated water solubility from TABLE 1.

d Log of measured LC50 values from TABLE 1.

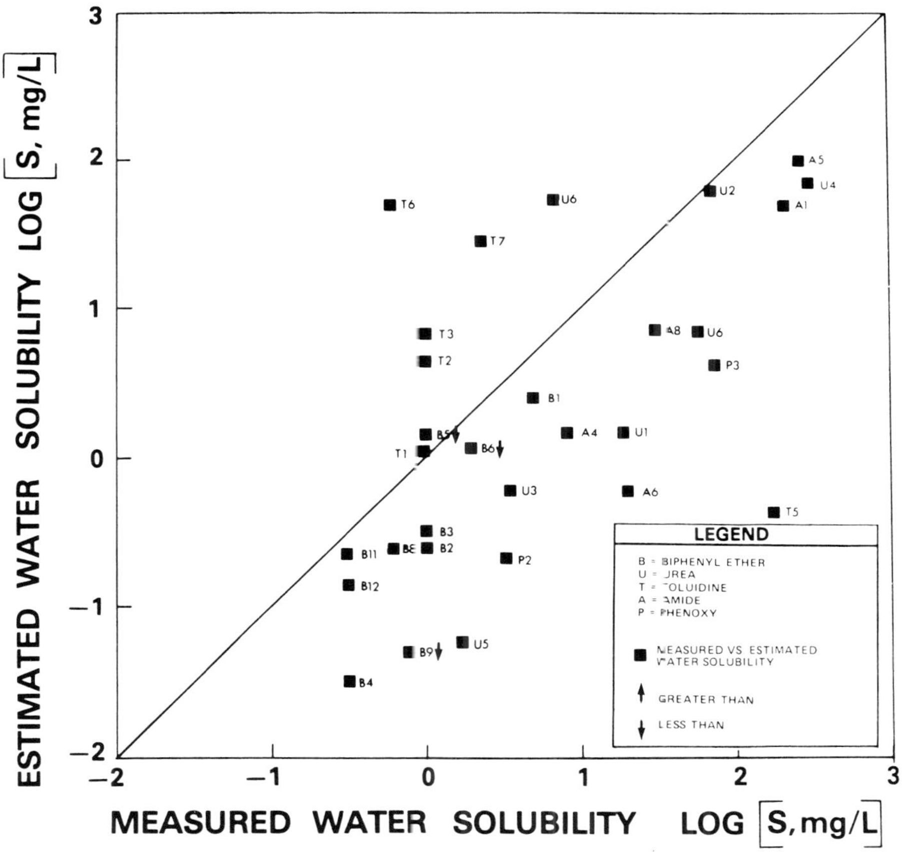

FIGURE 8: **Plot of measured versus estimated aqueous solubilities of selected non-reactive non-electrolyte herbicides**

ACKNOWLEDGEMENT

The authors are grateful to Dr. James H. Gilford, Chief of the Environmental Effects Branch, for his encouragement in carrying out this investigation.

REFERENCES

Albert, A. 1979. **Selective Toxicity,** 6th edition, Chapman and Hall, London, U.K.

Banerjee, S., Yalkowsky, S.H. and Valvani, S.C. 1980. Water solubility and octanol/water partition coefficients of organics. Limitations of the solubility-partition coefficient Correlations. **Env. Sci. Tech. 14:** 1227-1229.

Chou, J.T. and Jurs, P.C. 1979. Computer-assisted computation of partition coefficients from molecular structures using fragment constants. **J. Chem. Inf. Comput. Sci. 19:** 172-178.

Chou, J.T. and Jurs, P.C. 1980. Computation of partition coefficients from molecular structure by a fragment addition method. In **Physical Chemical Properties of Drugs,** S.H. Yalkowsky, A.A. Sinkula and S.C. Valvani (Eds.), Marcel Dekker, New York, pp. 163-198.

Ferguson, J. 1939. The use of chemical potentials as indices of toxicity. **Proc. R. Soc. London, Ser. B. 127:** 387-404.

Franks, N.P. and Lieb, W.R. 1982. Molecular Mechanisms of General Anesthesia, **Nature 300:** 487-493.

Hansch, C. and Leo, A.J. 1979. **Substituent Constants for Correlation Analysis in Chemistry and Biology,** Wiley Interscience, New York.

Hermans, J. 1983. The use of QSAR in toxicity studies with aquatic organisms. Correlation of Toxicity of Different Classes of Organic Chemicals with P_{oct}, pK_a and Chemical Reactivity. Proceedings Fourth European Symposium on Chemical Structure-Biological Activity Quantitiative Approaches, Bath, England, September 6-9, 1982. In **Quantitative Approaches to Drug Design,** J.C. Dearden (Ed.), Elsevier, New York, pp. 263-264.

Hisada, T. and Imamura, S. 1982. **Short Review of Herbicides.** 4th edition, Hodogaya Chemical Co. Ltd., Tokyo.

Könemann, H. 1981. Quantitative structure-activity relationships in fish toxicity studies. **Toxicology 19:** 209-221.

Lipnick, R.L. and Dunn, W.J. III. 1983. An MLAB study of aquatic structure toxicity relationships. Proceedings Fourth European Symposium on Chemical Structure-Biological Activity Quantitiative Aproaches, Bath, England, September 6-9, 1982. In **Quantitative Approaches to Drug Design,** J.C. Dearden, (Ed.), Elsevier, New York, pp. 265-266.

Magnuson, V.R., Harriss, D.K., Fulton, M.S. and Anderson, E.G. 1981. ISHOW, Information-System for Hazardous Organics in Water. 182nd Natl. Meet., Aug. 23-28, **Amer. Chem. Soc.,** paper no. CINF 22.

Perrin, D.D. 1972. **Dissociation Constants of Organic Bases in Aqueous Solution,** Supplement 1972, Butterworths, London, No. 4377.

Veith, G.D., Call, D.J. and Brooke, L.T. 1983. Structure-toxicity relationships for the fathead minnow, Pimephales promelas: Narcotic Industrial Chemicals. **Can. J. Fish. Aquat. Sci. 40:** 743-748.

WORKSHOP DISCUSSION

SCHULTZ:

We have looked at many carbazides, hydrazenes and hydrazides because they are groups of compounds that interact with connective tissue development in vertebrates and it is a very nice mode of action for teratogenesis. That would fit in very nicely to what we have seen in the past, that when you get that group, a C double bond with two nitrogens in a row, that seems to be the active centre.

McCARTY:

It intrigues me to find that the low log P compounds fall off your QSAR. There is something happening with this recipe and I will give you an idea that you may toss around. The drug transport model predicts that log P and BCF's of two and five are the break points. Also, remember that it was suggested for most of the QSAR relationships, it is not the inherent toxicity of the compound that is the problem, it is the accumulation that is changing the toxicity because the accumulation is changing faster than the toxicity. We thought that one thing that might be happening at levels below log P = 2, is that the compound doesn't accumulate anymore and therefore the inherent toxicity of the compound is being expressed without the modification through accumulation and therefore you expect more variation in your model at the bottom end. Just something we were tossing around and I think that suggests an explanation for what you are finding.

HODSON:

Then the possibility arises that given the range of ionization, accumulation is no longer dependent on log P, and other characteristics of the molecule become important. In this case we would still be looking at uptake limited toxicity but it would be irrelevant with respect to log P.

MACKAY:

Perhaps you should be careful about drawing diagrams where you have log P's as low as -4. That implies a substance to be ten thousand times more soluble in water than it is in octanol. Even water itself is only five or ten times more soluble in water than it is in octanol so I find it difficult to see what sort of molecule would have these properties. I suspect that there must be a minus sign missed out in the computer program somewhere. I have never seen an experimental value as low as that.

KAISER:

What kind of compound would be U7?

Comment:

Those are heterocycles. There is a problem with estimated log P in these compounds. The existing log P estimation methodology is limited by the availability of suitable fragment constants and correction factors for these compounds.

MACKAY:

I think there must be an error. When you add groups, normally, log P increases.

Comment:

It depends on which groups you are adding.

SCHULTZ:

If you look at fragment constants for nitrogen substituted into the ring you get a negative 1 and add an amino group to the ring you get a negative 1.23 It doesn't take long to add up a few nitrogens like that and you get a number like -3. I dcn't say that it is real, but I can see how you can calculate it. My question is how does a compound with that log P value ever get into the plant and be a herbicide?

MACKAY:

It just hates plants.

McCARTY:

You didn't edit your data all that well so is it a fair assumption that this is just a first cut of this thing?

Comment:

Those compounds for which appropriate fragment constants or proximity corrections are available to estimate log P will not be included in the final analysis.

McCARTY:

So you are going to go back and take a look at the data you have been using and make sure that it is all relatively comparable?

Comment:

We have done that, but we will go back and look at these compounds that fit the line. We pretty well know that these are outliers for one reason or another but we are going to go back and look at the data to see why these compounds closely fit the narcosis model.

THE RELATIONSHIP BETWEEN BIOCONCENTRATION FACTOR IN RAINBOW TROUT AND PHYSICAL-CHEMICAL PROPERTIES FOR SOME HALOGENATED COMPOUNDS

Barry G. Oliver

Environmental Contaminants Division
National Water Research Institute
PO Box 5050 Burlington, Ontario L7R 4A6
CANADA

ABSTRACT

The uptake of 13 chloro- and bromobenzenes and nine other chlorinated hydrocarbons by rainbow trout (Salmo gairdneri) from water has been studied. Equilibration between fish and water was achieved rapidly for most compounds, but some did not attain equilibrium over the 100-day study. For those chemicals which reached equilibrium, the bioconcentration factors (BCF) correlate well with the octanol/water partition coefficients (K_{ow}), and parachors (Pc). Chemicals which did not reach equilibrium have high partition coefficients ($K_{ow} > 10^5$) and high parachors ($P > 400$); their experimental bioconcentration factors are well below those predicted from the log BCF/log K_{ow} relationship.

INTRODUCTION

The determination of bioconcentration factors, BCF's, in fish is time consuming and expensive. For these reasons considerable efforts are currently being made to predict BCF's of chemicals from their physical and chemical properties. Earlier studies have shown that BCF correlates well with octanol/water partition coefficient

K. L. E. Kaiser (ed.), QSAR in Environmental Toxicology, 301–317.

(Veith *et al.* 1979; Neely *et al.* 1974; Chiou *et al.* 1977; Tulp and Hutzinger 1978; Oliver and Niimi 1983), water solubility (Kenaga and Goring 1980; Veith *et al.* 1980; Mackay 1982) and parachor (Tulp and Hutzinger 1978). In this paper I report BCF's for 22 halogenated compounds in rainbow trout (*Salmo gairdneri*) measured using low water exposure concentrations ($ng \cdot L^{-1}$), and attempt to correlate these BCF's with octanol/water partition coefficients and parachors of the chemicals. The uses and limitations of laboratory BCF's to predict residue levels in field fish are also discussed.

EXPERIMENTAL

Experimental procedures for this study have previously been described (Oliver and Niimi 1983). Briefly, hatchery reared rainbow trout (≈260 g initially) were maintained in 1 x 1 x 0.5 m^3 self-cleaning tanks and provided with 6 $L \cdot min^{-1}$ flow of activated charcoal filtered Lake Ontario water at 15°C. Fish were fed a dry commercial diet (≈2% body weight/day) and chemicals were spiked into the tanks using a multichannel peristaltic pump. Six fish were removed from the control and each exposure tank after each exposure period (≈0, 8, 20, 40, 60, 100 days). The fish were wrapped in aluminum foil and frozen until analysis.

The following abbreviations are used for the study chemicals: 1,2-dichlorobenzene (1,2-DCB); 1,3-dichlorobenzene (1,3-DCB); 1,4-dichlorobenzene (1,4-DCB); 1,3,5-trichlorobenzene (1,3,5-TCB); 1,2,4-trichlorobenzene (1,2,4-TCB); 1,2,3-trichlorobenzene (1,2,3-TCB); 1,2,4,5-tetrachlorobenzene (1,2,4,5-TeCB); 1,2,3,4-tetrachlorobenzene (1,2,3,4-TeCB); pentachlorobenzene (QCB); hexachlorobenzene (HCB); hexachlorobutadiene (HCBD); hexachlorethane (HCE); 1,3-dibromobenzene (1,3-DBB); 1,3,5-tribromobenzene (1,3,5-TBB); 1,2,4-tribromobenzene (1,2,4-TBB); 1,2,4,5-tetrabromobenzene (1,2,4,5-TeBB); 3,4-dichlorobenzotrifluoride (3,4-DCBTF); 2,4,5-trichlorotoluene (2,4,5-TCT); 2,3,4-trichloroanisole (2,3,4-TCA); 1,4-dichloronaphthalene (1,4-DCN); 2,4-dichlorodiphenylether (2,4-DCDPE); and 4,4'-dichlorobiphenyl (4,4'-PCB). The chemicals were tested in two batches. The chlorobenzenes, HCBD and HCE were tested at two concentration levels each. The other chemicals were tested at a later date using only a single water concentration.

The analytical procedure for water and fish using capillary gas chromatography with electron capture detectors has been described previously (Oliver and Nicol 1982a). Water samples (4 L) were extracted with 75 mL of hexane, the extract was concentrated to 1 mL, and cleaned up on a small H_2SO_4/silica column (collection volume 5 mL). Fish were homogenized and ground with Na_2SO_4 prior to soxhlet extraction for 24 hrs with acetone/hexane. The extract was back extracted with water to remove the acetone, shaken with concentrated H_2SO_4 to remove the bulk of the lipids, and (after evaporation to the appropriate volume) polished on a small H_2SO_4/silica column. The reproducibility of this technique on replicate samples was ±10%. All compound recoveries from spiked water and fish were greater than 80%.

RESULTS AND DISCUSSION

Bioconcentration factors and exposure water concentrations for the chemicals are tabulated in TABLE 1. Examination of the data for the chlorobenzenes in the table shows that BCF increases with increasing degree of chlorine substitution. It is also apparent from data set 1 that exposure concentration has an effect on BCF. On average, the BCF's from the higher exposure are 2.2 times those at the lower exposure. This observation indicates that the rate of elimination of the chemicals by the fish is probably concentration dependent. This result also emphasizes the importance of working at concentrations as close to environmental levels as possible if the data is to be used to predict behaviour in the natural environment.

A plot of the bioconcentration factor versus octanol/water partition coefficient (TABLE 2) for data set 1 at high and low exposure doses is shown in FIGURE 1. A good linear correlation is observed for the eleven compounds which reached equilibrium in this data set:

$$\log BCF = -0.63 + 1.02 \log K \qquad \text{High Dose} \qquad (1)$$
$$n = 11;\ r^2 = 0.986$$

and

$$\log BCF = -0.87 + 1.00 \log K \qquad \text{Low Dose} \qquad (2)$$
$$n = 11;\ r^2 = 0.986$$

TABLE 1: Bioconcentration factors and exposure chemical concentrations in water for the two compound sets

	Set 1				Set 2		
	High Dose		Low Dose				
Compound	Water Concentration ($ng \cdot L^{-1}$)	Mean log BCF	Water Concentration ($ng \cdot L^{-1}$)	Mean Log BCF	Compound	Water Concentration ($ng \cdot L^{-1}$)	Log BCF
1,2-DCB	940	2.75	47	2.43	1,3-DBB	5.4	2.82
1,3-DCB	690	2.87	28	2.62	1,3,5-TBB	0.32	3.70
1,4-DCB	670	2.86	28	2.57	1,2,4-TBB	0.58	3.63
1,3,5-TCB	45	3.61	2.3	3.26	1,2,4,5-TeBB	0.09	3.80[b]
1,2,4-TCB	52	3.51	3.2	3.11	3,4-DCBTF	1.0	3.18
1,2,3-TCB	72	3.42	4.3	3.08	2,4,5-TCT	0.13	3.68
1,2,4,5-TeCB	21	4.11	1.0	3.72	2,3,4-TCA	0.90	2.96
1,2,3,4-TeCB	26	4.08	1.4	3.72	1,4-DCN	1.7	3.75
QCB	9.3	4.30	0.34	4.11	2,4-DCDPE	2.7	3.97[b]
HCB	8.0	4.30[a]	0.32	4.08[a]	4,4'-PCB	19	3.58[b]
HCBD	3.4	4.23	0.10	3.76			
HCE	7.1	3.08	0.32	2.71			

a Highest value, did not equilibrate during 110-day.
b Highest value, did not equilibrate during 90-day study.

The slopes of the two lines are seen to be virtually identical but the intercepts are significantly different. It should be noted that HCB, which did not attain equilibrium in the study, plots well below and outside the 95% confidence interval of both lines.

If the BCF's of equilibrated compounds from data set 1 (low dose, eleven compounds) and data set 2 (also low exposure dose, seven compounds) are combined, a good linear correlation is still obtained but more data scatter is observed:

$$\log BCF = -0.68 + 0.94 \log K \quad (3)$$
$$n = 18;\ r^2 = 0.91$$

TABLE 2: Octanol/water partition coefficients, K_{ow}, and parachor for study compounds

Compound	Log K_{ow}	Parachor[a]	Compound	Log K_{ow}	Parachor[a]
1,2-DCB	3.40[b]	290	1,3-DBB	3.79[e]	310
1,3-DCB	3.44[b]	290	1,3,5-TBB	4.51[e]	360
1,4-DCB	3.37[b]	290	1,2,4-TBB	4.51[e]	360
1,3,5-TCB	4.15[c]	330	1,2,4,5-TeBB	5.13[e]	420
1,2,4-TCB	4.02[d]	330	3,4-DCBTF	4.43[f]	360
1,2,3-TCB	4.11[c]	330	2,4,5-TCT	4.82[f]	370
1,2,4,5-TeCB	4.52[c]	360	2,3,4-TCA	4.24[f]	380
1,2,3,4-TeCB	4.46[c]	360	1,4-DCN	4.79[f]	390
QCB	4.94[b]	400	2,4-DCDPE	5.62[f]	480
HCB	5.50[d]	440	4,4'-PCB	5.58[g]	460
HCBD	4.78[b]	410			
HCE	3.58[c]	350			

a Calculated by the method of Quayle (1953).
b Data from Banerjee *et al*. (1980).
c Data from Könemann *et al*. (1979).
d Data from Chiou *et al*. (1982).
e Data from Wataral *et al*. (1982).
f Calculated by the method of Hansch and Leo (1979).
g Data from Chiou *et al*. (1977).

The slopes of equations 1, 2 and 3 are similar and close to one. Using selected literature data, Mackay (1982) has recently demonstrated that a good correlation between log BCF and log K_{ow} can be obtained using a two parameter equation with slope equal to 1.0. The reason that earlier equations (Veith *et al*. 1979; Neely *et al*. 1974) had a significantly lower slope may have been due to the inclusion of compounds which had not attained equilibrium in the BCF test. As an example, if the four compounds (HCB, 1,2,4,5-TeBB, 2,4-DCDPE, 4,4'-PCB) that did not reach equilibrium in the experiments are included in the data analysis, the slope of the resulting equation is 0.68.

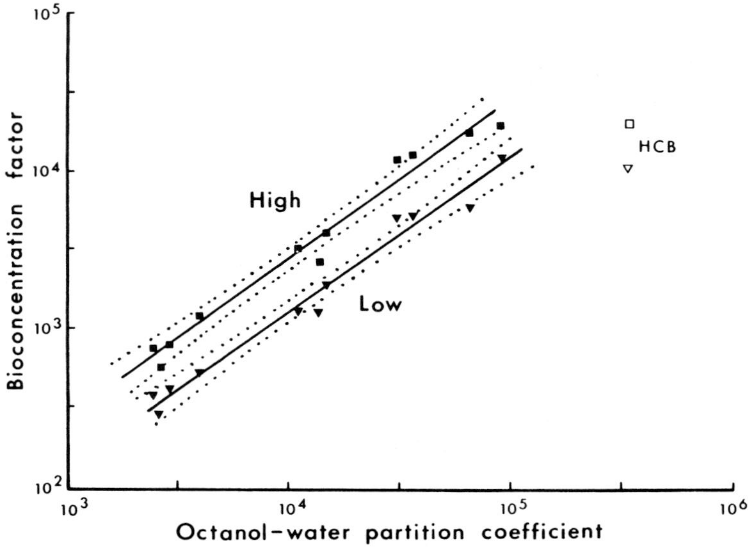

FIGURE 1: Bioconcentration factor, BCF, versus octanol/water partition coefficient, K_{ow}, for the chlorobenzenes, HCBD and HCE at low and high exposure doses

Another parameter which has been correlated to BCF is parachor (Tulp and Hutzinger 1978). Parachor, which can be thought of as the molar volume corrected for compressive effects of intermolecular forces, is represented by equation 4 (Quayle 1953):

$$P = \frac{M}{D-d} \cdot \gamma^{\frac{1}{4}} \qquad (4)$$

where M is the molecular weight, D is the density of the liquid, d is the density of the vapour, and γ is the surface tension, all measured at the same temperature. Parachor is a constitutive property and can be easily calculated from the molecular structure (TABLE 2) (Quayle 1953). A plot of log BCF versus parachor is shown in FIGURE 2 for the 18 compounds which reached equilibrium at the low exposure concentration in the two experiments. Fifteen compounds, all halogenated aromatics, fall reasonably well on a straight line. Three compounds HCBD, HCE and 2,3,4-TCA, fall well to the right of this line. HCBD and HCE are halogenated aliphatics and have a somewhat different shape than the rest of the compounds. Even though 2,3,4-TCA is also a halogenated aromatic, it has an oxygen substituent on the ring which renders it more water soluble and perhaps more readily excretable by fish than the other compounds in the series. These results show that parachor (molecular size) can be used as a BCF predictor for a closely related group of compounds, e.g., the chlorobenzenes. For predicting behaviour of diverse classes of compounds, the octanol/water partition coefficient appears to be a more useful parameter.

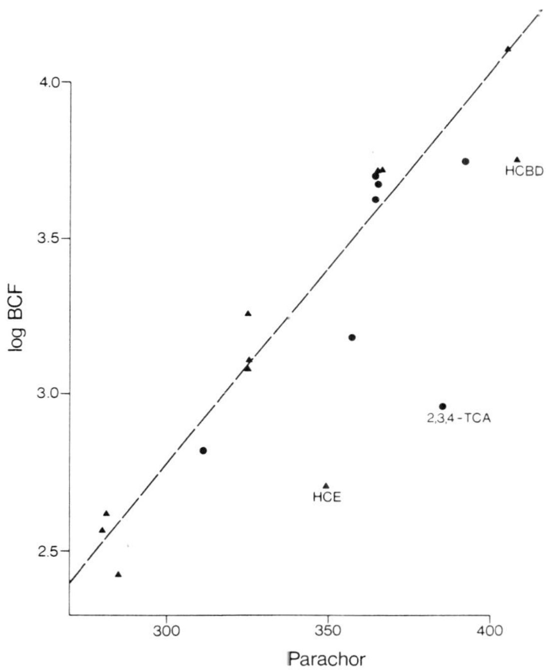

FIGURE 2: Log BCF versus parachor for 18 equilibrated compounds

Laboratory measured BCF's or those calculated from QSAR equations such as 2 or 3 can be used to predict residue levels in field populations of fish if the chemical concentrations in the waterbody is known. An example of such a prediction for some chlorobenzenes, CB's, is shown in TABLE 3. These results show that excellent agreement between predicted and measured CB residues is obtained for all compounds except HCB. The tri-, tetra- and pentachlorobenzenes all reached a steady equilibrium concentration in the fish in a fairly short time in the laboratory study. Because of this rapid equilibration between chemical concentration in the fish and in the water, the main controlling factor governing residue levels for these compounds in environmental fish should be the water concentrations of each chemical. The TABLE 3 results confirm this hypothesis.

TABLE 3: Predicted chlorobenzene, CB, residue levels in Lake Ontario rainbow trout compared to measured CB residues in ten rainbow trout from Lake Ontario

Compound	Predicted Concentration[a] in Fish ($ng \cdot g^{-1}$)	Measured Mean[b] Concentration in Lake Ontario Fish ($ng \cdot g^{-1}$)
1,2,4-TCB	0.8	0.6 ± 0.3
1,2,4,5-TeCB	0.4	0.5 ± 0.2
1,2,3,4-TeCB	0.4	1.0 ± 0.4
QCB	2.3	3.4 ± 1.3
HCB	2.6	33 ± 15

a Calculated by multiplying measured CB concentrations in Lake Ontario (Oliver and Nicol 1982b) by BCF's predicted using equation 2 in this paper.

b Data from Oliver and Niimi (1983).

On the other hand, HCB did not reach a steady concentration in the fish despite 110 days of exposure. Because HCB has a long half-life in rainbow trout (Niimi and Cho 1981), the concentration of this chemical continues to build up in the fish over the exposure period. The predicted HCB residue level from K_{ow} by the QSAR equation 2 is at least an order of magnitude below the observed environmental fish concentration (TABLE 3). HCB concentration in field populations appear to be mainly affected by HCB contamination of fish food and only minimally influenced by HCB water concentrations.

HCB was not the only compound which did not attain a steady state concentration in the fish. The chemicals 1,2,4,5-TeBB; 2,4-DCDPE and 4,4'-PCB also failed to equilibrate. From TABLE 2 it is apparent that these four compounds have similar properties - all have K_{ow}'s greater than 10^5 and parachors greater than 400. When the non-equilibrated BCF's for these compounds are included in the log BCF versus log K_{ow} plots they fall well below the linear correlation lines (equations 2 and 3). Spacie and Hamelink (1982) have recently also observed difficulties in linearly correlating BCF's to physical properties for large molecules with limited water solubility and high octanol/water partition coefficients, and have proposed the use of a drug transport model. Even if good predictions of the BCF's of such compounds could be made, the HCB discussion above illustrates that these BCF's would be of little use in predicting environmental concentrations because of the minimal influence of the direct contaminant uptake from water for these chemicals.

In conclusion, BCF is a useful predictive parameter for chemicals with $K_{ow} < 10^5$ and BCF correlates well with K_{ow} for these chemicals. BCF is of limited use for chemicals with $K_{ow} > 10^5$ (parachors > 400).

ACKNOWLEDGEMENTS

The author would like to thank Arthur Niimi of the Great Lakes Fisheries Research Branch for his many contributions to this research and to Karen Nicol for technical assistance.

REFERENCES

Banerjee, S., Yalkowski, S.H. and Valvani, S.C. 1980. Water solubility and octanol/water partition coefficients of organics. Limitations of the solubility-partition coefficient correlation. **Environ. Sci. Technol. 14:** 1227-1229.

Chiou, C.T., Schmedding, D.W. and Manes, M. 1982. Partitioning of organic compounds in octanol/water systems. **Environ. Sci. Technol. 16:** 4-10.

Chiou, C.T., Freed, V.H., Schmedding, D.W. and Kohnert, R.L. 1977. Partition coefficient and bioaccumulation of selected organic chemicals. **Environ. Sci. Technol. 11:** 475-478.

Hansch, C. and Leo, A. 1979. **Substituent constants for correlation analysis in chemistry and biology.** John Wiley and Sons, New York.

Kenaga, E.E. and Goring, C.A.I. 1980. Relationship water solubility, soil sorption, octanol/water partitioning, and concentration of chemicals in biota. In **Aquatic Toxicology,** J.G. Eaton, P.R. Parrish and A.C. Hendricks (Eds.), STP707, American Society for Testing and Materials, Philadelphia, pp. 78-115.

Könemann, H., Zelle, R., Busser, F. and Hammers, W.E. 1979. Determination of log P_{oct} values of chloro-substituted benzenes, toluenes and anilines by high-performance liquid chromatography on ODS-silica. **J. Chromatogr. 178:** 559-565.

Mackay, D. 1982. Correlation of bioconcentration factors. **Environ. Sci. Technol. 16:** 274-278.

Neely, W.B., Branson, D.R. and Blau, G.E. 1974. Partition coefficient to measure bioconcentration potential of organic chemicals in fish. **Environ. Sci. Technol. 8:** 1113-1115.

Niimi, A.J. and Cho, C.Y. 1981. Elimination of hexachlorobenzene (HCB) by rainbow trout (*Salmo gairdneri*), and an examination of its kinetics in Lake Ontario salmonids. **J. Fish. Aquat. Sci. 38:** 1350-1356.

Oliver, B.G. and Niimi, A.J. 1983. Bioconcentration of chlorobenzenes from water by rainbow trout: Correlations with partition coefficients and environmental residues. **Environ. Sci. Technol. 17:** 287-291.

Oliver, B.G. and Nicol, K.D. 1982a. Gas chromatographic determination of chlorobenzenes and other chlorinated hydrocarbons in environmental samples using fused silica capillary columns. **Chromatographia 16:** 336-340.

Oliver, B.G. and Nicol, K.D. 1982b. Chlorobenzenes in sediments, water, and selected fish from Lakes Superior, Huron, Erie, and Ontario. **Environ. Sci. Technol. 16:** 532-536.

Quayle, O.R. 1953. The parachors of organic compounds. **Chem. Rev. 53:** 439-589.

Spacie, A. and Hamelink, J.L. 1982. Alternative models for describing the bioconcentration of organics in fish. **Environ. Toxicol. Chem. 1:** 309-320.

Tulp, M.T.M. and Hutzinger, O. 1978. Some thoughts on aqueous solubilities and partition coefficients of PCB, and the mathematical correlation between bioaccumulation and physico-chemical properties. **Chemosphere 7:** 849-860.

Veith, G.D., Macek, K.J., Petrocelli, S.R. and Carroll, J. 1980. An evaluation of using partition coefficients and water solubility to estimate bioconcentration factors for organic chemicals in fish. In **Aquatic Toxicology,** J.G. Eaton, P.R. Parrish and A.C. Hendricks (Eds.), STP707, American Society for Testing and Materials, Philadelphia, pp. 116-129.

Veith, G.D., DeFoe D.L. and Bergstedt, B.U. 1979. Measuring and estimating the bioaccumulation factor of chemicals in fish. **J. Fish. Res. Board Can. 36:** 1040-1048.

Wataral, H., Tanaka, M. and Suzuki, N. 1982. Determination of partition coefficients of halobenzenes in heptane/water and 1 octanol/water systems and comparison with the scaled particle calculations. **Anal. Chem. 54:** 702-705.

WORKSHOP DISCUSSION

Question:

What equation did you use to predict BCF?

OLIVER:

The BCF's were measured. The predicted values were residue concentrations in the environment; we used laboratory bioconcentration factors, measured the lake water concentration and asked: If we put the fish in that lake what concentration would we predict? In this way we predicted the right values for most of the chemicals except hexachlorobenzene. For example, if the BCF of a chemical was 1000 and its concentration in water was one part per trillion, then you would have one part per billion in your fish - strictly a multiplication. This approach works well for chemicals that equilibrate rather quickly. Somebody actually phoned me up a while back from the United States, a consulting firm doing work for the EPA, and wanted to know how they could set loading levels for chemicals, to keep residue levels in fish below specified guidelines. You should be able to do this for chemicals such as the lower chlorobenzenes. For example, given a certain loading and flowrate, it should be possible to calculate the chemical concentration in the river. If the chemical's BCF is known, one can calculate expected residue levels.

HALLETT:

I was wondering about the problem with hexachlorobenzene when you were taking environmental residues versus what you predicted should happen. Did you go back to your aquarium experiment where you should have a better handle or should be able to measure what the hexachlorobenzene concentration was in that water much more accurately than trying to analyze all the water in Lake Ontario and get a mean level of hexachlorobenzene for that and then predict what the residue should be in the fish?

OLIVER:

The water in the aquarium was monitored. In other words, I could keep that fish in the tank at that exposure level, which is maybe ten times higher than you would get in Lake Ontario water and maybe still never attain the concentration of HCB in the fish that I would find if it were in Lake Ontario water because the fish is actually getting its HCB in Lake Ontario primarily from food and not from the water.

HALLETT:

I realize your argument but I guess the point is basically, how confident are you in your Lake Ontario water measurements for HCB which go against your relationship?

OLIVER:

I don't think it goes against the relationship.

HALLETT:

Well, it supports the argument you are telling us but if you don't accept the argument, what then? I am just a little concerned that you might be misleading yourself by some bad measurement or making an extrapolation.

OLIVER:

I have a lot of concentration data on the Niagara River and Lake Ontario for PCB's as well as chlorobenzenes. If one performs similar calculations for PCB's, again you find that predicted residue levels, from water concentrations and BCF's, are much lower than PCB residues actually found in the fish.

HALLETT:

My point was, when you did your experiment in your aquarium, you must have measured the concentration of hexachlorobenzene and you were getting a bio-concentration factor out of that. It wasn't equilibrating? How did you predict the level of hexachlorobenzene you should get in a fish? You must have derived some equation?

OLIVER:

We just took the highest value.

CULLEN:

Why didn't you just extrapolate the curves?

OLIVER:

Yes, you could do that if you took the K_{OW} relationship and take it off that; however, you only get a value a couple of times higher than the value that I have measured.

CULLEN:

BCF versus time and extrapolate for equilibrium so, why don't you take that?

OLIVER:

It does not matter whether you measure or calculate the BCF for chemicals such as HCB, using this BCF in calculations still results in a much lower residue value for field fish than what is observed.

KAISER:

I just wonder if we are getting into the argument of bio-concentration and direct uptake versus food-chain here. For chlorinated compounds, I think there is more experimental evidence now and if you look at field data and try to interpret them with respect to whether or not we have bio-concentrations with direct uptake or food-chain mechanism, I think it all indicates food-chain mechanism, at least that is my conviction.

ZITKO:

I agree with that. I think that with hexachlorobenzene we are getting into the area where relations should be revised. For compounds with low solubility the uptake rate is much slower. One might take this a step further and mix some hexabromobenzene in and you will find that it doesn't accumulate from water at all. The other thing is if you measure concentration of hexachlorobenzene in water there is no way of distinguishing between molecularly dissolved and particulate material because you extract both of them. You don't know whether you have single molecules or aggregates of colloids or what, because you probably work quite close to the solubility limit of the hexachlorobenzene and the fish may be able to take up, gradually, any dissolved hexachlorobenzene but not any particle.

OLIVER:

But these concentrations were less than a part per trillion and we are not even anywhere near the hexachlorobenzene solubility.

ZITKO:

We still don't know how it behaves in solution.

OLIVER:

We have done a fair bit of work actually in the Niagara River in filtering samples and actually most of the chemicals are in dissolved form in the water in the Niagara River. Also, in our aquariums we did some work on filtering samples and 99% of the hexachlorobenzene was dissolved, given the classical definition of a 0.45 micron filter.

SCHULTZ:

As an individual who works with low K_{ow} compounds, I have heard the term high K_{ow} compounds thrown out for the last couple of days. Is there any consensus on this? Is the cutoff at 5, is it at 6? When I left yesterday I felt it was at 6 and today it seems to be at 5.

KAISER:

The difference is whether you calculate it from log P of benzene and pi increments or measure it. If you measure log P of hexachlorobenzene, you get only 5. If you calculate it you get 6, so that is how these two figures come about.

MACKAY:

That may be a kinetic problem. With something like hexabromobenzene fish don't live long enough to reach equilibrium because of the slow uptake, not because of a low bioconcentration factor. The K_{ow} relationship is correct. The fish just don't live long enough to complete the experiment.

ZITKO:

I don't think so. You don't have to get equilibrium but you will get something in the fish.

McCARTY

I think I agree with Vlado Zitko. Perhaps one of the reasons that your result for hexachlorobenzene is different from what is predicted is for the very reason that the predicted values are based on water concentrations. If your fish are living for a long time in the lake and they are eating contaminated food, then your effective concentration is different. Therefore, it is going to accumulate to a higher level and if you compare your water level against the fish level, it isn't going to agree.

OLIVER:

That is what I am saying. Basically, you can't predict from the water concentrations what the residue level is because food is the most important source.

McCARTY:

You cannot predict it because you have to take into account another factor and if you consider that factor you can predict it. What you are saying is that the translation from the lab data to field observation seems to be incorrect for high log P compounds. That doesn't mean that it can't be predicted, it is just that we don't know, we haven't established what the correlation is in order to be able to predict it.

HODSON:

Do you know the reason for the relationships of BCF to log K_{ow} for different exposure concentrations? Why is it occurring?

OLIVER:

The only thing we can have is the possibility that the elimination process is concentration dependent. In other words, if you get very high concentrations in the tissues, the beast can't get rid of it as fast. I don't know why this is.

HODSON:

Could that be a primary sign of toxicity?

OLIVER:

Could be. The residue level was much higher in the high dose than in the lower dose, 3 ppm opposed to 70 ppb, so that you could be slowing it down, poisoning the system, or something. This shows the importance of working with concentrations close to environmental levels. You could never predict the concentrations in the environment if you were working with ppm models and trying to extrapolate to those $ng \cdot L^{-1}$ concentrations in the environment. At least from levels we see in the Niagara River, which is a contaminated river, one has to work in the part per trillion range.

McCARTY:

What you are saying is that these low levels appear to be safe levels and certainly any test that we would typically run, either acute test or chronic test, would indicate that those levels are safe. However, if you look at the environmental levels found in those fish, you find that they accumulated the toxics to a much higher level than you might otherwise predict on the basis of standard tests. Therefore, the standard tests are not adequate. Is that what you are saying?

OLIVER:

Yes.

WELLS:

There is an interaction between fish age and the duration of your experiment. Because you can't get a small fish for a long experiment, there is interaction with feeding rate and growth, inhibition and metabolic rate, etc.

OLIVER:

The fish grew from 260 g to 400 g over the study. The concentrations (for rapidly equilibrating compounds) reached a steady value, so "total" residues in the fish increased with fish size over the study.

TOXICITIES OF CHLOROANILINES TO PHOTOBACTERIUM PHOSPHOREUM AND THEIR CORRELATIONS WITH EFFECTS ON OTHER ORGANISMS AND STRUCTURAL PARAMETERS

Juan M. Ribo and Klaus L.E. Kaiser

Environmental Contaminants Division
National Water Research Institute
PO Box 5050, Burlington, Ontario L7R 4A6
CANADA

ABSTRACT

The toxicities of aniline and sixteen chloro-substituted anilines to Photobacterium phosphoreum have been determined using the Microtox test. The values obtained correlate well with the toxic effects of these compounds to four different species of yeast and with the octanol/water partition coefficients.

INTRODUCTION

The growing concern over the impact and toxicological effects of synthetic chemicals in the environment leads to the need for more information on their interactions with and effects on specific biota. Already, thousands of compounds have been identified in raw and potable water, in fishes, plants, soils, and sediments worldwide. In many instances, extremely little is known on the environmental fate and effects of such compounds. Moreover, new compounds are constantly being introduced and regulatory agencies are charged with assessing the environmental safety of such materials, frequently without having the necessary biological data. This lack of information has led to the successful application of quantitative structure-activity relationships (QSAR) to this field.

K. L. E. Kaiser (ed.), QSAR in Environmental Toxicology, 319–336.

Until a few years ago, QSAR has exclusively been used for the development of new compounds, in particular drugs and pesticides. Recently, this methodology has been applied to the correlation and prediction of the toxicities and effects of inorganic (Kaiser 1980) and organic compounds (Ribo and Kaiser 1983) to aquatic biota. We report here on further QSAR investigations, specifically on the toxicities of chloroanilines to *Photobacterium phosphoreum* in the Microtox^{TM} (Bulich and Isenberg 1981, Bulich *et al*. 1981) test. From our previous investigations on chlorobenzenes, chlorophenols, and para-substituted phenols (Ribo and Kaiser 1983) the Microtox test appears to be well suited to provide relevant toxicity data with good precision for comparative analysis with other biota and with structural parameters. This study on chloroanilines is another facet of the large mosaic on the biological effects of various functional groups and substituents on the benzene frame.

Apart from our own interest in this set of chloroanilines as a class of polychlorinated aromatic compounds, there are in the literature reports on aromatic amines in soil and sediment samples from the Buffalo River (Nelson and Hites 1980) and the Delaware River areas (Diachenko 1979). In both areas, these contaminants are thought to originate from industrial wastes from textile and dyestuff manufacturing plants. Several chloro- and methylanilines were also observed in the River Rhine (Wegman and DeKorte 1981).

METHODOLOGY

The flow chart of SCHEME 1 summarizes the process of our QSAR investigations.

The first step is the selection of a set of chemical compounds and the compilation of the toxicity values either from our own Microtox test, in which case the preparation of the sample and its purification is important, or from other toxicity tests reported in the literature. In a second stage we collect every possible information on structural parameters for those compounds from literature sources or from our own determinations. These include electronic, physico-chemical, quantum-chemical, and structure-information parameters.

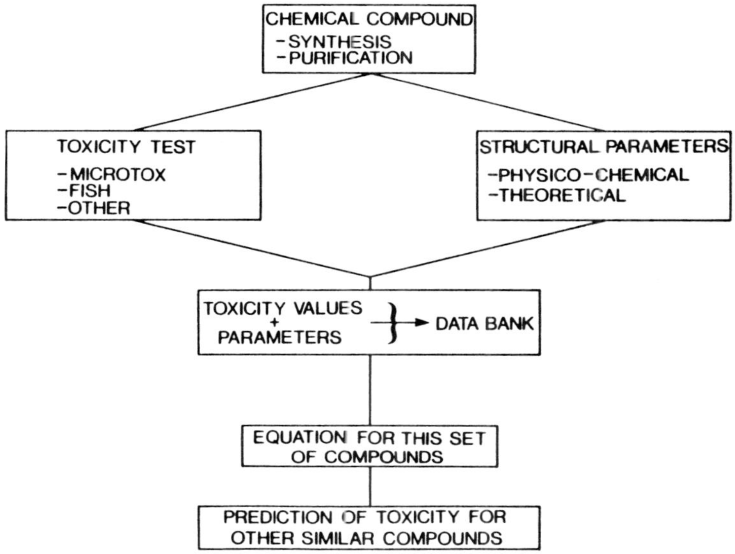

SCHEME 1: Methodology of our current QSAR investigations

The parameters and the toxicity values are entered in a computerized data bank after normalization of the toxicity values: all values are expressed as the negative logarithm of the molar concentration (i.e., $\log \frac{1}{C}$, where C is the concentration in $mmol{\cdot}L^{-1}$ or $mmol{\cdot}kg^{-1}$ depending on the nature of the test).

To calculate the QSAR equations, we use computerized statistical analysis of the toxicity data and structural parameters for the selected set of compounds to find a mathematical equation which models the activity of these compounds based on the values of the parameters chosen, and which would allow the prediction of the toxic activity of a compound not included in the training set or for which there is no experimental value.

This sequence is repeated for different sets of compounds and the next step will be to combine such data in terms of multi-dimensional QSAR equations to elucidate general and specific effects of certain funcional groups, such as the aryl-NH_2, and of common substituents, such as chlorine.

Toxicity Test

The test organism (Photobacterium phosphoreum) is a marine bacteria which emits light as part of its metabolism under normal conditions. The test uses the reduction of light emitted when the bacteria are exposed to a toxic substance, as measurement of the toxicity of that compound (Bulich *et al.* 1981). The reagent is prepared by reconstitution of the freeze-dried bacteria with ultrapure water and kept at 2 to 4°C until use. From this stock solution of the reagent five identical solutions are prepared for each test and kept at 15°C which has been found to be the optimum test temperature. After the temperature equilibration period of approximately 15 min, sample solutions ranging from 0 to 100% concentration (referred to the original sample) are added to the reagent and the light emmision is read before and at 5, 15 and 30 min after the addition of the investigated compound.

Data Reduction

TABLE 1 shows an actual example of the Microtox reading for 4-chloroaniline. The gamma function is the ratio of light lost to the

TABLE 1: Microtox reading for the 4-chloroaniline

	Concentration (ppm)	Final Readings				Gamma Function (Effects)		
		0 min	5 min	15 min	30 min	5 min	15 min	30 min
0	0 (blank)	94	81	76	69	-	-	-
1	1.17	86	55	53	49	0.35	0.31	0.29
2	2.33	86	45	44	42	0.65	0.58	0.50
3	4.67	88	33	33	31	1.30	1.16	1.08
4	9.33	87	22	21	20	2.41	2.35	2.19

Gamma Function Calculation:

$$\text{GAMMA (5 min)} = \frac{\frac{81}{94} * 86 - 55}{55} = 0.35$$

light remaining. The blank readings are normalized against the reading at time zero and every actual reading (light remaining) is subtracted from this normalized value to obtain the amount of light lost.

The dose-response curve (FIGURE 1) is the concentration versus gamma function plot and is normally a straight line when

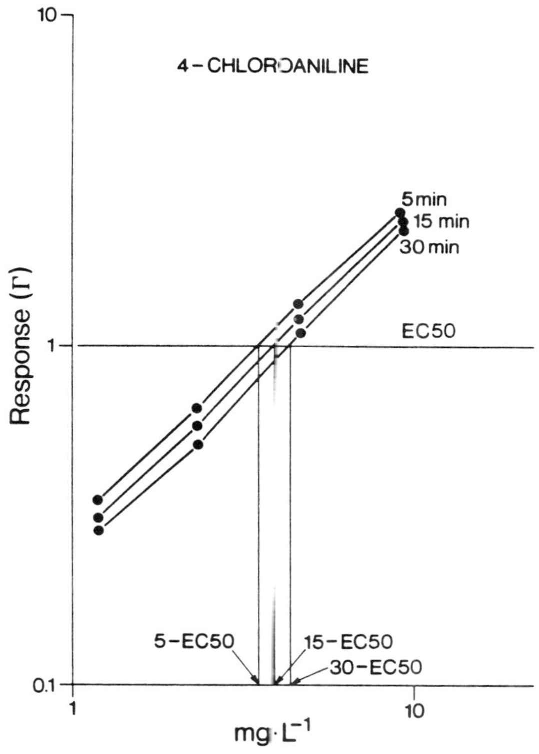

FIGURE 1: Dose-response curves for the 5, 15, and 30 min effects of 4-chloroaniline in the Microtox test

plotting the logarithm of concentration (ppm) against the logarithm of the effect observed (log C/log Γ). From this plot, the toxicity value EC50, the effective concentration of the chemical causing 50% of light reduction, i.e., the concentration at which Γ = 1, is then graphically determined. The EC50 values are a function of time and therefore are referred to as the 5-min EC50 (5-EC50), 15-min EC50 (15-EC50), or 30-min EC50 (30-EC50) concentrations. The EC50 values can also be calculated from the experimental data by least square regression analysis.

The toxicity values are expressed in the form p(t-EC50) (i.e., negative logarithm of the concentration in $mmol{\cdot}L^{-1}$ for the time t), in order to be able to compare these values with other previously determined toxicity data. Values from literature references are normalized in the same manner.

RESULTS AND DISCUSSION

Few toxicological data on chlorinated anilines are available in the literature. Adema _et al._ (1981) reported the toxicities of 3,4-dichloroaniline to several aquatic organisms and structure-activity correlations of acute growth inhibition of four strains of yeast by several chloroaniline congeners are reported elsewhere in this volume (Kwasniewska and Kaiser 1984). The carcinogenic activity of 4,4'-methylene-bis(2-chloroaniline) has been described by McQueen _et al._ (1981).

Structure-Activity Correlations

TABLE 2 shows EC50 values for aniline and 16 mono- and polychloro-substituted anilines together with the logarithms of the octanol/water partition coefficients.

As mentioned above, all toxicity values are always expressed as p(t-EC50) [-log(30-EC50)], where EC50 is the concentration causing the 50% of light reduction expressed in $mmol{\cdot}L^{-1}$ after 30 min of exposure. The log P (P = octanol/water partition coefficient) values used are taken from the compilation by Hansch and Leo (1979), except for those marked, which are either experimentally measured, or calculated from equation 1 (Kaiser 1983):

$$\log P_N = (N + 1)^b \log P_o \qquad (1)$$

where P_o is the octanol/water partition coefficient of the parent compound, b is a constant (b = 0.9110 for chloroanilines), and N is the number of chlorine substituents.

TABLE 2: Microtox toxicity values p(t-EC50), (EC50 in mmol L^{-1}) of chloroanilines and octanol/water partition coefficients (log P)

No.	Substituent	Microtox			log P
		p(5-EC50)	p(15-EC50)	p(30-EC50)	
1	-	0.16	0.13	0.12	0.98
2	2	0.95	0.93	0.91	1.92
3	3	1.01	0.98	0.96	1.90
4	4	1.60	1.53	1.40	1.83
5	2,3	1.81	1.77	1.77	2.78
6	2,4	1.61	1.55	1.54	2.66[b]
7	2,5	1.68	1.65	1.63	2.66[b]
8	2,6	2.04	1.98	1.97	2.66[b]
9	3,4	2.56	2.46	2.40	2.69
10	3,5	1.23	1.18	1.19	2.66[b]
11	2,3,4	2.02	1.96	1.92	3.46[b]
12	2,3,5				3.46[b]
13	2,3,6				3.46[b]
14	2,4,5	2.05	2.08	2.12	3.46[b]
15	2,4,6	1.66	1.65	1.63	3.42[a]
16	3,4,5	1.77	1.69	1.77	3.46[b]
17	2,3,4,5	2.31	2.29	2.37	4.24[b]
18	2,3,4,6				4.24[b]
19	2,3,5,6	2.20	2.15	2.16	4.24[b]
20	2,3,4,5,6	1.31	1.34	1.35	4.78[a]

Log P values from Hansch and Leo (1979) except for:

a Values measured.

b Calculated from Kaiser (1983).

The plot of partition coefficent values (log P) against Microtox toxicity values p(30-EC50) is shown in FIGURE 2 and the simple linear regression analysis gives the following equation 2:

$$p(30\text{-}EC50) = -0.041 + 0.570 \log P \qquad (2)$$
$$n = 15;\ r^2 = 0.80$$

The pentachloroaniline has been excluded from the regression, the basis being the low water solubility of the compound and therefore the probability of a wrong toxicity value.

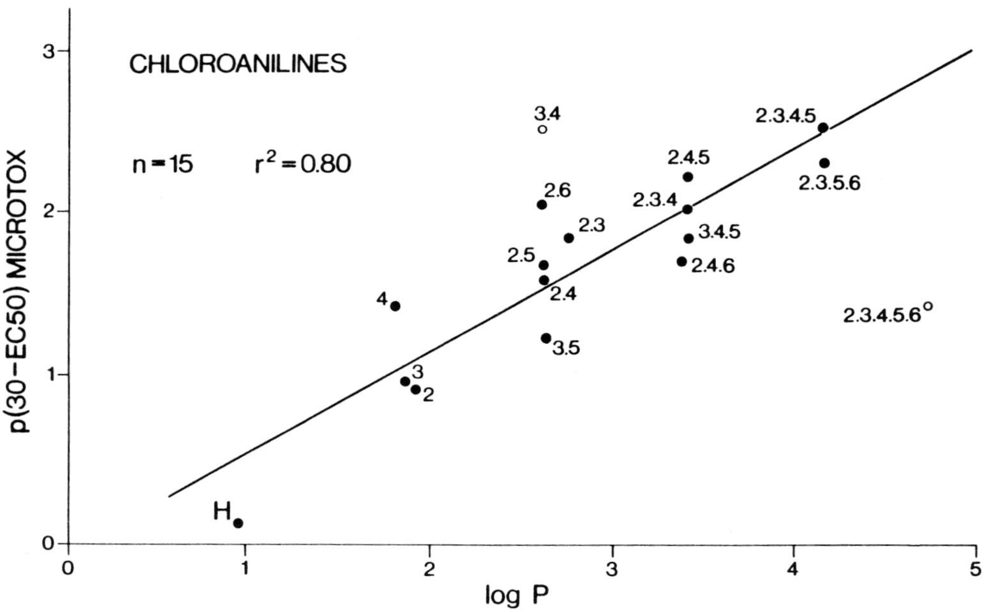

FIGURE 2: Plot of Microtox toxicity p(30-EC50) values of chloroanilines versus the octanol/water partition coefficients (log P); regression curve is given in equation 2

Our previous results (Ribo and Kaiser 1983) on chlorobenzenes and chlorophenols congeners show good linear correlations between toxicity values and log P as shown in the equations 3 and 4:

Chlorobenzenes: $p(30\text{-}EC50) = -0.979 + 0.631 \log P \qquad (3)$
$n = 11;\ r^2 = 0.76$

Chlorophenols: $p(30\text{-}EC50) = -0.679 + 0.678 \log P$ (4)

$n = 20;\ r^2 = 0.79$

In each of the three series, there is an increasing toxicity trend with increasing number of chlorine substitutions and, of course, with increasing octanol/water partition coefficients, which is not surprising in light of the proven dependence of log P on the number of chlorine atoms in the molecule (Kaiser 1983).

From regression analyses on the three sets of compounds it is evident that the slopes of the regression lines (FIGURE 3) are parallel but with different intercepts. This suggests that a similar

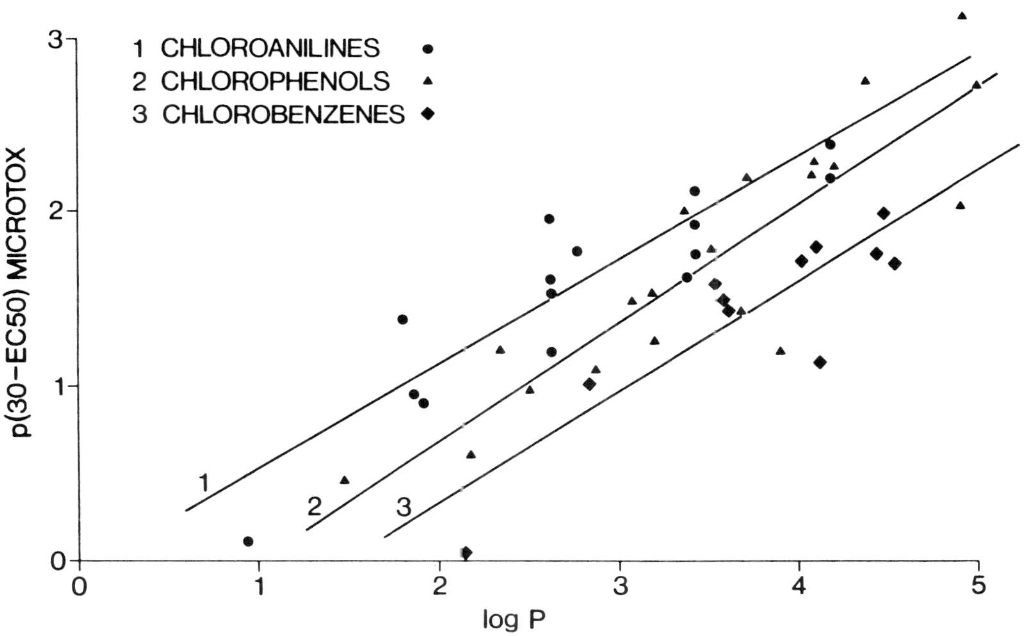

FIGURE 3: Plots of Microtox toxicity p(30-EC50) values of chloroanilines, chlorophenols, and chlorobenzenes versus the octanol/water partition coefficients

mechanism of action exists for these sets of compounds in terms of toxic activity to the test organism. As this slope is less than one, the influence of the chlorine substituents on the compound's toxicity appears to be greater than that of the functional groups. However, this observation cannot be generalized without rigorous testing of a variety of other sets both with different functional groups and nonfunctional substituents.

Correlation with Other Toxicity Tests

In order to investigate the Microtox toxicity test in terms of its ability to provide a measurement of the toxicity of hazardous or potentially hazardous chemicals, the above results were compared with the toxicities of selected chloroanilines to four different species of yeast (Kwasniewska and Kaiser 1984). TABLE 3 lists the concentration of the chemicals causing a 50% inhibition in the growth of the microorganisms. As usual this is expressed as the negative logarithm of the molar concentration in $mmol \cdot L^{-1}$.

TABLE 3: Toxicity values of selected chloroanilines to Microtox p(30-EC50) and four species of yeast p(IC50), expressed as negative logarithms of concentrations

Substituent	Microtox p(30-EC50)	Yeasts p(IC50)[a]			
		Pichia sp.	Sacch. cerev.	Rhod. rubra	Rhod. sp.
-	0.12	-0.03	-0.03	-0.63	-0.03
4	1.40	0.20	0.11	0.04	-0.07
2,4	1.54	0.49	0.13	0.97	0.91
2,6	1.97	0.27	0.40	0.48	0.39
3,5	1.19	0.78	0.99	1.10	0.75
2,3,4	1.92	1.21	0.98	1.21	1.09
2,3,5,6	2.16	-	1.36	1.36	1.20
2,3,4,5,6	1.35	0.16	0.80	0.36	0.36

a Yeast measurements from Kwasniewska and Kaiser (1984).

The regression analysis of these data provide the following equations:

1) Microtox - *Pichia* sp.

$$p(I50) = -0.195 + 0.529\ p(30\text{-}EC50) \quad (5)$$

$n = 6$; $r^2 = 0.49$ (excluding 2,6-Cl_2 chloroaniline)

This equation 5 does not show a significant correlation of the chloroaniline toxicities between the Microtox and the yeast species, even when excluding the 2,6-dichloroaniline.

2) Microtox - *Saccharomyces cerevisiae*

$$p(I50) = -0.0076 + 0.614\ p(30\text{-}EC50) \quad (6)$$

$n = 5$; $r^2 = 0.89$ (excluding 4-Cl; 2,4-Cl_2, and 2,6-Cl_2 chloroanilines)

Equation 6 shows a reasonably good correlation between *Saccharomyces cerevisiae* and the Microtox values. The slope of equation 6 is similar to that of equation 5, indicating similar overall mechanisms for the two yeast species.

3) Microtox - *Rhodotorula rubra*

$$p(I50) = -0.499 + 0.814\ p(30\text{-}EC50) \quad (7)$$

$n = 7$; $r^2 = 0.66$ (excluding 4-Cl chloroaniline)

The correlation of Microtox with *Rhodotorula rubra* values for seven of the chloroanilines is shown in equation 7. Compared to equations 5 and 6, both the intercept and the slope are significantly different for equation 7.

4) Microtox - *Rhodotorula* sp.

$$p(I50) = -0.128 + 0.610\ p(30\text{-}EC50) \quad (8)$$

$n = 6$; $r^2 = 0.86$ (excluding 2,6-Cl_2 and 4-Cl chloroanilines)

As with the correlations 5 and 6, equation 8 shows a smaller slope of the yeast toxicity as dependent on the Microtox values.

Plots of the p(30-EC50) Microtox values against p(IC50) for the four yeast species show that, within experimental errors, both toxicity tests are reasonably well correlated, as shown in FIGURE 4.

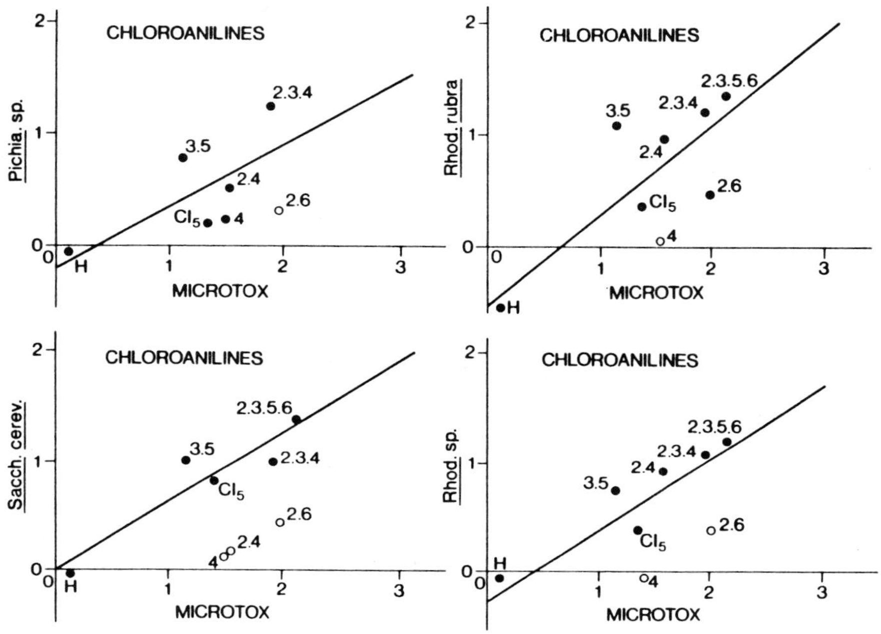

FIGURE 4: Plots of Microtox toxicities of selected chloroanilines versus their acute toxicities to four strains of yeast; regression curves are given in equations 5 to 8

CONCLUSION

Our previous results on chlorophenols and chlorobenzenes (Ribo and Kaiser 1983) indicate that the Microtox test is of great value in terms of cost, speed and accuracy. Furthermore, we showed that the toxicity results are well correlated with toxicity values for other organisms; this work confirms our previous conclusions. Furthermore, we have obtained some new toxicity values for another series of compounds that are potential aquatic contaminants and have preliminary conclusions showing similarity of mechanism of toxic action for the three series of compounds. Other series of polychloro-substituted aromatic compounds will be investigated in the near future.

ACKNOWLEDGEMENT

We thank the Great Lakes Water Quality Program for substantial support for this study.

REFERENCES

Adema, D.M.M. and Vink, G.J. 1981. A comparative study of the toxicity of 1,1,2-trichloroethane, dieldrin, pentachlorophenol and 3,4,-dichloroaniline for marine and fresh water organisms. **Chemosphere 10:** 533-554.

Bulich, A.A. and Isenberg, D.L. 1981. Use of the luminescent bacterial system for the rapid assessment of aquatic toxicity. **ISA Transactions 20:** 29-33.

Bulich, A.A., Greene, M.W. and Isenberg, D.L. 1981. Reliability of the bacterial luminescence assay for determination of the toxicity of pure compounds and complex effluents. In **Equatic Toxicology and Hazard Assessment: Fourth Conference,** D.R. Branson and K.L. Dickson (Eds.), ASTM STP 737, American Society for Testing and Materials, pp. 338-347.

Diachenko, G.W. 1979. Determination of several industrial aromatic amines in fish. **Environ. Sci. Tech. 13:** 329-333.

Hansch, C. and Leo, A.J. 1979. **Substituent Constants for Correlation Analysis in Chemistry and Biology.** John Wiley & Sons, New York.

Kaiser, K.L.E. 1980. Correlation and prediction of metal toxicity to aquatic biota. **Can. J. Fish. Aquat. Sci. 37:** 211-218.

Kaiser, K.L.E. 1983. A nonlinear function for the approximation of octanol/water partition coefficients of aromatic compounds with multiple chlorine substitution. **Chemosphere, 12:** 1159-1167.

Kwasniewska, K. and Kaiser, K.L.E. 1984. Toxicities of selected chloroanilines to four strains of yeast. In **QSAR in Environmental Toxicology,** K.L.E. Kaiser (Ed.), D. Reidel Publ. Co., Dordrecht, Holland, pp. 223-233.

McQueen, C.A., Maslansky, C.J., Crescenzi, S.B. and Williams G.M. 1981. The genotoxicity of 4,4'-methylene-bis(2-chloroaniline) in rat, mouse and hamster hepatocytes. **Toxicol. Appl. Pharmacol 58:** 231-235.

Nelson, C.R. and Hites, R.A. 1980. Aromatic amines in and near the Buffalo River. **Environ. Sci. Tech. 14:** 1147-1150.

Ribo, J.M. and Kaiser, K.L.E. 1983. Effects of selected chemicals to photoluminescent bacteria and their correlations with acute and sublethal effects on other organisms. **Chemosphere 12:** 1421-1442.

Wegman, R.C.C. and DeKorte, G.A.L. 1981. Aromatic amines in surface waters of the Netherlands. **Water Research 15:** 391-394.

WORKSHOP DISCUSSION

HART:

You say that you measure the light emission at 30 minutes. Did you find that the light emission decreases indefinitely?

RIBO:

It decreases so that if you come the next day, there is no light. The test has other, more important problems. The main problem we have found is with the solubility of the compound. For one, you do not have the choice of many solvents and methanol is one of those with very low toxicity. We found that concentrations of 5% or 10% methanol do not affect the final value, but in some cases, even that level of methanol may not be enough to get a compound into solution.

Question:

Going back to your chart where you have your log P's, my question is, how close is the correlation of the Microtox values with the LC50s which you discussed?

RIBO:

In absolute terms, the light output is decreasing with time. For example, if you measure at 60 minutes, you will find little or no light coming out. However, all values are determined relative to the control.

Comment:

You mentioned that you published the correlation between Microtox values and fish LC50's?

RIBO:

We didn't do that with certain compounds because we didn't find literature data for all of them but we did sc for the chlorobenzenes and the chlorophenols. Our results show that the LC50s and the 30 min EC50s correlate quite well with the correlation coefficients in the r^2 = 0.70 to 0.75 range.

McCARTY:

The slopes of those two regression lines are about the same. Is it correct to say that the fish toxicity line is going to have a steeper slope against log P than the Microtox one? I'm looking at some data by Veith for which the slope is found as LC50 = 0.94 log P. The Microtox test appears to give slopes around 0.5 to 0.6, so caused by the higher log P compounds which are not accumulating, i.e., not getting into the bacteria very fast. The two slopes are different and you are not going to get something parallel if you have different rates of uptake in that short time period.

CULLEN:

What kind of error do you get for the Microtox values?

RIBO:

We normally do the test three times and find the same value with errors in the 10% to 30% range.

SCHULTZ:

A few comments: First of all, we have found that we have the same problems with the 2,6 substitution and we're using a methyl substitution so, apparently whatever the group is, it doesn't interact. However, you do get a shielding effect that is somehow not constant.

RIBO:

We found that in chlorophenols, too.

SCHULTZ:

We have looked at a series of naphthalenes where we have looked at different substituents and then gone ahead and used Hansch and Leo's cluster group analysis with the 5 cluster group, whether you believe in the way they are clustered together or not, it doesn't matter, but if you then look at the ability to act as hydrogen acceptor or hydrogen donor, you'll find that cluster group 1 and cluster group 5 do not interact with water and cluster groups 2, 3, and 4 do and we get a very similar relationship. In the chlorobenzenes you have hydrogens and chlorines which do not interact with water. However, with the phenols and the anilines, you get a hydroxy group and amino group which will interact and we get the same sort of relationship for those that have substitutent groups that can interact with water and have a line that is above those that do not interact with water. It would be interesting to look at another series of chlorinated ring compounds, such as chlorinated pyridines to see if they overlap with the benzenes rather than with the other group.

RIBO:

That is precisely what we have in mind. The problem here though is that we have not found any other parameter which would allow a correlation between the groups.

HALLETT:

I'm just concerned, that you are starting to talk about fine lines of chemistry of these compounds without knowing whether things are happening as a result of this test or because of some different biochemical parameter.

KAISER:

No, I think we can differentiate that. One of the first things you get if you have physical limitations is you won't get the straight lines in your log of Gamma versus concentration. Now, we have had a few of these cases. We see, quite clearly, a case of either positive or negative slope and honestly, I am not quite sure why it can be either way. It may be a factor of the solubility, quite obviously you reach maximum rate for the transport of the chemical. The problem is that we don't know the solubility of the chemical in the final test solution with 5% to 10% methanol. They are hardly known in pure water, but then you add your other salts to achieve this saline solution because they are marine bacteria, so obviously the solubility is a little less than before and then you add the bacteria which likely adsorb physically. As a result, experimental conditions for lipophilic compounds are at the border line for solubility.

HALLETT:

In what concentration range of chemical is this test actually working? What solvents will work?

RIBO:

This is water and up to 5% to 10% methanol which we found not to interfere with the test. The concentrations we were using were in the $mg \cdot L^{-1}$ range, approximately 9 $mg \cdot L^{-1}$, maximum.

Question:

With the timeframe you are looking for at - half an hour, what if you ran it at a lower concentration for more time?

RIBO:

That could be a good idea, but then I wouldn't be able to compare both values. With a longer time, it would be difficult to distinguish the toxic effect from the normal loss of light emission.

Comment:

Things that are more soluble in water are probably taken up by the bacteria at a slower rate and therefore take longer to show up.

RIBO:

In traditional tests like LC50's you will have solubility problems, too.

Comment:

You will at a certain log P.

Question:

Do you have the same problems with quenching compounds since you are measuring light output and these compounds absorb light, some of the change in light readings could be simple quenching?

RIBO:

For coloured solutions there are special colour correction cuvettes. We have had experimental problems with some photosensitive compounds. So far we have no evidence for true quenching but it may occur in some cases.

Question:

At what wavelength are the bacteria emitting?

RIBO:

It is in the visible range. I don't have the exact wavelength here but the major emission band is in the 500 nm range.

Question:

Are chlorophenols more toxic than chlorobenzenes to the fish?

RIBO:

Usually yes.

Hallett:

The other thing I would like to mention, is that there has been a myth around that chemicals at the very low concentrations, in the waters of the Great Lakes, are adsorbed onto particulate matter. We did some very large volume sampling of some water, and find it very interesting to find concentrations, in the Niagara River, particularly, of PCB's, of chlorobenzenes, of dioxins and other compounds in a dissolved state. I think we should keep this in mind when we are looking at that. These compounds can occur dissolved in the water column, they can be at relatively high concentrations, especially when you are looking at groundwater, perhaps coming out of landfill.

STRUCTURE-ACTIVITY CORRELATIONS OF SELECTED AZAARENES, AROMATIC AMINES, AND NITROAROMATICS[1]

T. Wayne Schultz and Becky A. Moulton

Department of Animal Science
College of Veterinary Medicine
University of Tennessee
PO Box 1071, Knoxville, TN 37901
U.S.A.

ABSTRACT

Quantitative chemical structure - biological activity relationships between the molecular descriptors 1-octanol/water partition coefficient (log K_{ow}), the ${}^{1}X^{v}$ (one-chi-V) molecular connectivity index and (log K_{ow})^2 and the biological response measured as population growth impairment to the ciliate Tetrahymena pyriformis were investigated for a series of 21 nitrogen substituted aromatic molecules. The test molecules included seven each of the azaarenes, primary aromatic amines and nitroaromatics including single, double and triple ring compounds. Nitro-containing homologs are the most active followed by amino-containing ones and azaarenes. Within each analog series activity increases with alkyl substitution and ring addition. The QSAR model: log BR = 0.8255(log K_{ow}) - 1.6334 was determined to be an excellent predictive model which accounted for 86.9% of the variability in the observed biological activity. This model is thought to express non-specific toxicity due to membrane perturbation.

---

[1] Supported in part by Biomedical Research Support Grant NIH-RR05845-02.

K. L. E. Kaiser (ed.), QSAR in Environmental Toxicology, 337–357.

INTRODUCTION

The need for a time and cost effective means of predicting toxicity of organic compounds has spurred recent interest in quantitative structure activity relationships (QSAR) in ecotoxicology. QSAR is being shown to be a powerful research tool that permits the ranking of industrial chemicals on the basis of meaningful correlations between molecular structure descriptors and biological activity.

Molecular predictors include two basic types: each is a series of numerical indicators useful in the development of QSAR. One category consists of the free-energy related physicochemical properties including electronic, hydrophobic and steric parameters (Hansch and Leo 1979). The most often used physicochemical descriptor is the hydrophobic parameter log 1-octanol/water partition coefficient of log K_{ow} (Hansch 1969).

The second type of predictors includes molecular descriptors which are direct quantitations of molecular structure. This category includes: atom counts, molecular weight, molecular connectivity, molecular orbital theory and quantum mechanics. The most often used of these descriptors is molecular connectivity (Koch 1983; Kier 1980; Kier and Hall 1976). The molecular connectivity descriptor consists of a series of numerical indices which describe the topological characteristic of the molecule. The most common molecular connectivity index is the $^1X^v$ (one-chi-V). The $^1X^v$ term is the sum of the reciprocal square roots of the products of adjacent delta values (nonhydrogen bound valencies). This index numerically encodes information concerning number and types of atoms, branching, cyclization and saturation. The reciprocal square root transformation reduces isomeric redundancy.

The free-living ciliate *Tetrahymena pyriformis* has been shown to be a rapid, inexpensive and well characterized test system which lends itself to quantitative concentration-response analysis. Recently Schultz (1983) has reviewed and standardized its methodology as a short-term screen using population growth as the end-point. *Tetrahymena* is not only appropriate as a zooplankter aquatic toxicity test system but also as an eucaryotic cytotoxicity model.

Previous QSAR studies of nitrogen containing aromatic compounds have centered on neutral and basic nitrogen heterocyclic molecules (Schultz *et al.* 1980; Schultz and Cajina-Quezada 1982; Schultz *et al.* 1982; Schultz 1983). In this investigation are reported correlations between the molecular descriptors log K_{ow}, ${}^1X^v$, and $(\log K_{ow})^2$ and acute toxicity monitored as population growth impairment of *Tetrahymena* for 21 selected one, two and three ring azaarenes, primary aromatic amines and nitroaromatics.

MATERIALS AND METHODS

Axenic cultures of the amicronucleated strain GL-C of *Tetrahymena pyriformis* were tested at 28°C under acute static conditions. The ciliates were cultured in a proteose peptone medium (Schultz *et al.* 1980). The tested chemicals were secured from a commercial source (Aldrich Chemical Co., Milwaukee, WI, U.S.A.) and were not repurified. The compounds selected form a series of aromatic analogs and homologs which can be subdivided into three classes: (1) azaarenes; (2) primary aromatic amines; (3) nitroaromatics. Test chemicals were initially dissolved in reagent grade dimethyl sulfoxide (DMSO) to form 10,000, 50,000 or 500,000 $mg \cdot L^{-1}$ stock solutions. Subsequently, aliquots of stock solutions no greater than 300 μL were added to the culture medium. Previously, Schultz and Cajina-Quezada (1982) noted that DMSO in concentrations as high as 0.75% (375 μL/50 mL) does not affect population growth of *Tetrahymena*.

Based on the investigations of Schultz *et al.* (1981), the biological response of population growth was selected to quantify activity. Population growth was monitored using the method of Cooley *et al.* (1972) with cell density being measured spectrophotometrically as optical density (OD) at 540 nm. Cameron (1973) noted that the number of *Tetrahymena* in culture is directly proportional to the OD. Cultures were grown for 60 hours in 250 mL Erlenmyer flasks containing 50 mL of test solutions. Flasks for each test molecule were prepared in a five step graded concentration series. Flasks without test chemicals served as controls and each flask was inoculated with 0.2 mL of log-growth-phase *Tetrahymena* culture. Two flasks for each concentration were conducted simultaneously for a minimum of three replicates. Each replicate used freshly prepared stock solutions. For analysis, OD values were transformed to percent control OD. For each test compound the best concentration-response line was fitted by least-squares linear regression. The 60-hour IGC50 value (concentration which inhibits 50% growth) and 95% confidence limits were determined for each test compound.

For structure activity correlations, the IGC50 values ($mg \cdot L^{-1}$) were converted to $mmol \cdot L^{-1}$ prior to taking the log of the inverse, hereafter referred to as the log BR (Biological Response). The log K_{ow} values were secured from Appendix II of Hansch and Leo (1979). The first order valence ${}^1X^v$ molecular connectivity index values were determined using the one bond fragment algorithm

$$C_{ij} = (\delta_i^v \delta_j^v)^{-\frac{1}{2}}$$

with the δ^v values being based upon the equation

$$\delta^v = Z^v - h$$

where Z^v is the number of valence electrons and h is the number of bonded hydrogen atoms (Kier 1980). Correlations were examined for the 21 alkyl and ring addition analogs of pyridine, aniline and nitrobenzene using log K_{ow}, (log $K_{ow})^2$ and ${}^1X^v$ as the independent variables with log BR as the dependent variable. The structure activity models were developed using regression analysis with goodness of fit tested by the variance. Analyses were conducted using Statistical Analysis System software.

RESULTS

Presented in TABLE 1 are the Chemical Abstract Service (CAS) registry number, 60-hour IGC50 ($mmol \cdot L^{-1}$) and 95% confidence limits for each compound assayed. Biological activity varies between the three main chemical groups: azaarenes (compounds 1 to 7), primary aromatic amines (compounds 8 to 14) and nitroaromatics (compounds 15 to 21). Within each homologous series, the nitro derivative is the most active followed by the primary amine and lastly the in-ring nitrogen substituted molecule. Within each analog series, activity increases with increased alkyl substitution. Similarly within each main chemical group, activity increases with an increase in aromatic

TABLE 1: Growth inhibition to Tetrahymena of selected nitrogen-containing aromatic molecules

	Molecule	CAS Number	60-hour[a] IGC 50	95% Confidence[a] Limits
1	Pyridine	110-86-1	15.32[b]	11.8 - 21.0
2	3-Picoline	108-99-6	10.41[c]	6.27 - 17.27
3	4-Picoline	108-89-4	7.80	6.04 - 10.09
4	3,4-Lutidine	583-58-4	3.02	2.16 - 4.21
5	Quinoline	91-22-5	0.97[c]	0.67 - 1.41
6	4-Phenylpyridine	939-23-1	0.22	0.15 - 0.32
7	Acridine	260-94-6	0.04[c]	0.03 - 0.06
8	Aniline	62-53-3	1.66[b]	1.34 - 2.06
9	3-Toluidine	108-44-1	2.59	1.93 - 3.46
10	4-Toluidine	106-49-0	1.34	0.46 - 2.33
11	3,4-Xylidine	95-64-7	1.94	1.37 - 2.75
12	1-Naphthylamine	134-32-7	0.60[d]	0.48 - 0.76
13	4-Aminobiphenyl	92-67-1	0.15	0.09 - 0.25
14	1-Aminoanthracene	610-49-1	-	-
15	Nitrobenzene	98-95-3	1.16	0.67 - 1.99
16	3-Nitrotoluene	99-08-1	0.49	0.39 - 0.58
17	4-Nitrotoluene	99-99-0	0.58	0.43 - 0.79
18	4-Nitro-o-xylene	99-51-4	0.23	0.11 - 0.46
19	4-Nitronaphthalene	86-57-7	0.10[d]	0.09 - 0.12
20	4-Nitrobiphenyl	92-93-3	-	-
21	9-Nitroanthracene	602-60-8	-	-

a In $mmol \cdot L^{-1}$.
b From Schultz and Allison (1979).
c From Schultz *et al.* (1980).
d From Schultz *et al.* (1983).

ring addition. The exceptions are molecules 14, 20 and 21, 1-aminoanthracene, 4-nitrobiphenyl and 9-nitroanthracene, respectively, which due to their limited aquatic solubility do not elicit the measured response at saturation.

The log K_{ow}, ${}^1X^v$, $(\log K_{ow})^2$, (observed and predicted) log BR and residual values for each chemical tested are listed in TABLE 2. A plot of log BR versus log K_{ow} shows a direct linear

TABLE 2: Comparison of molecular descriptors and biological response of selected nitrogen-containing aromatic molecules

	Molecule	log K_{ow}^{a}	$^{1}\chi^{v}$	log BR obs.	log BR[b] pred.	Residual[b]	95% Confidence Limits
1	Pyridine	0.64	1.850	-1.185	-1.105	-0.080	-1.725 - -0.486
2	3-Picoline	1.24	2.260	-1.017	-0.610	-0.407	-1.198 - -0.231
3	4-Picoline	1.22	2.260	-0.892	-0.626	-0.266	-1.216 - -0.370
4	3,4-Lutidine	1.68	2.677	-0.480	-0.247	-0.233	-0.823 - 0.330
5	Quinoline	2.04	3.264	0.013	0.051	-0.038	-0.523 - 0.624
6	4-Phenylpyridine	2.55	3.920	0.658	0.471	0.186	-0.109 - 1.052
7	Acridine	3.40	4.679	1.398	1.173	0.225	0.554 - 1.792
8	Aniline	0.90	2.199	-0.219[e]	-0.890	0.671	-1.495 - -0.286
9	3-Toluidine	1.42	2.611	-0.414	-0.461	0.047	-1.044 - 0.121
10	4-Toluidine	1.40	2.611	-0.127	-0.478	0.351	-1.060 - 0.105
11	3,4-Xylidine	1.98	3.028	-0.288	0.001	-0.289	-0.572 - 0.574
12	1-Naphthylamine	2.23	3.610	0.222	0.207	0.015	-0.367 - 0.782
13	4-Aminobiphenyl	2.83[c]	4.335	0.824	0.703	0.121	0.113 - 1.292
14	1-Aminoanthracene	3.22	5.016	-	-	-	- - -
15	Nitrobenzene	1.84	2.500	-0.064	-0.114	0.050	-0.689 - 0.480
16	3-Nitrotoluene	2.43	2.912	0.310	0.373	-0.063	-0.205 - 0.950
17	4-Nitrotoluene	2.41	2.912	0.237	0.356	-0.119	-0.221 - 0.933
18	4-Nitro-o-xylene	2.96	3.324	0.638	0.810	-0.172	0.215 - 1.405
19	4-Nitronaphthalene	3.19	3.911	1.000	1.000	0.000	0.393 - 1.607
20	4-Nitrobiphenyl	3.78[d]	4.552	-	-	-	- - -
21	9-Nitroanthracene	4.17	5.317	-	-	-	- - -

a From Hansch and Leo (1979).
b Based on equation 1.
c Calculated by substituent method log K_{ow} = 4.06 (log K_{ow} biphenyl) -1.23 (π - NH_2).
d Calculated by substituent method log K_{ow} = 4.06 (log K_{ow} biphenyl) -0.28 (π - NH_2).
e Outlier based on equation 1.

correlation (FIGURE 1). Linear regression of these data by the least-squares method yields the model:

$$\log BR = 0.8255 (\log K_{ow}) - 1.6334 \quad (1)$$
$$n = 18;\ r^2 = 0.869$$

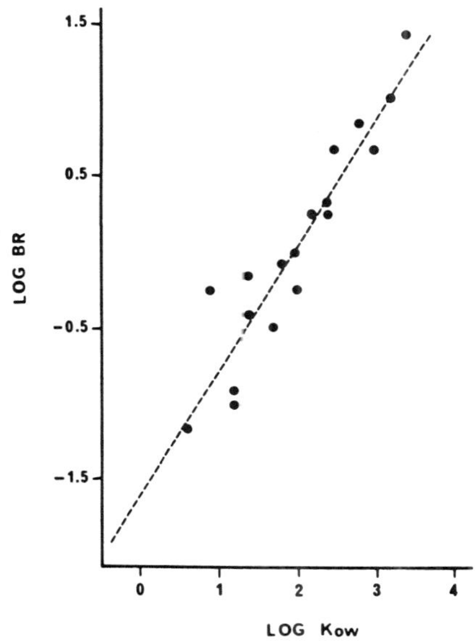

FIGURE 1: Linear regression analysis of log BR versus log K_{ow} for 18 azaarenes, aromatic amines and nitroaromatics

Similar linear QSAR between $^1X^v$ and $(\log K_{ow})^2$ and log BR are presented in models 2 and 3, respectively.

$$\log BR = 0.8152\ {}^1X^v - 2.4505 \quad (2)$$
$$n = 18;\ r^2 = 0.828$$

$$\log BR = 0.1996 (\log K_{ow})^2 - 0.8995 \quad (3)$$
$$n = 18;\ r^2 = 0.874$$

Attempts to develop higher order polynomial models for either log K_{ow} or $^1X^v$ did not increase the coefficent of determination (r^2) values of the models.

A matrix of squared linear model correlation coefficients for the three molecular descriptors is presented in TABLE 3. All three descriptors are highly colinear with all probability values ≥0.0001.

TABLE 3: Matrix of linear correlation coefficients (r^2)[a]

	log K_{ow}	$^1X^v$	$(\log K_{ow})^2$
log K_{ow}	1.000	0.822	0.964
$^1X^v$		1.000	0.816
$(\log K_{ow})^2$			1.000

a Based on the values of 18 molecules eliciting the measured response (see TABLE 2).

Comparisons of the type 1 sums of squares and the r^2 values adjusted for degrees of freedom of the various combinations of two independent variable models are presented in TABLE 4. In all cases the first parameter is highly significant to the model. However, only in models 5 and 8 does the addition of the second independent variable

TABLE 4: Comparison of two independent variable model statistics

Ordered Parameters	P > F for Type 1 SS For 1st var.	P > F for Type 1 SS For 2nd var.	Adjusted r^2	Equation Number
log K_{ow}, $^1X^v$	0.0001	0.0906	0.878	(4)
$^1X^v$, log K_{ow}	0.0001	0.0091[a]	0.878	(5)
log K_{ow}, $(\log K_{ow})^2$	0.0001	0.2657	0.863	(6)
$(\log K_{ow})^2$, log K_{ow}	0.0001	0.4154	0.863	(7)
$^1X^v$, $(\log K_{ow})^2$	0.0001	0.0062[a]	0.883	(8)
$(\log K_{ow})^2$, $^1X^v$	0.0001	0.0849	0.883	(9)

a Significant at $P < 0.05$ level.

significantly ($P < 0.05$) improve the model. The addition of ${}^1\chi^v$ to log K_{ow} (i.e., model 4 or 5) or ${}^1\chi^v$ to (log $K_{ow})^2$ (i.e., model 8 or 9) causes a slight increase in the adjusted r^2 when compared to the single descriptor models 1 and 3, respectively. However, comparison of the adjusted r^2 for the (log $K_{ow})^2$, log K_{ow} model (i.e., model 6 or 7) with that of either model 1 or 3 reveals a slight reduction in the quality of fit.

The results in TABLES 3 and 4 suggest that there is a great deal of overlap in the quantitation of molecular structure provided by the descriptors log K_{ow} and ${}^1\chi^v$, and (log $K_{ow})^2$ and ${}^1\chi^v$, but even more overlap between log K_{ow} and (log $K_{ow})^2$. In addition, log K_{ow} or (log $K_{ow})^2$ is a better single predictor than ${}^1\chi^v$.

The results of earlier studies including the investigations of Zitko _et al._ (1976), Schultz _et al._ (1978), McLeese _et al._ (1979), Könemann (1980), Rogerson _et al._ (1983), Veith _et al._ (1983) suggested _a priori_ that log K_{ow} is an excellent descriptor of toxic response. Thus, it is surprising that the quadradic model (log $K_{ow})^2$ is a better predictor of the present biological activity than the linear log K_{ow} model (see models 1 and 3). However, analysis of residual values (Anscombe 1960) from TABLE 2 shows compound 8, aniline, to be an outlier and more toxic than predicted. Because this molecule is at the lower end of the range of descriptor values, it exerts a biased influence when included in the analysis, suggesting a curvilinear relationship and favoring the (log $K_{ow})^2$ model over log K_{ow}. In the residual value analysis based on model 3, aniline is not an outlier. If aniline is removed from the analyses, models 1 and 3 become:

$$\log BR = 0.9099 \log K_{ow} - 1.8490 \qquad (10)$$
$$n = 17;\ r^2 = 0.933$$

and

$$\log BR = 0.2121 (\log K_{ow})^2 - 0.9915 \qquad (11)$$
$$n = 17;\ r^2 = 0.910$$

respectively, and log K_{ow} becomes the better predictor.

DISCUSSION

The 21 molecules selected for this study represent a test set with similar chemical structures, that is they are nitrogen-substituted aromatics. The nitrogen takes the form of an alicyclic amine, primary amine, or nitrogroup. Compounds 1 to 7 are characterized by weak hydrogen bonding. The other chemicals have functional groups (i.e., NH_2 or NO_2) which act as hydrogen acceptors and thus interact with water to varying degrees.

Ecotoxicological QSAR analyses have evolved from investigations in medicinal chemistry. These studies have shown the value of describing molecular structure in a form suitable for correlation analyses with biological activity. Prediction of toxicity, especially within a series of congeners, has been related to a number of structural and physical parameters including number of carbon atoms, alkyl-chain length and alkyl-ring addition, molecular weight, molecular connectivity indices, molecular volume, partition coefficient, boiling point, as well as various substituent constants. The log K_{ow} as indicated by the studies of Zitko *et al.* (1976), McLeese *et al.* (1979), Könemann (1980), Saarikoski and Viluksela (1982), Rogerson *et al.* (1983), Veith *et al.* (1983) and Schultz *et al.* (1983) is at present the most universal predictor. It is approximately proportional to the rate of movement of organic molecules through biological materials (Collander 1954) and a good index of biological partitioning (Leo *et al.* 1971). Thus log K_{ow} indicates passive transport of the molecule to its intracellular site of action.

While aniline, compound 8, statistically is an outlier from the relationship described by model 1, biologically it is very similar to the other test molecules. Schultz and Allison (1979) in studying the toxic interaction of aniline and pyridine showed this interaction to be slightly less than additive suggesting the same mode of action. One reason aniline appears to be an outlier in this investigation is that the aniline and pyridine data used here were taken from Schultz and Allison (1979). Schultz and Allison (1979) used a different formulation of medium, one lacking Fe EDTA. In addition, they read their assays at 72 rather than 60 hours. As noted by Levy (1973), the exact growing conditions, including medium and age, have an effect on the population growth and density of *Tetrahymena*. It has also been shown that iron content, which varies between batches of proteose-peptone and yeast extract (Hunter et al. 1973), affects population growth (Conner and Cline 1964). Therefore, aniline should be included in this QSAR.

Previous log K_{ow}-dependent QSAR studies have centered on alkyl substituted mononitrogen heterocyclic aromatics (Schultz et al. 1980), alkyl substituted dinitrogen heterocyclic aromatics (Schultz and Cajina-Quezada 1982), and multinitrogen heterocyclic aromatics (Schultz 1983). These data demonstrate similar models to that described in equation 1. Comparison of observed toxicity, i.e., log BR values, for the compounds assayed by Schultz et al. (1980), Schultz and Cajina-Quezada (1982) and Schultz (1983) with the 95% confidence interval for model 1 reveals three outliers: aniline, phenazine and phthalazine. Aniline toxicity has previously been discussed. Pthalazine is an outlier because of the poor estimation of its log K_{ow} (Schultz and Cajina-Quezada 1982). Based on molecular structure there is no reason to suspect phenazine to be so toxic. An explanation of its toxicity is wanting at this time. Following the removal of phthalazine, phenazine and duplicate values the data sets were pooled (FIGURE 2). The resulting combined regression analysis based on 49 chemicals yields the model:

$$\log BR = 0.7230 (\log K_{ow}) - 1.3459 \quad (12)$$

$$n = 49;\ r^2 = 0.903$$

Only one of these 49 compounds, 3-picoline, resides outside its 95% confidence limits of the observed log BR value. It, like aniline, is not a true outlier for there is no reason to believe that substitution of pyridine with a methyl group in the three position should alter the toxicity so drastically.

Schultz et al. (1978) suggest that quantitative changes in biological activity caused by structural differences are of two categories: (1) non-linear modifications caused by the addition or replacement of a major functional substituent, and (2) linear activity alterations caused by variation in the number and/or position of similar substituents. The present results necessitate a re-evaluation of these statements. Linear alterations in biological activity suggest the same mode of action. Such linear changes are related to number, type and position of functional groups only as they affect physicochemical properties and thus the ability of the toxicant to reach the active site. Non-linear alterations in biological activity suggest a change in mode of action, usually but not always, from a physical, non-specific toxicity (e.g., membrane perturbation), to a chemical or specific toxicity (e.g., blocking DNA synthesis).

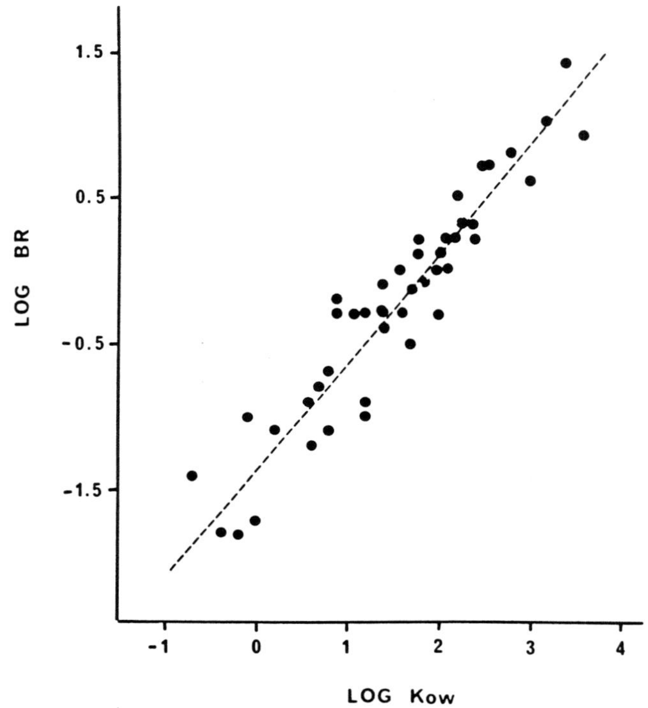

FIGURE 2: Linear regression analysis of log BR versus log K_{ow} for 49 nitrogen containing aromatic molecules

Specific toxic effects will be diverse because they are due to toxicant interactions with specific receptors or enzymes. Molecules which cause such specific toxic effects will in most cases be analogs or homologs of native biochemical molecules. Predicting such specific toxic effects may not be obvious from universal molecular descriptors. An example of such specific toxicity is demonstrated in the data of Saarikoski and Viluksela (1982) where 2,5-dinitrophenol is significantly more toxic than predicted by any of the 1-octanol/water partition coefficient and/or ionization term (ΔpKa) regression models based on the other phenols. The models of Saarikoski and Viluksela (1982) represent base line toxicity due to non-specific narcosis, while 2,5-dinitrophenol is a classic uncoupler of oxidative phosphorylation, a specific toxic effect.

Molecular connectivity indices such as $^1X^v$ are dimensionless numbers which are computed from nonempirical assignments of atom adjacency. However, while not physicochemical properties, they do correlate well with such properties (Kier and Hall 1976), especially for structurally simple molecules. The main advantage of the $^1X^v$ parameter is that it can be easily calculated. The strong colinearity between log K_{ow} and $^1X^v$ exhibited in these studies would be expected to become markedly reduced when more structurally complex molecules (e.g., multiple substituted nitroaromatics, aromatic amines and azaarenes) are investigated. In the latter cases, it is expected that substituent interactions will be expressed better by log K_{ow} than by $^1X^v$.

Linear QSAR have been demonstrated between log K_{ow}, $^1X^v$ and (log $K_{ow})^2$ and log toxicity measured in cell culture. These data suggest that alkyl substituted nitroaromatics, primary aromatic amines and azaarenes have a common mode of action, i.e., membrane perturbation. The model, log BR = 0.7230 (log K_{ow}) - 1.3459, expressed in equation 12 is a good one with which to begin looking at the toxicity of more structurally complex (e.g., 5-amino-6-nitro quinoline), compounds.

REFERENCES

Anscombe, F.J. 1960. Rejection of outliers. **Technometrics 2:** 123-147.

Cameron, I.L. 1973. Growth characteristics of _Tetrahymena_. In **Biology of Tetrahymena,** A.M. Elliott (Ed.), Dowden, Hutchinson and Ross, Stroudsburg, Pennsylvania, pp. 199-226.

Collander, R. 1954. The permeability of _Nitella_ cells to non-electrolytes. **Physiol. Plant 6:** 420-445.

Conner, R.L. and Cline, S.G. 1964. Iron deficiency and the metabolism of _Tetrahymena pyriformis_. **J. Protozool. 11:** 486-491.

Cooley, N.R., Keltner, J.M. Jr. and Forester, J. 1972. Mirex and Aroclor 1254: Effects on and accumulation by Tetrahymena pyriformis strain W. **J. Protozool. 18:** 636-638.

Hansch, C. 1969. A quantitative approach to biochemical structure-activity relationships. **Acc. Chem. Res. 2:** 232-239.

Hansch, C. and Leo, A. 1979. **Substituent Constants for Correlation Analysis in Chemistry and Biology.** John Wiley & Sons, New York.

Hunter, S.H., Baker, H., Frank, O. and Cox, D. 1973. Tetrahymena as a nutritional pharmacological tool. In **Biology of Tetrahymena,** A.M. Elliott (Ed.), Dowden, Hutchinson and Ross, Stroudsburg, Pennsylvania, pp. 411-434.

Kier, L.B. 1980. Molecular connectivity as a description of structure for SAR analysis. In **Physical Chemical Properties of Drugs,** S. Yalkowsky (Ed.), Marcel Dekker, New York, pp. 277-319.

Kier, L.B. and Hall, L.H. 1976. **Molecular Connectivity in Chemistry and Drug Research.** Academic Press, New York, 257 p.

Koch, R. 1983. Molecular connectivity index for assessing ecotoxicological behaviour of organic compounds. **Toxicol. Environ. Chem. 6:** 87-96.

Könemann, H. 1980. Structure-activity relationships and additivity in fish toxicities of environmental pollutants. **Ecotoxicol. Environ. Safety 4:** 415-421.

Leo, A.J., Hansch, C. and Elkins, D. 1971. Partition coefficients and their uses. **Chem. Rev. 71:** 525-616.

Levy, M.R. 1973. Effects of some environmental factors on the biochemistry, physiology and metabolism of Tetrahymena. In **Biology of Tetrahymena,** A.M. Elliott (Ed.), Dowden, Hutchinson and Ross, Stroudsburg, Pennsylvania, pp. 227-258.

McLeese, D.W., Zitko, V. and Peterson, M.R. 1979. Structure-lethality relationships for phenols, anilines and other aromatic compounds in shrimp and clams. **Chemosphere 2:** 53-57.

Rogerson, A., Shiu, W.Y., Huang, G.L. Mackay, D. and Berger, J. 1983. Determination and interpretation of hydrocarbon toxicity to ciliates protozoa. **Aquatic Toxicol. 3:** 215-228.

Saarikoski, J. and Viluksela, M. 1982. Relation between physio-chemical properties of phenols and their toxicity and accumulation in fish. **Ecotox. Environ. Safety 6:** 501-512.

Schultz, T.W. 1983. Aquatic toxicology of nitrogen heterocyclic molecules: Quantitative structure activity relationships. In **Aquatic Toxicology,** J.O. Nriagu (Ed.), John Wiley & Sons, New York, pp. 579-612.

Schultz, T.W. and Allison, T.C. 1979. Toxicity and toxic interaction of aniline and pyridine. **Bull Environ. Contam. Toxicol. 23:** 814-819.

Schultz, T.W. and Cajina-Quezada, M. 1982. Structure-toxicity relationships of selected nitrogenous heterocyclic compounds II. Dinitrogen molecules. **Arch. Environ. Contam. Toxicol. 11:** 353-361.

Schultz, T.W., Cajina-Quezada, M. and Dumont, J.N. 1980. Structure-toxicity relationships of selected nitrogenous heterocyclic compounds. **Arch. Environ. Contam. Toxicol. 9:** 591-598.

Schultz, T.W., Dumont, J.N., Sankey, F.D. and Schmoyer, R.L. Jr. 1983. Structure activity relationships of selected naphthalene derivatives. **Ecotox. Environ. Safety 7:** 191-203.

Schultz, T.W., Kier, L.B. and Hall, L.H. 1982. Structure-toxicity relationships of selected nitrogenous heterocyclic compounds III. Relationships using molecular connectivity. **Bull. Environ. Contam. Toxicol. 28:** 373-378.

Schultz, T.W., Kyte, L.M. and Dumont, J.N. 1978. Structure-toxicity correlations of organic contaminants in aqueous coal-conversion effluents. **Arch. Environ. Contam. Toxicol. 7:** 457-463.

Schultz, T.W., Richter, C.S. and Dumont, J.N. 1981. Cytotoxicity of acridine, a synfuel component, to Tetrahymena. **Environ. Pollut., Ser. A, 26:** 215-226.

Veith, G.D., Call, D.J. and Brooke, L.T. 1983. Structure-toxicity relationships for the fathead minnow, Pimephales promelas: Narcotic industrial chemicals. **Can. J. Fish Aquat. Sci. 40:** 743-748.

Zitko, V., McLeese, D.W., Carson, W.G. and Welch, H.E. 1976. Toxicity of alkyldinitrophenols to some aquatic organisms. **Bull. Environ. Contam. Toxicol. 16:** 508-515.

WORKSHOP DISCUSSION

DUNN:

When you said the model breaks down; do you mean that the compounds are more toxic than log P would predict?

SCHULTZ:

In all cases, the compounds are more toxic, so these models tend to be baseline toxicity. Whatever you are having after that is always more toxic than predicted but whether that toxicity results from a different mode of action or the incorrect molecular descriptor value, is unclear at this time."

DUNN:

We have seen the same thing with other data, for example with fish toxicity data that we've been looking at. We were looking for suitable descriptors and we finally found one which we think might be helpful; it is not easy to get but I would like to comment on it.

One thing we have noticed, is that those compounds that are more toxic than log P predicts, are usually small molecules that have a high charge concentration and they have a number of polar groups.

What we've done is a CNDO, a charge-density calculation on the molecule. We calculate the charge on the atoms and take the sum of the absolute values of the charge on the atoms and then divide that by some parameter which can measure the size of the molecule. For that, we are using molar refractivity right now because it is very easy to calculate and it is highly correlated with the size or area of the molecule. This charge density parameter accounts for, at least we think, a very high charge concentration of the molecule and may prove to provide another component to toxicity which log P doesn't account for. I would say that in about 100 small organic molecules, that are more toxic than log P predicts, we improve prediction of toxicity considerably so that we think that this might be an approach, at least to describe molecules that are heterogeneous and provide another indicator of their toxicity. I would like to have any comments on this if you have any. We haven't published this yet. We are still working on calculating charges, which has it's problems too but as a descriptor, so far, with the molecules we have looked at, we have had some luck.

KAISER:

How long does it take you to calculate a molecule like that?

DUNN:

We can do eight or ten calculations a night, but you have to input the data in advance and form the connection table and you need the coordinates, so that it is not real easy to do but it is the only thing that we have had any luck with at all and it does seem to bring the very small molecules with high charge-density into line.

KAISER:

Along the lines of your first question, my question was, is the toxicity higher than what you would predict on the basis of your molecular connectivity as well?

SCHULTZ:

Yes it is. We looked at those molecules that are outliers like 4-nitropyridine and dinitrobenzene, 4-nitroaniline and of course what mainly jumps out is that you have a 2-nitrogen substitution on the aromatic ring which is just like the classic uncoupling agent 2,4-dinitrophenol. So we tested 2,4-dinitrophenol and it fell right in line with those other molecules. But then, we went ahead and made the fatal flaw of going one step further and tested the 4-vinylpyridine. It inhibits protein synthesis and lo and behold that one is more toxic than predicted and since we have only got four or five compounds, it basically falls on the same line as the other ones so I don't really know if it is a change in the mode of action or if it actually is a problem with the molecular descriptor. You can start playing with some of those molecular descriptors if you are using a rather heterogeneous set like that and you can get lines to fall out based on whether the molecular connectivity is for single ring compounds or double ring compounds. The trouble here is that you have a limited number of data points and when you start to do regressions you don't know what is real and what is not. Right now we are going back to look at a larger number of these multiple nitrogen-substituted compounds. We anticipate that by the end of the year we will have about forty of these nitrogen-substituted compounds as opposed to the forty-nine we have for our other model and we should start to be able to shake some things out.

KAISER:

Are you eventually going to be able to determine the log P experimentally?

SCHULTZ:

That is a good question. I got Rogerson's paper and was interested in your determination of some of the log P values based on the retention time with high pressure liquid chromatography and that would be interesting. My problem, of course, is that as a biologist, I am really not set up to do those things and even if I had the money to get set up, I don't know what the time lag would be before I would get the expertise to feel comfortable in doing those things. From what I have heard from talking to other people, it seems to be an inherent problem to come up with molecular descriptors for the complex molecules. That is the one thing that has made molecular connectivity so beautiful, you can always calculate the value and know what it is.

ZITKO:

Which is more toxic, aniline or nitrobenzene?

SCHULTZ:

Nitrobenzene is, consistently.

ZITKO:

That really means then that the toxic compound is some product of reduction on the nitro group?

SCHULTZ:

That is probably true. It is interesting to note that the nitro-substituted pyridines are more toxic than what would be predicted by log P values, however, amino-substituted pyridines are not. They fall on the baseline, so apparently it is a little more than just reactivity.

ZITKO:

That would make sense because the rate of reduction on the nitro groupd would be slower meaning that the intermediate would be generated more slowly and have more chance to react.

SCHULTZ:

Also, if you want to make the problem even worse, when you start to look at 1-2 versus 1-3 versus 1-4 substitutions of the isomers - you start to see some things - the ortho effect that we have commonly referred to where you get the 1-2 more toxic than the other one doesn't always hold true. However, it seems to hold true with the very reactive groups.

MACKAY:

If we accept that, what we are seeing here is basically a partitioning of the chemical, between the aqueous solution and some phase within the organism which is related to K_{ow}. Wouldn't it be worth trying to regress BR (Biological Reactivity) against K_{ow} or trying the equation

$$BR = constant \times K_{ow}?$$

That is a one-constant equation. When we examine the bio-concentration factors and we find that the one constant equation does almost as well as the two constant equation of

$$\log BR = constant \times \log K_{ow} + constant$$

SCHULTZ:

Does this work for the small molecules?

MACKAY:

I don't know.

SCHULTZ:

We have a problem that once we get the partition coefficient around 1, which you get with a pair of molecules, it really starts to spread at the bottom of the scale.

MACKAY:

We are all brought up to do linear regressions and of course you can't do a linear regression of K_{ow} against BR. You have to take logarithms in order to normalize and express the errors reasonably. So rather than do a linear regression, of log K_{ow} against log BR, why not start doing regressions of K_{ow} divided by BR. If it is constant, then you could have a one constant equation. If it is not, then you have got a second constant to play with and you can try something else. We tend to be obsessed with plotting log something against log something. There is no a priori reason for doing this except that is what the statistics books suggest.

SCHULTZ:

We are playing with some planar models where we try to take into consideration both log P and $^{1}X^{v}$ and of course you can do that with the computer and you can come up with some beautiful planes and you can turn them around on your screen and take pictures of them in different positions which really look nice when you go to publish them. But, I really don't know if it is giving us any more information than what we had in the past and the problem has been that the molecules that we have looked at have been basically so simple and the log P value just jumps out at you with a beautiful value that there hasn't been any point in the past in pursuing it any further. It is only now that we have this model with sufficient numbers of compounds that we can start to compare that to new, more complex things, and we start to open up whole new avenues of approaches; at least a lot of questions!

LANDE:

Further to what you were talking about on mechanisms, am I correct in saying that the toxicity is a function of the rate of uptake such as the crossing of a membrane, which is the controlling kinetic effect?

SCHULTZ:

Log K_{ow} is just related to the partitioning of the material, the greater movement of the material to the site of action. Now, in baseline toxicity, in a cell system like this, toxicity is probably just a disruption of the outer cell membrane. However, if you have a specific site of action such as you would have with the uncoupling agent 2,4-dinitrophenol, we know that this agent also has to get to the inner mitochondrial membrane. So, that is really partitioning it into three different compartments. We don't know that this is the case for every compound but one of the advantages you would have with QSAR once you generate the data base with a number of diverse compounds is that you could start to let the organism tell you something about the mode of action. That may tell you differences between compounds and obviously, the biggest problem we have with QSAR right now is how do you lump and split the groups.

Comment:

At the very least, it has provided baseline toxicity, so you know what the bottom line toxicities are with respect to a narcotic mechanism.

SCHULTZ:

Yes, you could say this compound is going to be no more toxic than this, but it may be considerably more toxic than that if it has a different mode of action.

RIBO:

Do you have any data on di-amino compounds?

SCHULTZ:

Yes. We have looked at phenylenediamine, however, we want to look again at the data because it is substantially different than what you would predict on the basis of log P.

DATA EVALUATION IN THE DEVELOPMENT OF QSAR

Vladimir Zitko

Fisheries and Environmental Sciences
Department of Fisheries and Oceans
Biological Station, St. Andrews, N.B. EOG 2X0
CANADA

ABSTRACT

Descriptors of environmental distribution, structure, properties and effects of chemicals are discussed. It is stressed that the quality of quantitative structure-activity correlations depends on the information content of the descriptors.

INTRODUCTION

It has been just over eight years since the Symposium on Structure-Activity Correlations in Studies of Toxicity and Bioconcentration with Aquatic Organisms took place in Burlington, Ontario (Veith and Konasewich 1975) and I would like to offer my assessment of the progress achieved by structure-activity correlations in environmental sciences during this period and to mention a few current priorities.

Quantitative structure-activity relationships (QSAR) have gained general recognition as one of the important tools of environmental sciences and their scope has been expanded to include relationships between structure, properties, and environmental fate as well. The progress in these two fields has been particularly spectacular.

K. L. E. Kaiser (ed.), QSAR in Environmental Toxicology, 359–368.

DATA SETS AND DESCRIPTORS

Success of QSAR depends on two factors: data sets, obtained by testing selected chemicals, and descriptors, describing the chemicals, their properties and effects. Properites determine the effects and behaviour of chemicals. QSAR attempt to recognize and model these relationships by using structural descriptors, models, and measures of biological activity, as outlined in SCHEME 1.

Environmental distribution	←	Properties	→	Effects
		Descriptors		
Models		Structural descriptors		Biological descriptors
Environmental	←	Topology	→	Toxicity
Compartmental	QSAR	Geometry	QSAR	(EC50; etc.)
		Pharmacophores		

SCHEME 1: Generalized scheme of the relation of environmental distribution, properties, and effects of contaminants, and the corresponding QSAR

QSAR, in the bottom part of SCHEME 1, are relationships between structural descriptors, characterizing the structure of the chemicals and descriptors of environmental distribution (models) or descriptors of biological effects (such as EC50). The relationships are obtained by testing a number of chemicals (data sets). The degree to which these relationships reflect correctly the data sets is affected by the quality of the descriptors, and will improve with improvement in the descriptors (vertically in the above scheme). Thus for example, the accuracy with which QSAR predict effects depends also on the extent with which the effects are reflected by EC50, or whatever other effect descriptor has been selected.

Structural Descriptors

A considerable degree of sophistication has been achieved in expressing structure of chemicals by structural descriptors in pharmaceutical and in chemical information research. Topological descriptors are mostly based on connectivity indices (Hall and Kier 1977). Geometrical descriptors include molecular surface, volume, principal axes of rotation, etc. (Hasan and Jurs 1983). Systems using pharmacophores and their relative positions in the molecule are available as well (Avidon _et al._ 1982). Environmental QSAR are not exploiting fully all the possibilities available in the field of structural descriptors and descriptors other than log P, solubility in water and possibly a few "classical" substituent constants remain the exception rather than the rule.

If one descriptor is good, two are not necessarily better, particularly when they reflect the same property (or, in other words, are highly correlated). This so-called collinearity of descriptors requires a fairly specialized mathematical treatment (Mager 1982). For example, the tight relationship between solubility in water and log P is well known. Similarly, molar refractivity, volume, Traube's rule volume and parachor are significantly correlated with the van der Waals volume (Charton and Charton 1979).

Structure-Properties Relationships

QSAR have been used successfully in relating structure and properties of chemicals. The techniques of calculating log P from contributions of molecular fragments are well known and relationships between log P and solubility in water, and log P and adsorption coefficient have been established. Many other structure-physical properties relationships are available, particularly in the chemical-engineering literature. Vapor pressure correlations developed recently (Mackay _et al._ 1982) are very useful for predictions of the distribution of organic chemicals in the environment. Other properties are predictable as well. One may, for example, predict chromatographic retention times (see for example Hasan and Jurs 1983) and assistance in identification of chemicals may soon become another environmental application of QSAR.

Environmental Models

There has been dramatic progress in the development of environmental models. The simpler, fugacity-based assessment models (Mackay and Paterson 1982) are more comprehensible and are very helpful in understanding the principles of the movement of organic chemicals in the environment. More complex models, such as the NRC model (Roberts et al. 1981) and EXAMS (Burns et al. 1982), have been developed, but do not appear to be widely used. This is likely to change once these models become more readily available.

Compartmental Analysis

Closely related to environmental models are compartmental models of uptake and excretion of chemicals in fauna and flora. The research in this field has concentrated on the relationship between log P or solubility in water and the bioconcentration factors, primarily in aquatic fauna, and good correlations have been obtained. However, the mathematical background of the bioconcentration factors, if used at all, is based on the one-compartment model, which is a considerable oversimplification. On the other hand, even with this simple model, additional relationships could be obtained if one looked at the uptake and excretion rate constants and not limited the investigations to their ratio, the bioconcentration factor. Limited data available in the literature indicate that the uptake rate constant increases, and the excretion rate constant decreases with increasing log P (Zitko 1980). It may be that in this case two descriptors are better than one. Three or more compartment models may have to be used in certain instances. The effect of the organism's size on the uptake, distribution and excretion of chemicals has not been studied to any detail.

Progress in compartmental analysis should clarify the relationships between concentration in organs and tissues and toxicity, and determine the role of transfer rates in relating various exposure regimes to toxicological effects. Good understanding of the toxicological significance of observed concentrations of contaminants is very important for the interpretation of biota-monitoring data.

Effect Descriptors

Closely related to compartmental analysis is also the subject of effect descriptors. The majority of QSAR effects studies uses just a single descriptor (for example, EC50 or LC50) even though much more information has been gathered to obtain these values. It is worth to consider the whole lethality curves and use their parameters (Zitko 1982; Kooijman 1983) as effect descriptors.

Mathematical Techniques

So far, mathematical techniques of data evaluation in environmental QSAR have been straightforward, mainly because of the relatively simple data structure. As the data become more complex, the techniques will have to follow suit. An excellent overview paper is available (Wold and Dunn 1983) and an example of multivariate pattern recognition analysis by the SIMCA method is also described in this volume (Dunn _et al_. 1983).

Data Sets for QSAR

The selection of chemicals for the development of aquatic toxicity QSAR leaves a lot to be desired. For example, at this workshop, the bulk of the data deals with chlorinated benzenes, phenols, and anilines! We seem to be, consciously or unconsciously, selecting series of closely related chemicals that differ mostly in log P and then are very pleased that their toxicity differences are explainable by the variation in log P. It was suggested a few years ago at another workshop that a set of about 2000 general organic chemicals should be selected for QSAR development. Unfortunately, nothing has been done about it.

Up till now we have been dealing with well-defined single chemicals. Many commercial chemical materials are complex, more often than not ill-defined mixtures, with composition either unknown or hidden behind a chemically meaningless tradename. QSAR will no doubt be applicable to such mixtures, but their composition must be known.

Pesticides and QSAR

Effects QSAR are not likely to play a leading role in the case of pesticides. Because of the widely varying nature of these compounds, detailed toxicity testing will always be required. On the other hand, environmental distribution QSAR will play an important role in hazard estimation and risk assessment (Zitko and McLeese 1980).

CONCLUSION

QSAR have undergone considerable progress during the last eight years, but are not yet able to provide predictions of effects for a wide variety of chemicals to the extent expected and hoped for by regulatory agencies. Much more work is required to reach this goal. On the other hand, the path leading to it is now much better defined than it was eight years ago.

REFERENCES

Avidon, V.V., Pomerantsev, I.A., Golender, V.E. and Rosenblit, A.B. 1982. Structure-activity relationship oriented languages for chemical structure representation. **J. Chem. Inf. Comput. Sci. 22:** 207-214.

Burns, L.A., Cline, D.M. and Lassiter, R.R. 1982. Exposure Analysis Modeling System (EXAMS). User Manual and System Documentation. **Environ. Res. Lab. Athens,** GA. EPA 600/3-82-023.

Charton, M. and Charton, B.I. 1979. Significance of "Volume" and "Bulk" parameters in Quantitative Structure-Activity Relationships. **J. Org. Chem. 44:** 2284-2288.

Dunn, W.J., Stalling D. and Wold, S. 1983. The use of SIMCA pattern recognition in the analysis of complex chromatographic data. In **QSAR in Environ. Toxicol.** , K.L.E. Kaiser (Ed.), D. Reidel Publ. Co., Dordrecht, Holland, pp. 75-88.

Hall, L.H. and Kier, L.B. 1977. Structure-activity studies using valence molecular connectivity. **J. Pharm. Sci. 66:** 642-644.

Hasan, M.N. and Jurs, P.C. 1983. Computer-assisted prediction of liquid chromatographic retention indexes of polycyclic aromatic hydrocarbons. **Anal. Chem. 55:** 263-269.

Kooijman, S.A.L.M. 1983. Statistical aspects of the determination of mortality rates in bioassays. **Water Res. 17:** 749-759.

Mackay, D. and Paterson, S. 1982. Fugacity revisited. **Environ. Sci. Tech. 16:** 654A-660A.

Mackay, D., Bobra, A.M., Chou, D.W. and Shiu, W.Y. 1982. Vapor pressure correlations for low-volatility environmental chemicals. **Environ. Sci. Tech. 16:** 645-649.

Mager, P.P. 1982. Principal component regression analysis applied to break collinearities and multicollinearities in QSAR. **Zbl. Pharm. 121:** 115-126.

Roberts, J.R., Mitchell, M.F., Boddington, M.J., Ridgeway, J.M., McGarrity, J.T. and Marshall. W.K. 1981. A screen for the relative persistence of lipophilic organic chemicals in aquatic ecosystems - an analysis of the role of a simple computer model in screening. **National Research Council of Canada,** Publication No. NRCC 18570, 302 p.

Veith, G.D. and Konasewich, D.E., Eds. 1975. **Structure-Activity Correlations in Studies of Toxicity and Bioconcentreation with Aquatic Organisms.** International Joint Commision, Windsor, Ontario, 347 p.

Wold, S. and Dunn, W.J. III. 1983. Multivariate quantitative structure-activity relationships (QSAR): Conditions for their applicability. **J. Chem. Inf. Comput. Sci. 23:** 6-13.

Zitko, V. 1980. Relationships governing the behaviour of pollutants in aquatic ecosystems and their use in risk assessment. In **Proceedings of the Sixth Annual Aquatic Toxicity Workshop,** November 6-7, 1979, Winnipeg, Manitoba, J.F. Klaverkamp, S.L. Leonard and K.E. Marshall (Eds.), **Can. Tech. Rept. Fish. Aquat. Sci. 975:** 243-265.

Zitko, V. 1982. LETCUR, the lethality curve program. **Can. Tech. Rept. Fish. Aquat. Sci. 1134,** 10 p.

Zitko, V. and McLeese, D.W. 1980. Evaluation of hazards of pesticides used in forest spraying to the aquatic environment. **Can. Tech. Rept. Fish. Aquat. Sci. 985,** 21 p.

WORKSHOP DISCUSSION

KAISER:

Do you think that experimentally, two or three compartment models can be handled?

ZITKO:

They can be handled. There are, of course, some other things we have to know, the compartment size which we usually neglect, which would mean to determine the volume of blood, the weight of the gills or whatever. It can be handled, it will only be more difficult, it is more work. For the procedures of fitting constants to models, the more compartments you have the more difficult it is fitting the constants but it has been worked out in pharmacology, pharmokinetics so there are established programs to do this. The main thing is to get the data. Looking at the literature, it is amazingly hard to find what is the typical liver weight in trout, what is normal, nobody measures that.

CHU:

That is true, in mammals the data are known for a certain size of animal and they increase their organ weight in roughly a constant ratio to their body weight but the blood and the amount of muscle occupy approximately 70% of the body weight. I think that the data for the aquatic species are lacking.

ZITKO:

It doesn't require too much to do chemical analysis because of instrumentation but you need to have a skilled chemial technician to do it. With very little skill you can teach somebody how to disect aquatic species and to take a range of sizes and just have them disect the specimens and weigh the organs. This type of information is most valuable.

MACKAY:

I like your thought about not just using an LC50 but looking at the entire curve. Is there merit in trying to establish the nature of that curve and to develop equations which will enable us to extrapolate down towards lower concentrations and longer times which are closer to environmental reality?

ZITKO:

I think it is very important to have a model and describe it mathematically, then test whether we understand what is going on because there is also another factor, toxicity. We know for many compounds that you can expose fish to a chemical and kill it quickly but the residues will be fairly low and you can expose a fish to a lower concentration for a longer time without killing it and the residues will be higher. We don't quite understand quantitatively what is going on. I think it is very important to look at toxicity first, to have a working model from which we could extrapolate, and to also understand the mechanism of how the toxicity comes about. Obviously, if you kill a fish at the high concentration very quickly, you saturate something, somewhere. It would be nice to know what. Also, it would be nice if we go out into the field and collect trout from Lake Ontario and find PCB levels in the brain and in the liver, what does it mean? Does it mean that the fish are badly affected or does it mean that we are far away from some danger level to the fish, we don't really know. It would be nice to know and I think the place to start is that we should develop a relationship with the total system.

MACKAY:

Do you think there is any possibility of developing multi-compartment models and trying to validate them using the injection-type techniques that Peter Hodson described?

ZITKO:

I think it would be very good, you inject the fish and follow where it goes. You could get quite a few inter-compartment exchange rates out of it and that would be the way to do it. Of course, you could get the other part from water exposure because we still have to know how it gets in, the kinetics of the uptake but both ways are approaches to the problem.

SCHULTZ:

Would you care to make any statement concerning what you would think would be first time cut as far as the number of different groups or how would you go about doing that? After making that statement, if we all went out and did that, do you think it would be better if we went out separately or to try and come up with some master list with which to try and work with so we can start to compare between the different systems? That is one of the advantages we have seen the past couple of days with the chloroanilines, the chlorobenzenes and we are starting to see some trends that are reproducible.

ZITKO:

I think that a group of people should get together and develop a master list and this actually was a recommendation of the San Antonio workshop, to select about two thousand chemicals and then do the testing and this eliminates duplicity. As far as the groups go, we have the normal selection of amino, hydroxy, and carboxy groups, esters, etc., and we should probably limit it to aromatic compounds, to benzene derivatives. I think that the best thing to do would be to have a group of people and use their collective ideas on the matter.

HODSON:

I just wanted to comment on your idea about the toxicity curve. There are a couple of little problems in there that a lot of people don't realize. One is that the slope of the curve is directly related to the precision with which the final endpoint is measured. When you assemble this data, you just take the final endpoint without the slopes, even though a lot of the data may be estimated with completely different precisions. In the classical, statistical sense, these date are not strictly comparable. We may be comparing apples and oranges in some of our QSARs. The slope itself may then become a variable which could be used in a QSAR.

ZITKO:

I would like to see some theory behind it in order to model it somehow. For example, to kill a fish we have to reach a certain concentration of a chemical in a certain compartment, to develop an equation on this basis to test it - rather than to have the one point and slope in the classical toxicological sense, I would like to have some more complete information.

GENERAL DISCUSSION

KAISER:

We certainly need new tests. Whether it is now serum sorbitol dehydrogenase or intraperitoneal injection as a fast way to get around some uptake problems such as solubility in water or cost. New tests to determine mutagenic or other nasty properties of compounds. However overall, as a chemist, I do fear that we have too many tests already, so should we forget about all the old ones? I just wish, as a chemist and interested in structure-activity correlations, that the biologists would come up with a limited number - maybe five or maybe fifty - standard tests where all the chemicals they test would really be applied to. I understand the biologists are working on this and I hope they will succeed.

What do the biologists want from the chemists? I really have to ask the biologists. Let's keep the discussion along these lines. If there are any specific comments or generalized statements about specific papers or sessions, please let's hear them.

Comment:

We still have our standard tests that we are still doing.

SCHULTZ:

I think that once we start to compare the results of the different tests much like you alluded to in your talk, that although we think we have different endpoints, such as with the Microtox test, the yeast test, the Daphnia 48-hour or 96-hour fish test, that acute tests are basically measuring the same thing. We may be able to reduce the total number of tests we need to come up with a limited number of tests. On the question of chronic versus acute; I think that one is much more up in the air. Maybe we can draw correlations between bluegill and trout but I don't know if we can compare Daphnia to fish. Eugene Kenaga has opened that question but unfortunately with his data base you don't really know where some of his problems

K. L. E. Kaiser (ed.), QSAR in Environmental Toxicology, 369–377.

lie and there were a lot of apples and oranges thrown into one pot. On the other hand, the number of chemicals or maybe more correctly the number of chemical groups that were actually looked at, are very few and we may find that some of the problems we have is the fact that we have looked at chlorinated or alkyl-substituted aromatics for 90% of the data. Once we get into the more reactive compounds, hydrogen donors and acceptors or the really more complex molecules with a variety of substituents on the rings, we may need different descriptors and larger number of tests than we have now. So, some of these things will have to be addressed, I think.

KAISER:

So, what you are saying then, is that we should generally look and move towards testing of possibly less persistent and more reactive compounds?

SCHULTZ:

That is one possibility. What about anilines versus nitroaniline, for example? That is something that I recently attempted to address and a lot of things were not as clear and simple. I am going to use "MY" system to test reactive compounds. What about other systems?

HODSON:

My feeling about biological testing is that it is quite right that it is the weak area in this science, I would say that chemistry and modelling are miles ahead of biology in terms of sophistication. What is missing are data appropriate to the end use of QSAR. This whole business of looking at the short-term toxicity of essentially insoluble compounds, generates a lot of useless data. Also, in terms of the end use, it is certainly important to show whether a QSAR exists, or whether you can use certain parameters to predict toxicity. For this purpose, you can use nearly any organism in any test which is suitable and convenient. However, if we are really interested in estimating environmental impact, then many of our methods are irrelevant. For example, look at acute toxicity in fish, how many times do we ever see acute toxicity in the environment? If you had a test that you felt was a better indicator of long-term toxicity it would not necessarily have to be a chronic test. For example, if a certain molecular parameter was correlated with chronic toxicity, then you could ask the companies for data on that molecular parameter. I wouldn't let current practice get in our way at this point; if we continue to ask for relevant data then I think QSAR will really progress.

SCHULTZ:

I take exception to that. I think that QSAR has progressed to the point where we can now say we have got some results that are relevant. We certainly wouldn't have said that in 1975 and the first thing was the fact that we are sitting here now and talking about log P and throwing out a number of 4,5 or 6 depending on whether it is calculated or whether it is experimentally determined and we said "we are going to have problems". That was never addressed before. It was considered that that scale went on forever and the same problem exists at the other end. We are saying that when we have low log P's or log P's that are negative, such as for some heterocyclics, with a lot of substitutions on them, which are very hydrophylic, then we also have problems. I don't know what region is in between, but the point of it is that from several independent sources we are coming to that same general trend and it is opening our eyes and we are saying, "okay, now we know we have these problems, but we can use this very simple linear relationship and are able to compare between species and between acute and chronic effects as long as the log P values for our compounds are within this range. When they are not, then we are going to have problems". Those are points that need to be addressed.

KAISER:

What is the overall view of the usefulness of QSAR for legislative, administrative purposes? Can you use it for that purpose?

LANDE:

I like your comments also being one of the regulatees. I would like to add a comment about molecular descriptors, what they really mean and a type of question. Are people just using log K_{ow} because it has always been there or has it been questioned? I really question its application without going any further into what the molecular descriptors are. The same holds true for the use of the connectivity index. Are people using it without question and knowledge of the molecular basis of its activity?

Comment:

I would like to address that point for a moment. We have a structure-activity panel of experts that have been in the business for years on all aspects, from metabolism to carcinogenicity. They always want to know what is the octanol/water partition coefficient. They figure that if a thing is going to be bioavailable then you have a basis for looking at it further for its toxicity.

McCARTY:

It is the first approximation and the thing is, people must explore to see if and when it will break down. To start off with something completely new means you throw all the usefulness you might have gained out of all this effort. Somewhere, along the line someone might find a better descriptor and then you can proceed on that basis and we can refine the models that have been established rather than tear the whole building down and start with the ground floor again. You have to build on what you have done in the past and just can't come along and take a break and start all over again because then you have lost continuity, so just as this has been changing over the years it will continue to change. Just as we were in this workshop, questioning some of the log P's and their applicability and the accuracy, I wasn't aware that there were as many inaccuracies as there are, so therefore I have learned something new so I am going to be a little more careful the next time when using that and I have learned that there may be an error introduced which I didn't think was there.

LANDE:

How about then, at the limits to use it as a descriptor, will you know at that particular point that there are questions and what the log P actually does?

SCHULTZ:

It is only as we develop the data bases chemically and biologically that we are going to be able to address these questions. You can't do it on six points and two descriptors (i.e., log P and molecular connectivity) because until you get into a mixture where you have those compounds that have a negative effect on log P and have a positive effect on ${}^{1}X^{v}$ do you ever see that linearity break down and we are just now starting to scratch the surface. If you will notice in the literature in the last two years there have been four or five papers on molecular connectivity and toxicity. Those form the first wave and I would compare that to the 60's and early 70's and the papers with log P.

KAISER:

Unfortunately Rainer Koch couldn't be here or he would have been able to tell us more about it.

ZITKO:

I think we should select compounds that expand the applicability of log K_{ow} and do some very careful testing both at very high K_{ow}'s and at very low K_{cw}'s to gain some understanding of what is going on. In the meantime, let us do some more testing on a wider variety of compounds because we always select some homogogous group so that K_{ow} is going to apply because that is what we are trying to prove. Let's go into some other functional groups and on top of it, let's look at the ends of the regions.

LANDE:

I think Don Mackay very accurately presented the position that you have to understand the thermodynamic basis for these things. Sometimes when there is a question of fuzziness, if you go back to that thermodynamic basis, you don't need another parameter.

FREITAG:

I have another problem. We tried to correlate structure and activities and we made computer models and the effect is that very seldom is the input in the environment in form of a pure compound. In most cases, we use mixtures of compounds with small parts of other compounds. So, which of the structures will we correlate the activity with, the bio-concentration factors so that all the results would be measured? I think this is a big problem.

KAISER:

Undoubtedly we are working with very simplified systems and if we have difficulties with just one compound we will have much more with mixtures.

ZITKO:

Í think that we have just solved biochemistry if we assume that they all behave ideally and independently of each other.

KAISER:

I agree with you but what kind of solution do you foresee for this dilemma we have? How is this handled in National Health and Welfare where you are concerned for the safety? It concerns your getting an absolute number and deciding, can we use this?

CHU:

We are using an empirical approach. For instance, when we are dealing with interactions of environmental contaminants, what we do in order to answer the questions quickly, we just feed the animals with a bunch of contaminants, all together and feed them concentrations which somehow relate to the prevalent levels and then observe if there is any toxic effect. This will provide a quick answer. We have done this with organic contaminants and also with heavy metals. So far, we have found it is very safe.

HODSON:

That can be very dangerous if you have antagonistic materials and if one gets fractionated out in the environment, the antagonism can be removed and toxicity expressed.

CHU:

Could be if there is antagonism but I doubt very much that, at the levels which are presently in the environment, that would cause any effect. Usually the antagonism or the potentiation are effects which you observe only when you give the animal several hundred milligrams per kilogram but the levels in the environment are usually parts per billion levels.

SCHULTZ:

We made some stab at complex environmental mixtures where we have taken the complex environmental mixture and tested it in whole and then sub-fractioned it along classical chemical lines, acid base and neutral, and looked at those sub-fractions and then put them together in all possible combinations to see where it is additive and where it is non-additive. The problem, again, is that the chemistry is so far advanced over the biology. The chemists are able to separate that and come up with thirteen different sub-fractions and with all the different possible combinations, you simply run out of rats much less being able to reproduce it the next year or five years down the road.

WELLS:

Which is a very good argument for having some smaller organism-sensitive assays.

SCHULTZ:

With the Ames test, every time someone says they are going to take their money away, that is the flag they wave and with very good reason. We aren't able to look at all thirteen class-fractions and do that in a relatively short period time.

Comment:

The problem is that it tends to be qualitative rather than quantitative which shakes up people.

WELLS:

I think that the ultimate problem is mixtures but I think that in addition to looking at combinations we lack better models than the classical ones we have been playing with over the last few years to look at mixtures, mathematical models and the experimental design phase in particular.

McCARTY:

What intrigues me, out of the things we have seen so far, is what Alice Bobra showed us; that basically the toxic effect is a function of the concentration of the compound in the animal rather than the inherent toxicity. What that tells me is that for all the compounds that follow that kind of grouping which is all the compounds we have been talking about, that the toxicity of the mixture is a summation of those of the individual compounds.

Comment:

As long as you deal with compounds that have the same mode of toxic action!

SCHULTZ:

The problem is with the mode of action. You cannot continue to assume that. You can look at chlorophenols, nitrophenols and methylphenols and you are alright until you get to 2,4-dinitrophenol, which consistently, whether it is fish, *Tetrahymena*, *Daphnia*, or rats, is more toxic than predicted. This is fine until you get that one false negative and then it blows the whole thing.

Comment:

I agree with you but I am just saying that there has been a lot of work done on this thing and we still don't know why we get such outliers.

SCHULTZ:

We probably do in that case because that is one of the few compounds that has been really examined from the mechanism of action and things like that. The problem is with other ones, that may potentially come up, we may not be able to spot. There is nothing based on this log P value or a structure or anything like that, if they had not been tested and the mechanism of action demonstrated in the 1930's, they would have been able to say "that one is going to be an outlier". If you did your test, you would say, "gee, I did something wrong with that test or that log P value, there is a decimal point mistake in that log P literature", that is what you would say. You wouldn't flag it.

Comment:

Which one are you referring to?

SCHULTZ:

If you looked at phenols, methyl-substituted phenols, nitro-substituted phenols and chloro-substituted phenols and I say that because there is a fairly wide data base in the literature looking at those types of things and the 2,4-dinitrophenol is one that has been tested in all those groups and it consistently comes out to be more toxic than predicted and there is nothing to suggest it would be more toxic based on past findings.

KAISER:

I didn't show you that, but I plotted, for example, some Microtox toxicity data from the various Beckman Instrument Notes and the paper by Curtis et al. [Curtis, C., Lima, A., Lozano, S.J. and Veith, G.D. 1982. Evaluation of a bacterial bioluminescence bioassay as a method for predicting acute toxicity or organic chemicals to fish. In **Aquatic Toxicology and Hazard Assessment: Fifth Conference,** ASTM STP 766, J.G. Pearson, R.B. Foster and W.E. Bishop (Eds.), American Society for Testing and Materials, 1982, pp. 170-178]. I plotted those against the log P's. They do lie, relativley nicely, most of them on the one-to-one slope, particularly those that don't have any special functional groups. Others lie exactly on it but there are a few compounds, just two or three, which are really way, way out where the Microtox toxicity is about three to four orders of magnitude higher than you would expect from the log P and so, for example, hydroquinone. Now, phenol, itself, is just about right on the curve but hydroquinone is not. Obviously, there are reasons and plausible explanations but it would not be apparent from log P. However, in light of my earlier investigations on metals which showed correlations with redox potentials, if we were to look at all those, we would probably be able to flag those just by a computer listing right away without any chemical considerations.

Comment:

I think, also, we have to look at the structures, look at the molecular weight, look at the electron withdrawing groups, look at the reactivity of these compounds with the double bonds and the triple bonds where they have got the chlorines on one end, alpha to an unsaturated carbon. You may have to go back to your toxicology or pharmacology books to find out the mode of action from that respect but at least it will give you some type of idea of the toxicity mode other than just simple narcosis.

KAISER:

Then, I take it we are all agreed that QSAR in environmental toxicology is for real! Thank you very much for your contributions. I believe each of us has learned from the presentations and discussions. I hope we can meet again sometime in the future to advance further our knowledge of this important field of environmental research.

AUTHOR INDEX

FORMULA INDEX

$C_{12}H_8Cl_2O_1$	2,4,-dichlorodiphenyl ether
$C_{12}H_8Cl_6$	Aldrin
$C_{12}H_8N_2$	phenazine
$C_{12}H_9Cl_1$	2-chlorobiphenyl
$C_{12}H_9Cl_1$	3-chlorobiphenyl
$C_{12}H_9Cl_1$	4-chlorobiphenyl
$C_{12}H_9N_1O_2$	4-nitrobiphenyl
$C_{12}H_{10}$	biphenyl
$C_{12}H_{10}Cl_2N_2$	3,3'-dichlorobenzidine
$C_{12}H_{10}O_1$	biphenyl ether
$C_{12}H_{10}O_1$	p-hydroxybiphenyl
$C_{12}H_{10}O_2$	p-hydroxydiphenyl ether
$C_{12}H_{11}N_1$	4-aminobiphenyl
$C_{12}H_{11}N_1O_2$	Carbaryl
$C_{12}H_{12}N_2$	benzidine
$C_{12}H_{14}O_4$	diethylphthalate
$C_{12}H_{18}N_2O_1$	Isoproturon
$C_{12}H_{18}N_4O_6S_1$	Oryzalin
$C_{12}H_{26}$	dodecane
$C_{12}H_8Cl_6O_1$	Dieldrin
$C_{13}H_7Cl_1F_3N_1O_3$	Nitrofluorfen
$C_{13}H_7F_3N_2O_5$	Fluorodifen
$C_{13}H_9Cl_2N_1O_4$	Chlormethoxynil
$C_{13}H_9N_1$	acridine
$C_{13}H_{10}Cl_1$	2-methyl-2'-chlorobiphenyl
$C_{13}H_{10}F_1$	2-methyl-2'-fluorobiphenyl
$C_{13}H_{10}O_1$	benzophenone
$C_{13}H_{11}N_1O_3$	TOPE
$C_{13}H_{12}Cl_2N_2$	4,4'-methylene-bis(2-chloroaniline)
$C_{13}H_{16}F_3N_3O_4$	Benfluralin
$C_{13}H_{16}F_3N_3O_4$	Trifluralin
$C_{13}H_{18}Cl_1N_1O_1$	Solan
$C_{13}H_{19}N_3O_6S_1$	Nitralin
$C_{13}H_9Cl_2N_1O_4$	Chlormethoxynil
$C_{14}H_9Cl_2N_1O_5$	Bifenox
$C_{14}H_9N_1O_2$	9-nitroanthracene
$C_{14}H_{10}$	anthracene
$C_{14}H_{10}$	phenanthrene
$C_{14}H_{10}Cl_4$	p,p'-DDD
$C_{14}H_{11}N_1$	1-aminoanthracene
$C_{14}H_{13}N_1O_3$	Farmaid

$C_{14}H_{16}Cl_1O_5P_1S_1$	Coumaphos
$C_{14}H_{20}N_2O_1$	Siduron
$C_{14}H_{21}N_3O_3$	Karbutilate
$C_{14}H_{21}N_3O_4$	Butralin
$C_{14}H_{22}O_1$	2,6-di-tert-butylphenol
$C_{14}H_{30}$	tetradecane
$C_{14}H_9Cl_5$	DDT
$C_{15}H_{11}Cl_1F_3N_1O_4$	Oxyfluorofen
$C_{15}H_{13}Cl_2N_1O_2$	Mapica
$C_{15}H_{13}Cl_2N_1O_2$	N-methyl-2-methoxy-3,6-dichlorobenzanilide
$C_{15}H_{15}Cl_1N_2O_2$	Chloroxuron
$C_{15}H_{23}N_3O_4$	Isopropaline
$C_{16}H_{34}O_1$	hexadecanol
$C_{16}H_{17}N_1O_1$	Diphenamide
$C_{16}H_{22}O_4$	di-n-butylphthalate
$C_{16}H_{32}O_2$	palmitic acid
$C_{17}H_{15}Cl_1F_1N_1O_3$	Flampropmethyl
$C_{17}H_{20}N_2O_1$	Dymron
$C_{17}H_{21}N_1O_2$	Napropamid
$C_{18}H_{12}$	benzanthracene
$C_{18}H_{17}Cl_2N_1O_3$	Benzoylpropethyl
$C_{18}H_{36}O_2$	palmitic acid ethylester
$C_{19}H_{20}O_4$	butylbenzylphthalate
$C_{20}H_{12}$	benzo-a-pyrene
$C_{20}H_{12}$	perylene
$C_{21}H_{28}O_5$	cortisonacetate
$C_{22}H_6O_2N_4Cl_8$	Pigment TK 10466
$C_{22}H_{14}$	dibenzanthracene
$C_{22}H_{19}O_3N_1Cl_2$	Cypermetrin
$C_{22}H_{30}N_{10}Cl_2$	Chlorhexidin
$C_{22}H_{46}$	docosane
$C_{36}H_{34}O_4N_6Cl_2$	Pigment TK 10090
$C_{24}H_{38}O_4$	phthalic acid (2-ethylhexyl)ester
$C_{57}H_{110}O_6$	tristearin

SUBJECT INDEX